AF816148

Penicillium and *Acremonium*

BIOTECHNOLOGY HANDBOOKS

Series Editors: Tony Atkinson and Roger F. Sherwood

PHLS Center for Applied Microbiology and Research
Microbial Technology Laboratory
Salisbury, Wiltshire, England

Volume 1 *PENICILLIUM* AND *ACREMONIUM*
Edited by John F. Peberdy

A Continuation Order Plan is available for this series. A continuation order will bring delivery of each new volume immediately upon publication. Volumes are billed only upon actual shipment. For further information please contact the publisher.

Penicillium and *Acremonium*

Edited by
John F. Peberdy
University of Nottingham
Nottingham, England

Springer Science+Business Media, LLC

Library of Congress Cataloging in Publication Data

Penicillium and *acremonium.*

(Biotechnology handbooks; v. 1)
Bibliography: p.
Includes index.
1. Fungi—Biotechnology. 2. *Penicillium*—Biotechnology. 3. *Acremonium*—Biotechnology. I. Peberdy, John F., 1937- . II. Series.
TP248.27.F86P46 1987 660′.62 87-12328
ISBN 978-1-4899-1988-5

ISBN 978-1-4899-1988-5 ISBN 978-1-4899-1986-1 (eBook)
DOI 10.1007/978-1-4899-1986-1

Originally published by Plenum Press, New York in 1987
Softcover reprint of the hardcover 1st edition 1987

Contributors

B. L. Brady • CAB International Mycological Institute, Kew, Surrey TW9 3AF, England

B. W. Bycroft • Department of Pharmacy, University of Nottingham, Nottingham NG7 2RD, England

Paul F. Hamlyn • Biotechnology Group, Shirley Institute, Didsbury, Manchester M20 8RX, England

G. Holt • Biological Laboratory, University of Kent at Canterbury, Kent CT2 TNJ, England

P. G. Mantle • Department of Biochemistry, Imperial College of Science & Technology, London SW7 2AY, England

Maurice O. Moss • Department of Microbiology, University of Surrey, Guildford GU2 5XH, England

A. H. S. Onions • CAB International Mycological Institute, Kew, Surrey TW9 3AF, England

J. F. Peberdy • Department of Botany, University of Nottingham, Nottingham NG7 2RD, England

Brian F. Sagar • Biotechnology Group, Shirley Institute, Didsbury, Manchester M20 8RX, England

G. Saunders • School of Biotechnology, Faculty of Engineering and Science, The Polytechnic of Central London, London W1M 8JS, England

R. E. Shute • Department of Pharmacy, University of Nottingham, Nottingham NG7 2RD, England

David S. Wales • Biotechnology Group, Shirley Institute, Didsbury, Manchester M20 8RX, England

Foreword

Biotechnology is a word that was originally coined to describe the new processes which could be derived from our ability to manipulate, *in vitro*, the genetic material common to all organisms. It has now become a generic term encompassing all "applications" of living systems, including the more traditional fermentation and agricultural industries. Recombinant DNA technology has opened up new opportunities for the exploitation of microorganisms and animal and plant cells as producers or modifiers of chemical and biological products.

This series of handbooks deals exclusively with microorganisms which are at the forefront of the new technologies and brings together in each of its volumes the background information necessary to appreciate the historical development of the organisms making up a particular genus, the degree to which molecular biology has opened up new opportunities, and the place they occupy in today's biotechnology industry. Our aim was to make this primarily a practical approach, with emphasis on methodology, combining for the first time information which has largely been spread across a wide literature base or only touched upon briefly in review articles. Each handbook should provide the reader with a source text, from which the importance of the genus to his or her work can be identified, and a practical guide to the handling and exploitation of the organisms included.

It is perhaps apt that Volume 1 of the series should deal with a group of organisms involved in producing chemicals which have dominated the fermentation and pharmaceutical industries for the past forty years—the antibiotics. In terms of the degree to which molecular biology has made an impact on this genus to date, it is clear that the major rewards are still to come.

Tony Atkinson
Roger F. Sherwood

Wiltshire

Preface

Fungi have a long-standing importance in biotechnology. Of the filamentous fungi, species of *Penicillium* are probably the best known, having applications in three major areas of industrial activity. In this volume, I have taken the genus *Acremonium* as a "bookfellow," the link being the importance of these fungi as the source of two of the most commercially significant products, the β-lactams penicillin and cephalosporin.

In compiling the book, I have attempted to meet the different needs of its readers. For the industrial microbiologist, there are accounts of the fundamental aspects of the physiology and genetics of these fungi. For the academic microbiologist interested in their industrial use, there are contributions that provide the biochemical background to the various processes involved. There are no descriptions of the different commercial processes, but these are discussed in many other texts. The book opens with a chapter on taxonomy and throughout reflects the clear necessity for careful identification of organisms that are used or may be used in commercial processes.

The authors have working experience with the subject areas of their respective chapters. I am greatly indebted to all of them for taking part in the production of this book despite their busy schedules. They will join me, I am sure, in expressing the wish that all workers in the field of biotechnology find this a useful reference in their day-to-day work.

John F. Peberdy

Nottingham

Contents

Chapter 1

Taxonomy of *Penicillium* and *Acremonium*

A. H. S. Onions and B. L. Brady

Chapter 2

Morphology and Physiology of *Penicillium* and *Acremonium*

Maurice O. Moss

Chapter 3

Genetics of the *Penicillia*

G. Saunders and G. Holt

Chapter 4

Genetics of *Acremonium*

J. F. Peberdy

Chapter 5

Chemistry and Biosynthesis of Penicillins and Cephalosporins

B. W. Bycroft and R. E. Shute

Chapter 6

Secondary Metabolites of *Penicillium* and *Acremonium*

P. G. Mantle

Chapter 7

Extracellular Enzymes of *Penicillium*

Paul F. Hamlyn, David S. Wales, and Brian F. Sagar

Taxonomy of *Penicillium* and *Acremonium* 1

A. H. S. ONIONS and B. L. BRADY

1. INTRODUCTION

The fungi form one of the largest groups of organisms, comprising some 65,000 species. They vary from the larger bracket fungi and toadstools to the common soil fungi or recycling organisms. The latter play an important ecological role in breaking down dead plant and animal materials and returning them to the soil. However, from time to time, they attack materials grown or manufactured by man. In the process, they often damage the material either mechanically or enzymatically or by other chemical reactions. Interesting secondary metabolites including the useful antibiotics or undesirable mycotoxins are produced. This potential activity is harnessed in manufacturing processes. Two genera of particular interest in these respects, especially on account of their production of antibiotics and mycotoxins, are *Penicillium* and *Acremonium.*

1.1. Anamorphs and Teleomorphs

Penicillium and *Acremonium* by definition reproduce by vegetatively produced conidia (spores). They belong, therefore, to the class Hyphomycetes of the Deuteromycotina or Fungi Imperfecti. However, several genera of the Ascomycotina (Perfect Fungi) that reproduce sexually produce *Penicillium* or *Acremonium* vegetative asexual structures. The vegetative form is known as the *anamorph,* while the ascosporic form is called the *teleomorph.* According to the Botanical Code (Voss *et al.*, 1983), the teleomorph takes precedence and the teleomorphic name should be used.

A. H. S. ONIONS and B. L. BRADY • CAB International Mycological Institute, Kew, Surrey TW9 3AF, England.

However, some species do not produce a teleomorph, in which case the anamorphic name is applied. Since there are several teleomorphic genera that produce *Penicillium* anamorphs, it is convenient to treat them together, though they may not be truly related. The correct procedure is to use the teleomorph name, but many authors have found it convenient to classify all these organisms together in *Penicillium*. The situation is similar in *Acremonium*.

1.2. Priority of Nomenclature

Fungus names are controlled by the International Code of Botanical Nomenclature (Voss *et al.*, 1983), one of the provisions of which is priority of publication, whereby the earliest name under which a genus or species was described should be used (Hawksworth, 1984).

2. *ACREMONIUM*

2.1. The Names *Cephalosporium* and *Acremonium*

The name *Cephalosporium*, which has long been familiar to biotechnologists and which fathered that of the cephalosporin compounds, has been largely replaced by the name *Acremonium* since the monograph of Gams (1971). *Cephalosporium* was introduced by Corda (1839) for colorless molds with simple unbranched conidiophores and condiogenous cells bearing at the tip a group or "head" of unicellular conidia, from whence the name was derived. At that time, little was known about the interrelationships of these fungi, and the name came to be used for a wide variety of Hyphomycetes. However, there is a suspicion that Corda had applied the name to a member of the Mucorales, but none of his original material is now available for verification. Gams (1968) reexamined specimens of the original (type) material of *Acremonium* Link (1809), which has nomenclatural priority over *Cephalosporium*. This material of *A. alternatum* Link was well preserved and enabled Gams to redescribe the genus from living cultures of later isolates of that species. He applied this name to hyaline fungi that form numerous conidia enteroblastically in basipetal succession from erect conidiogenous cells on sparsely branched conidiophores and included most of the species formerly referred to *Cephalosporium*. However, many authors had come to use the name *Cephalosporium acremonium* for all such molds, and since the history of this name is so confusing, Gams (1971) described a new species *A. strictum* to encompass those fungi, retaining *A. kiliense* Grütz (the earlier name) for very similar forms that additionally produce hyaline chlamydospores, are frequently isolated from the soil, and have been associated with skin infections of man.

At first sight, the morphology of *Acremonium* is simple compared with that of *Penicillium*. It is only when the arrangement, form, and function of the conidiogenous cells (often referred to in earlier work as "phialides") are compared in the individual species that an underlying complexity is revealed.

The conidiogenous cells that generate conidia in both *Penicillium* and *Acremonium* do so by an active wall-building region at the tip of the cell budding off the first conidium, both inner and outer wall layers being involved; a basal cross wall is laid down, and the process is described as "holoblastic." Thereafter, in *Acremonium*, this conidium becomes detached by the separation of the two layers of wall between conidiogenous cell and conidium (schizolysis), and a further conidium is formed at the tip of the conidiogenous cell, the original inner wall layer becoming the outer wall and a new inner wall layer being laid down within it. These and subsequent conidia are said to be formed "enteroblastically," and after they secede, they may remain either clustered in the typical "head" of a "*Cephalosporium*" or loosely connected end-to-end in a "false chain." In *Penicillium*, the second and subsequent conidia produced by a conidiogenous cell are formed from the same two original wall layers, cross walls forming at intervals, but the outer wall layers of the whole chain of conidia remain continuous, and the chains are known as "true chains." The different forms of conidiogenesis are described in detail by Minter *et al.* (1982, 1983a,b).

2.2. Problems of Genus and Species Concepts

In his monograph of the *Cephalosporium*-like fungi, Gams (1971) transferred 18 *Cephalosporium* species to *Acremonium* and described many more *Acremonium* species in the genus, adding others later (Gams and Lacey, 1972; Gams, 1975). He also transferred most of the species of *Paecilomyces* Bainier with solitary conidiogenous cells to the genus *Acremonium* and included many of *Gliomastix* Guéguen, despite their dark-pigmented conidia. Subramanian (1972), recognizing that many of the species first described in *Paecilomyces* and *Gliomastix* formed their conidia in true chains as in *Penicillium*, transferred these to a new genus, *Sagrahamala* Subramanian, while Gams (1978) similarly placed these true chain-formers in another new genus, *Sagenomella* W. Gams. While Subramanian transferred *G. luzulae* to *Sagrahamala*, Gams considered it better left in *Acremonium*, to which he had transferred it from *Gliomastix* in 1971. This dark-spored species is probably best left in *Gliomastix* and is so treated here. Several *Cephalosporium* species, including *C. nordinii*, were transferred by Gams (1971) to *Monocillium* Saksena, a genus in which the wall at the base of the conidiogenous cell is thickened, the distal part expanding in outline before narrowing again at the tip. Some species of *Cephalosporium* were transferred to *Verticillium*, which typically has conidiogenous cells that arise together at one level to form whorls or verticils, but in which these cells are sometimes found singly or in "whorls" of only two; *C. aphidicola* Petch was transferred by Gams to *V. lecanii* (Zimm.) Viégas. On account of the distinct collarette at the mouth of the conidiogenous cell in *C. gregatum*, Gams transferred this species to *Phialophora.* Thus, although *Cephalosporium* species are generally speaking now renamed *Acremonium*, there are many exceptions. Furthermore, some combinations into *Acremonium* have never been made, usually because the taxonomic information on the fungus is too scanty. For instance, *C. caerulens* Matsumae, Kamio, and Hata was inadequately described, and the authors made no mention of type material; Gams (1971) suggested that it was a microconidial state of a species of *Fusarium* Link. Gams considered *C. gramineum* to be synonymous with *Hymenula cerealis* Ell. and Ev., a fungus that in nature is sporodochial. *Cephalosporium mycophilum* was never moved to *Acremonium*, since there was no type material available for study; a suggestion by Tubaki (1955) that it is the same as *A. butyri* was discounted by Gams. These controversial species are most appropriately still referred to as "*Cephalosporium*" so long as their position and status are in doubt.

All the fungi dealt with here are conidial states belonging to the class Hyphomycetes. Many Hyphomycetes are anamorphs (conidial states) of members either of the ascomycetes or of the basidiomycetes. *Acremonium* anamorphs are found in many ascomycete orders including the Clavicipitales, Eurotiales, Hypocreales, and Sordariales; "*Cephalosporium salmosynnematum*" is the *Acremonium* anamorph of *Emericellopsis salmosynnemata* Groskl. and Swift, considered by Gams to be synonymous with *E. minima* Stolk in the Eurotiales. *Acremonium* stages are found especially in the younger or juvenile phases of development of other genera. Microconidial stages of *Fusarium* resemble *Acremonium*, and similar simple conidiogenous cells bearing unicellular conidia are found in *Cylindrocarpon*, *Gliocladium*, *Sarocladium*, and *Verticillium* and constitute additional conidial states in *Humicola* and *Thermomyces*.

2.3. Genus Description: *Acremonium* (Figs. 1 and 2)

Colonies on culture media more or less slow-growing, attaining a diameter of less than 25 mm in 10 days on malt extract or oatmeal agars at 20°C. Hyphae thin-walled and hyaline. Typically, the erect conidiogenous cells are formed singly, but conidiophores with simple or verticillate branching occur in some species. Conidiogenous cells usually delimited by a basal septum, occasionally continuous with the hypha from which they are formed, narrowing in shape gradually toward the tip. First conidium formed holoblastically, secession schizolytic, sometimes leaving a short collarette; later conidia produced enteroblastically without further elongation of the tip

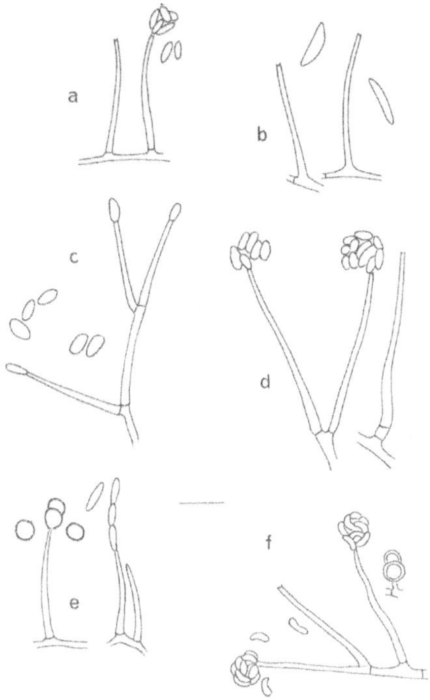

Figure 1. *Acremonium* species. (a) *A. strictum*; (b) *A. coenophialum*; (c) *A. chrysogenum*; (d) *A. crotocinigenum*; (e) *A. fusidioides*; (f) *A. recifei*. Scale bar: 10 μm.

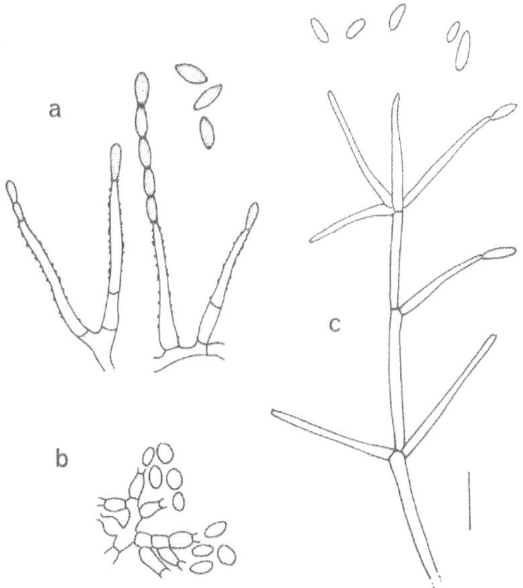

Figure 2. (a) *Gliomastix luzulae*; (b) *Phialophora gregata* (after Gams, 1971); (c) *Verticillium lecanii*. Scale bar: 10 μm.

of the conidiogenous cell. Conidia hyaline, usually consisting of a single cell, rarely two-celled, globose, ovoid, ellipsoidal, cylindrical, or fusiform, collecting after secession at the tip of the conidiogenous cell in a head or a false chain. Chlamydospores sometimes formed. Sclerotia sometimes formed. Some species have a teleomorph; these are in *Cordyceps*, *Emericellopsis*, *Nectria*, *Torrubiella*, *Wallrothiella*, and several other genera.

2.4. Identification of Species (see Section 4.1)

Gams (1971) divides the genus into three sections: *Simplex* (which includes *A. strictum* and *A. fusidioides*), *Gliomastix* (including *A. luzulae*, here kept in the genus *Gliomastix*), and *Nectrioidea* (including *A. chrysogenum*, *A. crotocinigenum*, and *A. recifei*). The characters employed in identification of the species are: appearance of the culture on malt extract or oatmeal agars; presence and form of the conidiophore or its absence; form of the conidiogenous cell; presence or absence of a collarette after the first conidium has seceded; way in which the conidia collect at the tip of the conidiogenous cell, i.e., in "heads" or "false chains"; presence and form of chlamydospores and of sclerotia.

2.4.1. *Acremonium chrysogenum* (Thirum. and Sukap.) W. Gams (Fig. 1c)

Colonies on malt extract agar after 10 days, 8–15 mm in diameter, yeast-like, slimy, chrome yellow or paler yellow, underside an intense chrome yellow, mycelium largely submerged in the medium. Sporulation sparse, conidiophores indistinguishable from the vegetative mycelium. Conidiogenous cells simple, upright, smooth, without an apical collarette, 25–50 μm long, 1.5–2.5 μm at the base, narrowing to 0.6–1.2 μm. Conidia in slimy heads, short or long ellipsoidal with a flattened base, occasionally slightly curved, 4–7.5 × 1.5–3 μm. Chlamydospores absent, although swollen hyphal segments occasionally occur. This species is placed by Gams (1971) in his section *Nectrioidea*. Teleomorph unknown.

2.4.2. *Acremonium coenophialum* Morgan-Jones and W. Gams (Fig. 1b)

Colonies on malt extract agar very slow-growing, 3 mm in diameter in 10 days, 20 mm in diameter in 2 months, white and cottony on malt extract agar, cream-colored and waxy on cornmeal agar. Conidiophores lacking, conidiogenous cells arising directly from the aerial mycelium, usually not delimited by a basal septum from the subtending hypha, 12–34 μm long, 1.5–2 μm wide at the base, tapering to 0.5–0.8 μm at the tip, without an apical collarette. Conidia formed singly, collecting in small groups at the tip of the conidiogenous cell after secession, elongate, slightly curved with a truncate base and obtuse tip, hyaline, smooth, 7.5–10 × 1.5–3 μm. Chlamydospores absent. Teleomorph unknown. From seeds and culms of *Festuca arundinacea*, North America.

2.4.3. *Acremonium crotocinigenum* (Schol-Schwarz) W. Gams (Fig. 1d)

Colonies on malt extract agar after 10 days 30–40 mm in diameter, white to cream-colored with a raised rim, reverse colorless or pale orange. Sporulation profuse, above the surface of the agar. Conidiophores septate, occasionally branched from the base, which is often thick-walled and warty. Conidiogenous cells with a wavy appearance at the tip and with a short collarette, 20–65 μm long, 2–3 μm at the base, narrowing to 1–1.3 μm at the tip. Conidia produced singly, collecting in slimy heads, long-ovoid to cylindrical, rounded at each end, 4–7.5 × 1.6–2.8 μm, sometimes larger, 10–16 × 3 μm, often with a cross wall. Chlamydospores present after 2 weeks, especially on oatmeal agar, in short intercalary chains, 5–8 μm in diameter, smooth or with fine warts. This species is placed by Gams (1971) in his section *Nectrioidea*. Teleomorph unknown. In nature, this fungus is found in soil and leaf-litter and on the fruiting bodies of larger fungi.

2.4.4. *Acremonium fusidioides* (Nicot) W. Gams (Fig. 1e)

Colonies on malt extract agar after 10 days 8–10 mm in diameter, white, ochraceous to gray-brown, often with a reddish tint, reverse dark brown. Sporulation profuse, conidiophores occasionally branched, but conidiogenous cells mostly solitary, 10–20 μm long, 2–2.5 μm at the base, 0.8–1 μm at the tip, no collarette. Conidia of two types, mostly (1) spindle-shaped, slightly pointed at each end, forming long false chains, 5–6.5 × 1–2 μm or (2) spherical in heads or false chains, slightly pigmented, sometimes slightly rough, 3.5–4.5 μm in diameter. Chlamydospores absent. Teleomorph unknown. This species is classed in the *Simplex* section of Gams (1971). It has been isolated from soil, animal droppings, and sputum of man.

2.4.5. *Acremonium recifei* (Leão and Lôbo) W. Gams (Fig. 1f)

Colonies on malt extract agar in 10 days 30–40 μm in diameter, white to pale salmon or gray, reverse colorless to ochraceous becoming brown with age. Conidiophores often branched, with two or more conidiogenous cells arising at one level. Conidiogenous cells 20–55 μm long, 1.5–2.5 μm at the base, 0.8–1 μm at the tip, the distal end of the cell often undulating; an inconspicuous collarette is present. Conidia produced singly, collecting in slimy heads after secession, elongate and strongly curved with rounded ends, smooth, hyaline, 4–6 × 1.2–2 μm. Chlamydospores occasional in old cultures, thin-walled, 3–5 μm in diameter. Teleomorph unknown. This species is classed in the *Nectrioidea* section of Gams (1971). First described from a mycetoma of the human foot; most other strains have been isolated from plant sources, mainly nuts.

2.4.6. *Acremonium strictum* W. Gams (Fig. 1a)

Colonies on malt extract agar after 10 days 16–25 mm in diameter, of variable color from white to cream, pink, or orange, and appearance from powdery to velvety or synnematous, reverse remaining colorless or turning pink to orange. Conidiophores simple, occasionally branching. Conidiogenous cells long and narrow, but of varying length according to the strain. Conidia ellipsoidal, short, cylindrical with rounded ends, occasionally curved, sometimes almost spherical, 3–7 × 1–2 μm, remaining in a "head" at the tip of the conidiogenous cell after secession. Chlamydospores absent. This species, which includes those strains formerly called *Cephalosporium acremonium*, is very common almost everywhere and on account of its simple morphology probably embraces many different strains that cannot be distinguished using current methods.

2.4.7. *Gliomastix luzulae* (Fuckel) Mason ex Hughes (Fig. 2a)

Colonies on malt extract agar after 10 days 5 mm in diameter, white to pink becoming greenish black in areas of sporulation, reverse pink to brown. Sporulation sparse to profuse; conidiophores lacking, or consisting of a few short cells; conidiogenous cells erect, hyaline, and slightly roughened, slightly pigmented at the tip, 20–35 μm long, 2.5–3.5 μm at the base, 1–1.5 μm at the tip. Conidia in very long false chains or in heads, dark olive green with a darker median band, ellipsoidal with ends truncate, 5–8 × 2–2.8 μm. Chlamydospores absent. Teleomorph unknown. Gams (1971) classes this species in *Acremonium*, Murorum section, but it is better kept in the genus *Gliomastix* with other dark-spored relatives (see Dickinson, 1968).

2.4.8. *Phialophora gregata* (Allington and Chamberl.) W. Gams (Fig. 2b)

Colonies on malt extract agar in 10 days 14–16 mm in diameter, white to dark gray. Sporulation sparse. Conidiophores branched several times bearing conidiogenous cells in groups. Conidiogenous cells bottle-shaped, narrowing at the tip with a distinct collarette, 5–9 μm long, 2–3 μm at the base, 1.2–1.5 μm at the tip. Conidia collecting in slimy heads, ovoid, 3.5–5 × 2.3–2.5 μm. Chlamydospores absent. Teleomorph unknown. This species is known as the cause of a brown stem rot of soybeans, but no material was available for study and the description and Figure 2b are based on Gams (1971).

2.4.9. *Verticillium lecanii* (Zimmerm.) Viégas (Fig. 2c)

Colonies on malt extract agar after 10 days 18–22 mm in diameter, white to pale yellow, velvety to cottony, reverse colorless or pale to chrome yellow. Conidiophores differing little from the vegetative mycelium, more or less branched, and with conidiogenous cells either solitary or in whorls of two or more. Conidiophores of variable size depending on the strain, those of *C. aphidicola* collections being 17–37 × 1.5–3 μm. Conidia ellipsoidal to cylindrical with rounded ends, very variable in size in the different strains, in *C. aphidicola* 4.5–8 × 1.3–2.3 μm. Chlamydospores absent. Teleomorph unknown, although Evans and Samson (1982) suggest a teleomorph association with *Torrubiella confragosa* Mains that has not been proved by ascospore germination. *Verticillium lecanii* strains are active parasites of a wide range of insects and mites and are also hyperparasites of fungal plant pathogens. Gams (1971) classes the species in his Prostrata section of the genus.

2.4.10. Other Species Mentioned in This Book

Only those species that produce industrially important metabolites and the universal *A. strictum* are given descriptions above. There follows a list of others mentioned in the book, with their authorities; full descriptions of all are given by Gams (1971) except where otherwise stated. *Acremonium alternatum* Link; *A. butyri* (v. Beyma) W. Gams; *A. diospyri* (Crandall) W. Gams; *A. fungicola* (Sacc.) Samuels (1973); *A. furcatum* (F. and R. Moreau) ex W. Gams; *A. kiliense* Grütz; *A. sordidulum* W. Gams and D. Hawksw. (see Gams, 1975) [syn. *Sagrahamala sordidula* (W. Gams and D. Hawksw.) Subramanian]; *A. terricola* (Miller, Giddens, and Foster) W. Gams [syn. *A. implicatum* (Gilman and Abbott) W. Gams (1975)]; *A. rutilum* W. Gams; *A. zonatum* (Sawada) W. Gams; *Cephalosporium aphidicola* Petch [see *Verticillium lecanii* (Zimm.) Viégas]; *C. caerulens* Matsumae, Kamio, and Hata [probably *Fusarium* sp. (see Gams, 1971)]; *C. gramineum* Nisikado and Ikata (syn. *Hymenula cerealis* Ell. and Ev.); *C. mycophilum* (Corda) Tubaki, of uncertain affinity; *C. nordinii* Bouchier [anamorph of *Monocillium nordinii* (Bouchier) W. Gams]; *C. salmosynnematum* Roberts (anamorph of *Emericellopsis salmosynnemata* Groskl. and Swift).

3. *PENICILLIUM*

3.1. The Name *Penicillium*

The generic name *Penicillium* was first introduced by Link (1809). As a result of changes including the starting date in the International Code of Botanical Nomenclature (Voss *et al.*, 1983), the generic name and *Penicillium expansum* can now be attributed to "Link" alone, and the latter is acceptable as lectotype for the generic name (see Hawksworth, 1985).

Notable works on identification started with Dierckx (1901) and Thom (1910), followed by Biourge (1923), Vuillemin (1904), Westling (1911), Thom (1930), and Raper and Thom (1949). These were succeeded by postwar monographs, stimulated by the discovery and use of antibiotics, by Abe (1956) and Pidoplichko (1972). Recently, with increased interest arising from mycotoxin production, further works have come from Fassatiová (1977), Udagawa and Horie (1973), Samson *et al.* (1976), Pitt (1979b), and Ramirez (1982).

3.2. Problems of Species Concepts, Application of Names, and Misidentifications

This wealth of literature on *Penicillium* systematics does not appear to have greatly clarified the situation, since each monographer has somewhat differing species concepts. For example, Pitt (1979b) accepted about 150 species, while Ramirez (1982) used over 400 species names to cover the same series of organisms. In some parts of the genus, such as the fasciculate species, the literature has become particularly confused, and the identity of isolates is therefore often questionable. Workers concerned with the biochemistry of metabolites do not always deposit their isolates with a culture collection, and although a strain may have been given a name, this may depend on the opinion of the identifier, who may follow one of several monographs, have his own interpretation of their work, or be confused and not have sound concepts of the species.

Since it is the overlapping or lack of clear-cut morphological characters that causes the confusions, attempts are now being made to separate these organisms by other means, such as the physiological tests used in bacteriology (Cowan, 1974) and yeasts (Kreger-van Rij, 1984). This is not a new concept, since Thom (1910) had already examined a few physiological and biochemical characteristics. Most such workers have tended to concentrate on one type of technique, for example, secondary metabolites (Frisvad and Filtenborg, 1983), electrophoresis (Clare *et al.*, 1968), production of enzymes (Bridge and Hawksworth, 1984; Bridge, 1985), serology (Polonelli *et al.*, 1985), and hybridization (Anné, 1985). The results have been interesting, but so far have not produced a consistent and compatible system.

A multidisciplinary approach is now being taken at the CAB International Mycological Institute in which over 300 strains are being used to produce a matrix made up of 200 characters from studies of morphology, physiology, primary and secondary metabolites, enzyme production, and scanning-electron-microscopic (SEM) microtopography of surfaces. These characters will be subjected to cluster analysis to see whether species concepts can be clarified (Onions *et al.*, 1984; Bridge *et al.*, 1985; Paterson and Hawksworth, 1985). The results of this major study will not be available until 1987, and in the interim, species separations remain almost entirely based on morphology.

3.3. Genus Description: *Penicillium*

Vegetative mycelium made up of narrow septate hyaline (pale or brightly colored) hyphae as usually seen in cultures grown on agar forms a loose to closely woven mat; fertile branches (conidiophores) arising more or less perpendicular to submerged or aerial hyphae; the basal portion or main stalk (stipe or conidiophore) septate, relatively narrow, bearing a branched fruiting structure at the apex; the latter (conidiophore, penicillus) may consist of conidiogenous cells (phialides) borne directly at the apex of the conidiophore or on one or more levels of branches (rami, ramulae); the conidiogenous cells are usually flask-shaped with a swollen base narrowing gradually or abruptly to a tube from within which conidia (spores) are produced in basipital unbranched chains (i.e., with the oldest at the top); the conidia are aseptate, hyaline, small, variously shaped from globose to cylindrical, smooth to strongly roughened.

Dense mycelial masses or sclerotia may be produced. Some species may produce a teleomorph consisting of hard or soft nonostiolate ascomata (cleistothecia).

3.3.1. Anamorph (Figure 3)

3.3.1a. Branching Systems. When dealing with a large genus such as *Penicillium*, it is helpful to divide it into groups of similar or apparently similar species. The arrangement of the branches in the conidiophore (penicillus) has been one of the main criteria of separation.

The branch arrangement arises from the development of the conidiophore, which begins with the formation of a conidiogenous cell (phialide), then produces further conidiogenous cells and branches. This process does not appear to follow a set pattern and is complete at various stages according to species. However, a few conidiophores in any isolate may

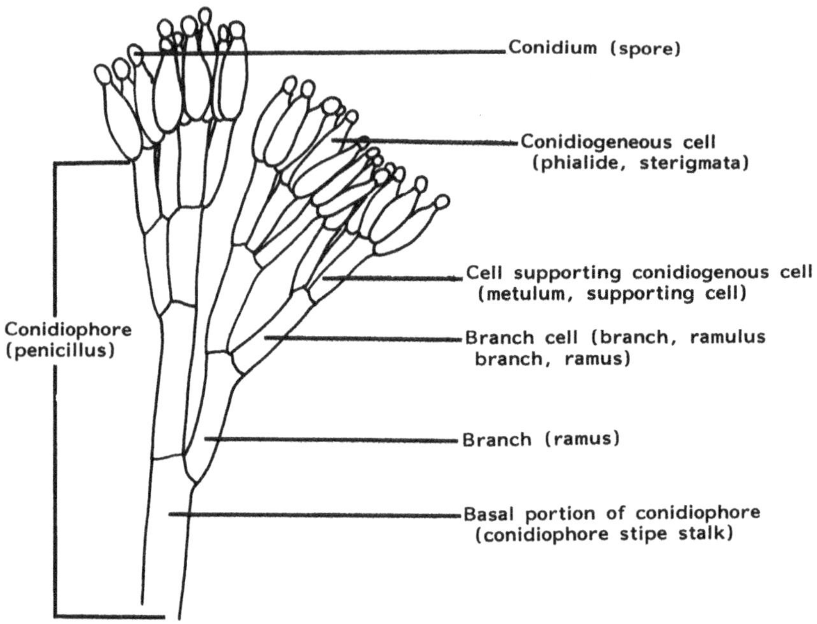

Figure 3. Conidiophore (penicillus). The elements are labeled and alternative terminology is given in parentheses.

not progress to full development, while some may progress further. Thus, in ascribing an isolate to a branching pattern, the majority appearance is used.

Interpretation of the branching pattern varies from author to author. All agree in counting levels of branching from the top down. Figures 4 and 6 illustrate the various forms.

3.3.1b. Characters Used in Describing *Penicillium* Isolates. Colony texture is often used in the description of cultures, e.g., velvety, floccose, funiculose, fasciculate. Figure 5 illustrates the arrangements of hyphae that give rise to these textures. The differences among the textures are not always easy to recognize, or several states may be seen in one colony or during its development. Raper and Thom (1949) used texture as a main means of separation, while Pitt (1979b) avoided it as far as possible.

Growth Characters. Morphological descriptions are based on a combination of growth characters and morphology. Since growth characters vary with cultural conditions, to obtain consistent results, it is necessary to grow the culture using standard conditions. In our laboratory, it is usual to grow *Penicillium* cultures on Czapek's agar and malt agar + Czapek's salts at 25°C in the dark for 7–10 days. Raper and Thom (1949) used Czapek's and malt agar, but Pitt (1979b) used a modified Czapek + yeast agar, malt

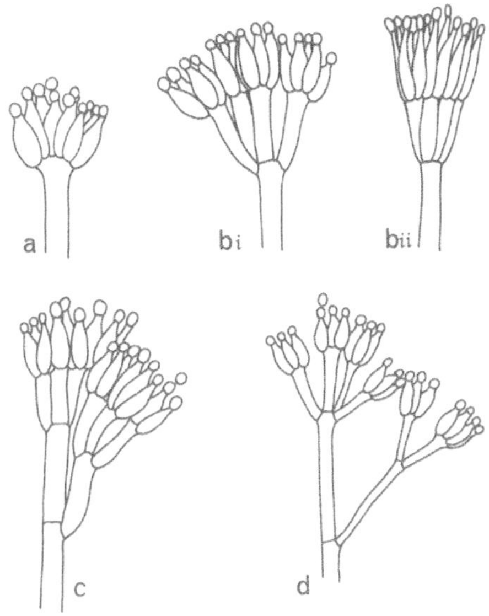

Figure 4. Branching patterns. (a) Monoverticillata; (b) Biverticillata: (i) asymmetrica, (ii) symmetrica; (c) Terverticillata; (d) Divaricata.

agar, and a low-water-activity medium, G25N. Pitt also used three temperatures, 5, 25, and 37°C.

Characters used in describing *Penicillium* cultures include growth rate on standard media, depth, profile, wrinkling, color of colony, production of drops, pigment in the agar, colony reverse, and smell. Smelling cultures, though often helpful, is not currently "in fashion," since the odors may well be produced by, or associated with, toxic metabolites.

Microscopically, the arrangement of conidia in columns, separate, tangled, or otherwise, the length, breadth, and surface texture of the basal portion of conidiophores, branches, cells supporting conidiogenous cells (metulae), conidiogenous cells (phialides), and conidia are also used in descriptions.

3.3.2. Teleomorphs

Sexual stages of species that produce hard sclerotiumlike ascomata are placed in *Eupenicillium* and those that produce soft ascomata in *Talaromyces*.

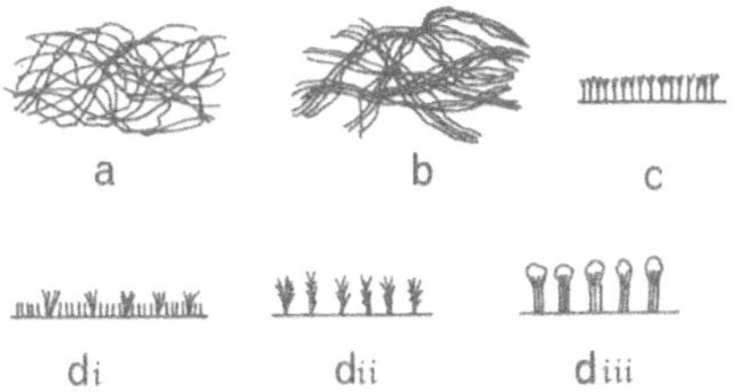

Figure 5. Colony textures. (a) Floccose (lanose); (b) funiculose; (c) velvety; (d) fasciculate: (i) small bundles, (ii) feathery bundles, (iii) synnemata.

3.4. Identification of Species

A key to genera and subgenera with "*Penicillium*-like" anamorphs is given in Section 4.2.

3.4.1. *Eupenicillium*

Species that produce *Penicillium* anamorphs and hard sclerotium-like ascomata are placed here. They occur in soil and are seldom of major economic significance.

Taxonomic studies of this group have been made by Stolk and Scott (1967), Pitt (1979b), and Udagawa and Horie (1973). Stolk and Samson (1983) made a comprehensive study of the genus, reducing many species to synonymy. Many well-known anamorphic *Penicillium* species produce sclerotia, which in many cases have been shown on aging or under special conditions to produce asci. Stolk and Samson (1983) included in their monograph species with *Penicillium* anamorphs that they regarded as being related to species of this genus but that had lost the ability to produce ascomata and reduced many of them to synonymy with *Eupenicillium* species. It would be interesting to investigate the validity of these synonymies based on morphology by the alternative methods now available. The monograph simplifies the identification of these organisms.

The most common species are *Eupenicillium javanicum* and members of the *E. javanicum* series. Stolk and Samson (1983) made many synonyms in this series, including *Penicillium simplicissimum* (Oudemans) Thom and *P. janthinellum* Biourge. Other species treated in Stolk and Samson's monograph include *P. thomii* Maire, *P. sclerotiorum* van Beyma, *P. lividum* Westling, and *P. citreonigrum* Dierckx.

3.4.2. *Talaromyces*

Species usually with symmetrical biverticillate branching systems that produce soft nonostiolate ascomata are referred to *Talaromyces*. These were described by Raper and Thom (1949) as part of the genus *Penicillium*, but Benjamin (1955) and later Stolk and Samson (1972) treated them separately: Pitt (1979b) has adopted most of the classification of Stolk and Samson (1972), though with some additional species and name changes.

Two species are fairly common in soil, *Talaromyces flavus* (Klöcker) Stolk and Samson (*P. vermiculatum* Dangeard) and *T. wortmannii* (Klöcker) C. R. Benjamin (*P. wortmannii* Klöcker; *P. kloeckeri* Pitt). The two species are very similar when grown in culture, producing bright yellow ascomata without a true wall and bound by a soft hyphal web. Growth of *T. wortmannii* is more compact and restricted. The ascospores are almost identical,

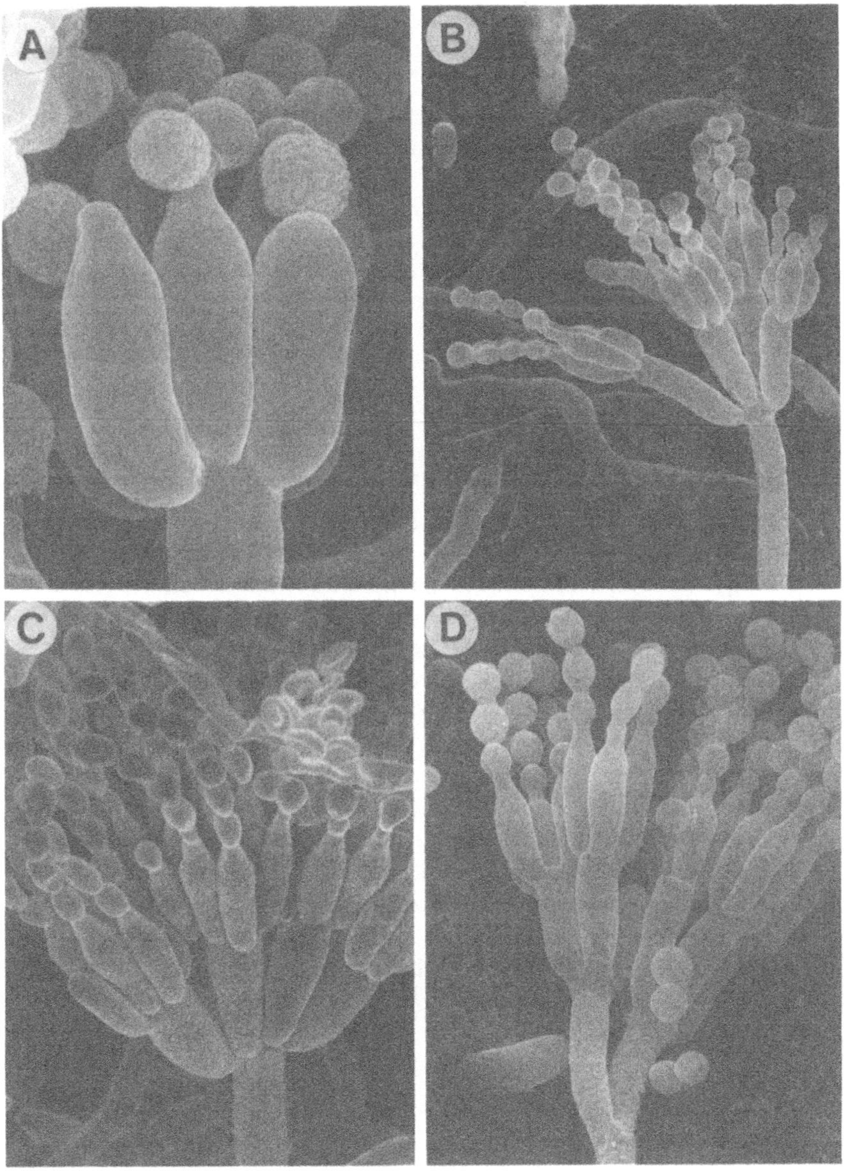

Figure 6. Scanning electron micrographs of branching patterns in *Penicillium*. (A) *P. glabrum* (Monoverticillata); (B) *P. citrinum* (Biverticillata, Asymmetrica); (C) *P. funiculosum* (Biverticillata, Symmetrica); (D) *P. cyclopium* (Terverticillata). Scale bars: 4 μm (A, C); 10 μm (B, D). Micrographs courtesy of Z. Lawrence (Ministry of Agriculture, Fisheries and Food). Crown copyright.

ovate, without a furrow, and finely spinulose, 3.5–5 × 2.5–3.5 μm. The easiest method of separation is by their different ascomatal initials. In *T. flavus*, these consist of a thick, deeply staining, club-shaped hypha around which thinner hyphae are tightly coiled, while in *P. wortmannii*, the initials are smaller, difficult to find, and consist of irregular knots of thickened hyphae. In order to see either, it is necessary to make mounts from the yellow edge of actively growing cultures.

Other species of *Talaromyces* mentioned later in this book include *T. galapagensis* Samson and Mahoney, *T. gossypii* Pitt, *T. helicus* (Raper and Fennell) C. R. Benjamin, *T. panasenkoi* Pitt, *T. stipitatus* (Raper and Fennell) C. R. Benjamin, *T. trachyspermus* (Shear) Stolk and Samson, *T. luteus* (Zukal) C. R. Benjamin, and *T. thermophilus* Stolk (*P. dupontii* Griffin and Maubl.). They are described in Pitt (1979b).

3.4.3. *Penicillium*

3.4.3a. Monoverticillata. A key to the Monoverticillata is given in Section 4.2.1.

Many monoverticillates have a simple form that when it deteriorates or varies in culture easily loses any recognizable character. The species concept is difficult to define. Raper and Thom (1949) and Pitt (1979b) adopted a conservative number of species (20–40), whereas Ramirez (1982) described many intermediates, and Stolk and Samson (1983) included many monoverticillate species in their monograph on *Eupenicillium*, taking a very broad species concept. Only a few species are common, of economic importance, or distinctive. The main criteria of Pitt (1979b) for separation are probably the most satisfactory. He divides the species on the presence or absence of a swollen apex to the conidiophore and on the basis of growth rate, length of conidiophore, and shape and ornamentation of conidia.

Pitt accepts two main sections, Aspergilloides, with a swollen apex to the conidiophore, and Exilicaulis, with nonvesiculate conidiophores.

Aspergilloides

Series *glabra*—Spreading Colonies

Penicillium chermesianum Biourge. Growth spreading, velvety to matted floccose, occasionally producing red pigment in the colony reverse. Conidiophores swelling at the apex, borne on aerial hyphae; conidia elliptical, smooth. Occurs in soil. Sclerotia sometimes produced. Stolk and Samson (1983) regard this as a synonym of *E. meridianum* Scott (*P. decumbens* Thom).

Penicillium glabrum (Wehmer) Westling (*P. frequentans* Westling). Growth spreading, velvety, rich green with at first a yellow then deep brown reverse. Conidiophores long, smooth, swelling at the apex, conidiogenous cells (phialides) ampulliform; conidia globose, smooth to slightly rough, about 3 μm in diameter, produced in solid columns, often crusting or breaking away to form a cloud of dust. Very common.

Penicillium spinulosum Thom. Similar to *P. glabrum*, but has a loose felt or is slightly floccose, blue-green to gray-green, reverse white to cream or purplish; conidia globose, spinulose, 3–3.5 μm in diameter, and borne in loose columns.

Penicillium purpurescens (Sopp) Thom. Similar to *P. glabrum* with spreading velvety colonies, green, providing reddish purple to brown shades in reverse; conidia globose, distinctly rough, 3.5–4.5 μm in diameter.

Penicillium thomii Maire. Growth spreading, velvety, gray-green, but dominated by pink sclerotia; reverse yellow to brownish; conidiophores long, slightly rough, swelling at apex; conidia elliptical, 3–3.5 μm long, finely to coarsely roughened, in long irregular columns. Sclerotia profusely produced in fresh isolates. Common in soil. Stolk and Samson regard it as a synonym of *P. lividum* Westling.

Penicillium implicatum Biourge. Growth restricted, rich blue-green, with yellow or orange-red reverse; conidiophores smooth, swelling at apex, not very long, strictly monoverticillate, conidia elliptical to subglobose, smooth, 2.5–3 × 2–2.5 μm, so numerous as to form crusts that break when the petri dish is knocked. Common in soil-contaminated materials. Stolk and Samson (1983) place this in *P. citreonigrum* Dierckx.

Series *implicata*—Restricted Growth

Penicillium lividum Westling. Produces restricted colonies, a distinctive blue in color; conidiophores long, rough, swelling at apex; conidia elliptical, rough, 3–4 × 2.5–3 μm.

Penicillium fellutanum Biourge. Growth restricted, bluish green to gray-green, reverse pale; conidiophores smooth from matted aerial hyphae, not very long, usually swelling at apex, often with one or more branches, conidiogenous cells noticeably tightly packed; conidia subglobose to elliptical, smooth to finely roughened, 2.5–3 × 2–2.2 μm. Stolk and Samson place this in *E. cinnamopurpureum* Stolk and Scott (*P. dierckxii* Biourge). *Penicillium waksmanii* Zaleski has more spreading growth than *P. fellutanum* and globose conidia.

Exilicaulis

Series *restrictum*

The most common species here is *P. restrictum* Gilman and Abbott. Growth is very restricted, conidia very dark green or gray in color, but may be very sparse; reverse colorless; conidiophores very short; conidiogenous cells few, short ampulliform with well-swollen base tapering to a narrow neck; conidia globose and distinctly rough to echinulate, about 2.5 μm in diameter. Common in soil.

Other species, rare but rather similar, include *P. vinaceum* Gilman and Abbott and *P. roseopurpureum* Dierckx, both of which produce reddish pigment. *Penicillium dimorphosporum* Swart produces both smooth and rough conidia. Stolk and Samson place *P. restrictum*, *P. vinaceum*, and *P. roseopurpureum* in *E. euglaucum* (van Beyma) Stolk and Scott (syn. *P. citreonigrum*) and *P. dimorphosporum* in *E. stolkiae* Scott (syn. *P. velutinum* van Beyma).

Series *citreonigrum*

Penicillium citreonigrum Dierckx (syn. *P. citreoviride* Biourge = *E. euglaucum*). This species is important because it produces the mycotoxin citreoviridin on rice, which is considered to be the cause of cardiac beriberi in Japan. Raper and Thom regard it as a common spoilage organism.

Growth is fairly restricted, strongly wrinkled, velvety to floccose with yellow mycelium and pale gray-green conidia; reverse bright yellow to brown; conidiophores borne on aerial hyphae moderately long, 50–100 μm, smooth, not swelling at apex; conidia globose, 2–3 μm in diameter. Stolk and Samson regard it as a very wide species and include many monoverticillate species as synonyms.

Other monoverticillate species referred to elsewhere in this book are *P. capsulatum* Raper and Fennell, *P. turbatum* Westling, and *P. cyaneum* (Bainier and Sartory) Biourge.

3.4.3b. Biverticillata. This is a general term used to include all species with one level of branching below the conidiogenous cells. Species that show this branching form fall into two distinct groups, one with slightly irregular branching and the other with a very definite funnel-shaped pattern. The former produces conidiogenous cells with swollen bases tapering fairly abruptly to short necks. This is typical of most *Penicillium* species except for the group with regularly branched heads called Symmetrica by Raper and Thom, which have lance-shaped conidiogenous cells with a narrow base tapering gradually to the conidium-bearing tip. This latter group of species should perhaps be regarded as distinct from the rest of the genus.

Asymmetrica. Interpreting this in the narrow definition as used here, there are relatively few species and only two of importance.

Penicillium citrinum Thom is a very common species in soil and spoilage situations and is well documented as producing citrinin and other metabolites. Growth is restricted, dense, velvety, with a narrow margin and radial wrinkles, green to dull gray-green producing pale dull yellow drops; reverse yellow to dull orange with agar colored yellow or occasionally pinkish; conidiophores smooth with a cluster of three to five branches (cells supporting conidiogenous cells, metulae) of more or less equal length; conidia globose to subglobose, smooth, 2.5–3 μm in diameter, packed in solid divergent columns one to each branch. More floccose isolates with less yellow pigment were formerly referred to *P. steckii* Zaleski.

Penicillium corylophilum Dierckx. Growth somewhat restricted, blue-green to gray-green with reverse pale becoming dirty brown; on malt agar, reverse almost black; conidiophores smooth, bearing two or three unequal branches; conidia globose to elliptical, 2.5–3 μm in diameter, in roughly parallel tangled chains. The almost black reverse color on malt agar is distinctive. An occasional extra branch is seen at a lower level.

Some other species could be placed in this group, such as *P. paxilli* Bainier, which is common on decaying large basidiomycetes and has large heads with five more short, swollen appressed branches. Some species previously referred to the Biverticillata Symmetrica have ampulliform conidiogenous cells and might be better placed in this group, such as *P. herquei* Bainier and Sartory, *P. diversum* Raper and Fennell, and *P. verruculosum* Peyronel. *Penicillium raistrickii* G. Smith might also belong here.

Symmetrica (*Biverticillium*). A key to the Biverticillata Symmetrica (*Biverticillium*) is given in Section 4.2.2.

A large number of species can be placed here. Pitt (1979b) refers the ascosporic species to *Talaromyces* and refers the rest to his subgenus *Biverticillium*, which he divides into species that produce synnemata (section *coremigenum*) and nonsynnematous species (section *simplicium*).

Section *coremigenum*

These are interesting taxonomically and generally attractive to grow, but rare, the most common being *P. duclauxii* Delacroix and *P. claviforme* Bainier.

Section *simplicium*

Pitt (1979b) separates these on the rate of growth, length of conidiophores, and shape and marking of conidia. As in other large *Penicillium* sections, the species tend to overlap, but a few are fairly distinctive and common. Species confines are difficult to establish, but biochemical and other studies may eventually clarify the situation.

Series *miniolutea*

These species with rapid growth are some of the most common *Penicillia*.

Penicillium funiculosum Thom. Raper and Thom (1949) described this in very broad terms. As redescribed by Pitt and from our experience, this is a well-defined species.

Colonies spreading with a tough basal felt and deep aerial growth consisting of definite tufts and ropes of mycelium; conidia gray-green; reverse pink to reddish, but some colonies are colorless and some show yellow mycelium; conidiophores are very short, smooth, borne perpendicular to the ropes of hyphae; conidiogenous cells are lanceolate; conidia elliptical to subglobose with thick walls, smooth to finely roughened, 2.5–3.5 × 2–2.5 μm. It is common in soil.

Penicillium pinophilum Hedgcock is also very common. Colonies spreading, loose floccose to slightly tufted; yellow mycelium dominates the colonies; conidia dark green. Conidiophores often produced from the basal mat and overgrown by yellow hyphae; reverse colorless to golden brown or red; conidiophores long with numerous closely packed cells supporting the conidiogenous cells forming a wide funnel-shaped head; conidia globose to elliptical, smooth to finely roughened, 2.5–3 μm long.

Penicillium minioluteum Dierckx is similar to *P. pinophilum* but lacking in well-defined character with less pigment, more elliptical conidia, and narrower heads.

Penicillium purpurogenum Stoll has spreading, usually velvety colonies, sometimes floccose, heavy sporing, dark green, with yellow mycelium; reverse and agar strongly pigmented in red; long, smooth conidiophores bearing narrow conidial heads, with narrow cells supporting the conidiogenous cells and narrow conidiogenous cells; conidia heavy-walled, elliptical, usually rough. Some more delicate isolates, heavily pigmented and with smooth conidia, have been referred to *P. rubrum* Stoll. Common in soil and various other substrates.

Penicillium verruculosum Peyronel has spreading colonies, floccose, funiculose, or loose in texture, with mixed yellow and white mycelium; conidia yellow-green; reverse greenish or pale brown; conidial heads rather short and broad, conidiogenous cells ampulliform; conidia globose and very rough. It is very similar to *P. aculeatum* Raper and Fennell, which has less yellow mycelium, and to *P. herquei* Bainier and Sartory. The presence of ampulliform conidiogenous cells suggests that these three species should perhaps be placed with the asymmetric species, though the yellow mycelium is typical of the symmetric species.

Series *islandica*

Penicillium piceum Raper and Fennell produces thick, matted yellow colonies dominated by dark yellowish green conidia, with reverse in dull orange-brown. The mature heads are very characteristic, forming solid conical conidial masses like a spruce tree; conidiophores are short with conidiogenous cells incurved; conidia subglobose to elliptical, rough, 2.5–3 × 2.2–2.8 μm.

Penicillium islandicum Sopp produces restricted, thick to tufted colonies of orange and red mycelium with dark green conidia; reverse dull orange to red to red-brown; conidiophores often short, heavy-walled, conidiogenous cells some lanceolate but some tending to ampulliform; conidia heavy-walled, smooth, elliptical, 3–4.5 × 2–3 μm, borne in tangled chains. The species is common in soil and occurs in cereals, where it produces mycotoxins.

Other interesting species include *P. primulinum* Pitt, with large heads of many divergent cells supporting conidiogenous cells, much yellow mycelium, and poor growth on Czapek's agar; *P. variabile* Sopp, with distinctive gray-green velvety colonies, but conidial heads like *P. islandicum*; and *P. rugulosum* Thom, with restricted colonies, sometimes irregular branching, and very rough ellipsoidal conidia.

3.4.3c. Terverticillata. Three levels of branching including the conidiogenous cells. This is the largest group of *Penicillium* species, designated by Pitt (1979b) as the subgenus *Penicillium*. It includes many species of economic importance on account of their metabolites, spoilage activity, and toxin production. The best known is *P. chrysogenum* Thom, which produces penicillin. Some species are well defined, others less so.

Raper and Thom (1949) divided them according to their colony texture; Pitt (1979b) divided them according to growth rate and conidiophore and conidial shape and surface, but he also used other characters. A combination of the two approaches is taken here.

Since there is at present particular interest in the species that provide fascicles (bunches of conidiophores), these are described together, while important species of the rest of the subgenus are dealt with separately. This is admittedly an arbitrary separation, but proper relationships of the species remain uncertain.

A key to common species of the subgenus *Penicillium* is given in Section 4.3.

Nonfasciculate Species. *Penicillium chrysogenum* Thom has many synonyms (see Samson *et al.*, 1977a), the most notable of which is *P. notatum* Westling, which was the first reported to produce penicillin. It produces broadly spreading colonies, blue-green to green, with broad white margins, smooth velvety with yellow drops; reverse yellow diffusing into the agar; conidial heads branched once, twice, or more times, although often complex, usually with a divergent pattern; branches and conidiophores smooth, conidiophores long, produced from basal felt; conidiogenous cells ampulliform with a short neck; conidia globose to elliptical, smooth, 3–4 μm long, arranged in long, separate, divergent columns that may be twisted. It is a large species, often showing considerable variation from this basic description, being floccose, lacking yellow pigment, or producing pinkish colors. It is common as a spoilage organism.

Penicillium brevicompactum Dierckx (syn. *P. stoloniferum* Thom) grows very restrictedly, producing blue to gray-green velvety colonies, often wrinkled with a narrow white margin; reverse dull yellow to greenish to brown; a few pale to brownish drops; the surface may look almost granular under low magnification due to the very large conidial heads; conidial heads are complex, two or three times branched, with branches short, swollen especially at the apex, and very tightly appressed; conidiophores coarse, smooth to slightly rough; conidia globose, slightly rough, 3.5–4 μm in diameter, arranged in complexly tangled chains.

Penicillium digitatum Saccardo grows very thinly on Czapek's agar, but spreading, smooth velvety, and dusty on malt agar; dull yellowish green; conidial heads simple and fragmentary, all parts being large, irregular, often with only one branch; conidia ovate to cylindrical, smooth, 6–8 × 4–7 μm or larger. The heads are very fragile and difficult to mount, but a successful mount may look like the skeleton of a human hand. It occurs on citrus fruit, causing olive to yellowish green rot. The fruit shrivels up.

Penicillium italicum Wehmer causes blue or soft rot of citrus fruit and is seldom found elsewhere. Growth is restricted, pale gray-green, usually with definite fascicles or even synnemata at colony margins; produces a sweet odor; reverse pale brown to yellowish brown; three-stage branched heads are common, with branches narrowly arranged, large, and relatively few in number; conidiophores smooth; conidia elliptical, smooth, 4–5 × 2.5–3.5 μm or more. This differs from *P. digitatum* in rate of growth, color, more complex conidial heads, and production of a soft rot in fruit.

Penicillium expansum Link (the type species of the genus; very common). Growth is spreading, dull green with a white margin; reverse colorless in some strains to deep brown in others; conidial heads long and compact with branches appressed; conidiophores usually smooth; conidiogenous cells cylindrical with a short neck; conidia elliptical at first, but may become globose, 3–3.5 μm or longer. The elongate character of the heads in a good isolate is very distinctive. The species is very common and quite variable, occurring ubiquitously on many substrates, but it is well known for rotting apples and other fruit in storage.

Penicillium oxalicum Currie and Thom produces spreading growth, dull green, velvety with dense masses of conidia that break off in crusts when the culture is disturbed and give the cultures a silky appearance at low magnifications. The smooth conidia are elongate and large, 5–5.5 × 3–3.5 μm. Colonies may become floccose in culture. It is common in soil, especially from the tropics.

Penicillium pallidum G. Smith. Pitt (1979a) transferred this to the genus *Geosmithia.* It is a floccose funiculose species lacking green color, with elongate heads and large cylindrical conidia.

Penicillium camembertii Thom and *P. roquefortii* Thom. The *Penicillia* that grow on cheese were the first ones to be studied by Thom (1906) and initiated his interest in the genus. Recently, Samson *et al.* (1976, 1977b) have made a close study of these fungi and concluded that there are two main species with many synonyms. *Penicillium camembertii* (syn. *P. caseicola* Bainier) produces floccose white or slightly greenish colonies, with colorless to cream-colored reverse; conidial heads are irregular with several levels of branching; conidiophores rough; conidiogenous cells elongate, 9–14 × 2.5 μm; conidia elliptical to subglobose, smooth, produced in tangled chains up to $5 \times 4.5\ \mu m$. *Penicillium roquefortii* has broadly spreading colonies, very thin on Czapek's agar, blue green to dark green, smooth, velvety, reverse colorless to dark green or black; conidial heads large, asymmetric with three-stage branching; conidiophores rough, 4–6 μm in diameter; conidia smooth, globose or nearly so, large for *Penicillium*, 4–6 μm in diameter, in loose columns or tangled chains. Both these species are used in the ripening of cheeses (see Chapter 7). *P. roquefortii* is also found in spoilage situations such as silage and compost heaps where similar conditions prevail. It has recently been reported as producing mycotoxins (Wei and Liu, 1978).

Fasciculate Species. A few species are distinct in this area, but the morphology of the majority overlaps, and the various workers with the genus hold different species concepts. Since they are interesting biochemically as producers of toxins and other metabolites, efforts are being made to separate them by alternative means (Frisvad and Filtenborg, 1983; Bridge, 1985).

Penicillium expansum (see above).

Penicillium claviforme Bainier produces definite synnemata, with distinct pink stalks and gray-green heads. It is common on dung and often isolated from soil.

Penicillium corymbiferum/hirsutum and related species. Opinions differ as to the species concept in this area. Pitt (1979b) places them together as *P. hirsutum* Dierckx. Samson *et al.* (1976) have two species, *P. verrucosum* var. *corymbiferum* (Westling) Samson, Stolk, and Hadlock and *P. hordei* Stolk. Raper and Thom (1949) used the name *P. corymbiferum* Westling.

Growth is spreading yellow-green to blue-green, very fasciculate to very deeply fasciculate, with yellow-brown to blood-red drops; reverse deep reddish brown with the color diffusing into the agar; conidial heads are compact three-stage branched, with rough conidiophores; conidia smooth globose to subglobose, about 3 μm in diameter.

There appears to be some difference in growth pattern. Many isolates come from liliaceous bulbs, others from cereals. They are also isolated from soil. It is hoped that biochemical and SEM studies may clarify the situation.

Penicillium gladioli McCulloch and Thom. Isolates that attack gladiolus corms and produce pink sclerotia at 20°C with fasciculate growth and "terverticillate" conidial heads are placed in *P. gladioli.*

Penicillium griseofulvum Dierckx produces restricted colonies with abrupt margins, pale gray-green to light gray; granular with numerous fascicles; colorless drops, reverse pale dull yellow to brownish pink; color diffusing; conidial heads large, irregularly divergently branched; conidiophores smooth, sinuate; conidiogenous cells short with swollen base, 4.5–6 μm long, crowded; conidia smooth elliptical to subglobose, 2.5–3 μm long in divergent chains. Common in soil, food, and cereals. Produces patulin, which is now regarded as an important mycotoxin. Well-known synonyms are *P. patulum* Bainier and *P. urticae* Bainier.

Penicillium cyclopium/viridicatum/verrucosum group. These species constitute some of the most industrially interesting species in the genus *Penicillium.* They are all very similar, and there is an extensive literature on their secondary metabolites.

They all produce somewhat restricted fasciculate colonies, often variable or showing sectors. This variation within the isolates contributes to the confusion. Conidial heads tend to be similar, consisting of rough conidiophores, somewhat appressed branches, ampulliform conidiogenous cells with short necks, and globose to subglobose conidia borne in medium-size tangled heads. Conidia are usually smooth to very finely roughened as seen under the light microscope. They occur mostly in soil and cause food spoilage, especially of cereals. Samson *et al.* (1976) and Pitt (1979b) produced different synonymy.

There follow a few species concepts that might be applied but that may require modification as a result of work currently in progress.

Penicillium echinulatum Raper and Thom ex Fassatiová has darkish, somewhat spreading colonies and very rough globose conidia.

Penicillium puberulum Bainier is on the edge of the group with very fine fasciculate, low, dense colonies often appearing velvety, conidia bluish gray to greenish gray, clear exudate, and only very finely roughened conidiophores. Some workers consider it near to *P. brevicompactum* or even a colorless *P. chrysogenum.*

Penicillium cyclopium Westling as placed by Raper and Thom (1949) = *P. aurantiogriseum* Dierckx according to Pitt (1979b) and *P. verrucosum* var. *cyclopium* Samson, Stolk, and Hadlock by Samson *et al.* (1976). It is typical of the group. It has moderately rapid growth and blue-green conidia.

Penicillium viridicatum Westling as described by Raper and Thom (1949) = *P. viridicatum* Westling for fast-growing isolates and *P. verrucosum* Dierckx for slow-growing isolates according to Pitt (1979b), while Samson *et al.* (1976) place it in *P. verrucosum* Dierckx. These isolates are slower growing than the typical *P. cyclopium* and produce yellow-green conidia.

Frisvad and Filtenborg (1983) and Ciegler *et al.* (1981) have studied the secondary metabolites and divided these latter accordingly into distinctive groups *P. viridicatum* I, II, III, and IV and *P. viridicatum* I and II, respectively.

Penicillium olivicolor Pitt. Isolates that lack a green pigment and have olive or ochraceous conidia are variously referred to as *P. olivicolor* Pitt, *P. ochraceum* Bainier, or *P. verrucosum* var. *ochraceum* (Bainier) Samson, Stolk, and Hadlock.

Penicillium crustosum Thom. Some isolates produce darkish green, more spreading colonies, spore very profusely to such an extent that they appear almost velvety, and form crusts that break away in irregular pieces when the petri dish is tapped. It seems likely that these constitute a good separate species.

3.4.3d. Divaricata. Some isolates are branched in a spreading, divergent, and irregular pattern. They have been variously placed within the schemes of classification. Many of them are certainly not related. Pitt (1979b) placed most of them in his subgenus *Furcatum,* many of which have one level of branches below the conidiogenous cells.

Series janthinellum of Section divaricatum. Conidial heads of an irregular pattern and growth broadly spreading.

Penicillium janthinellum Biourge [syn. *P. simplicissimum* (Oudemans) Thom]. Stolk and Samson (1983) regard these as synonyms, while Pitt (1979b) places them far apart. This is a very variable group with broadly floccose, spreading colonies, radially furrowed, pale gray-green, with reverse varying from colorless to yellowish greenish or showing reddish pigments; conidiophores finely to coarsely rough, conidial heads irregular, often consisting of a simple whorl of cells supporting the conidiogenous cells; conidiogenous cells characteristic, very divergent, tapering abruptly to a long, slender tip; conidia mostly elliptical to subglobose, delicately roughened, arranged in delicate divergent twisted chains.

Penicillium ochrochloron Biourge is a common spoilage organism that appears in unusual situations such as copper-treated military equipment or chemical solutions. It is very similar to *P. janthinellum,* though with very reduced conidial heads.

Penicillium lilacinum Thom is somewhat similar to these, but the colonies lack green color, being dull pinkish. It has been transferred to *Paecilomyces lilacinus* (Thom) Samson.

Series P. canescens. Growth is restricted, with divaricate branching; conidiophores or conidia or both are rough; conidiogenous cells ampulliform with a short, broad neck; conidia are rough. Two species are common, *P. canescens* Sopp and *P. janczewskii* Zaleski.

Penicillium canescens Sopp produces gray-green colonies; rough conidiophores and globose, smooth conidia in distinct columns. The name is accepted by Stolk and Samson (1983).

Penicillium janczewskii Zaleski (syn. *P. nigricans* Bainier) has restricted thick, velvety colonies, almost pure gray to dark gray in age; orange-brown reverse; divaricate heads, often with tetrads of cells supporting the conidiogenous cells (metulae); conidiophores smooth and conidia globose and very rough, 3–3.5 μm in diameter.

Penicillium melinii Thom, which may or may not belong here, has both rough conidiophores and rough conidia.

4. KEYS

4.1. Key to the Species of *Acremonium* and Related Species of Industrial Importance

1		Conidiogenous cells disposed in whorls	*Verticillium lecanii*
		Conidiogenous cells disposed singly or in pairs	2
2	(1)	Conidiogenous cells with a distinct apical collarette	*Phialophora gregata*
		Conidiogenous cells without a distinct collarette	3
3	(2)	Conidiogenous cells occurring mainly singly on the vegetative mycelium without a distinct conidiophore, chlamydospores absent	4
		Conidiogenous cells occurring mainly on the distal parts of a simple branched conidiophore, chlamydospores present	8
4	(3)	Conidia after secession remaining at the mouth of the conidiogenous cell in a cluster or "head"	5
		Conidia after secession remaining attached to each other end-to-end in a "false" chain	7
5	(4)	Cultures very slow-growing (3 mm in diameter in 10 days)	*Acremonium coenophialum*
		Cultures faster-growing (8–25 mm in diameter in 10 days)	6
6	(5)	Colony white to pink or orange	*A. strictum*
		Colony yellow, the reverse and surrounding agar intensely so	*A. chrysogenum*
7	(4)	Conidia in "false" chains, of two shapes, fusiform and spherical	*A. fusidioides*
		Conidia in long chains, dark-pigmented	*Gliomastix luzulae*
8	(3)	Conidia strongly curved	*A. recifei*
		Conidia long-ellipsoidal	*A. crotocinigenum*

4.2. Key to Genera and Subgenera with "*Penicillium*-like" Anamorphs

1		Producing ascomata	2
		Lacking ascomata	3 *Penicillium*
2	(1)	Hard ascomata	*Eupenicillium*
		Soft ascomata	*Talaromyces*
3	(1)	Lacking branches	Monoverticillata (Aspergilloides)
		Branches present	4
4	(3)	One level of branching	5 Biverticillata
		More than one level of branching	6
5	(4)	Branches of uneven length and irregular arrangement	Asymmetrica (*Furcatum*)
		Branches of even length in regular funnel-shaped arrangement	Symmetrica (*Biverticillium*)
6	(4)	Definite branched head	Terverticillata (*Penicillium*)
		Irregularly arranged branching system	Divaricata (part of *Furcatum*)

4.2.1. Key to Monoverticillata

1		Apex of conidiophore swollen	2 Aspergilloides
		Apex of conidiophore not swollen	11 Exilicaulis

4.2.1a. Aspergilloides

2	(1)	Rapid growth	3 Series *glabra*
		Restricted growth	8 Series *implicata*
3	(2)	Long conidiophores, globose conidia	4
		Short conidiophores, elliptical conidia	*P. chermesianum*
4	(3)	Velvety growth	5
		Floccose growth, rough conidia	*P. spinulosum*
5	(4)	Smooth conidia	*P. glabrum*
		Rough conidia	6
6	(5)	Conidia blue in mass	*P. lividum*
		Conidia green in mass	7
7	(6)	Conidiophores rough, conidia elliptical, pale sclerotia	*P. thomii*
		Conidiophores smooth, conidia globose	*P. purpurescens*
8	(2)	Long conidiophores, conidia elliptical, blue in mass, rough-walled, strictly monoverticillate	*P. lividum*
		Not so	9
9	(8)	Moderate conidiophores, conidia elliptical, smooth-walled, strictly monoverticillate	*P. implicatum*
		Short, sometimes branched conidiophores	10
10	(9)	Growth restricted, conidia elliptical to sub-globose	*P. fellutanum*
		Growth spreading, globose conidia	*P. waksmanii*

4.2.1b. Exilicaulis

11	(1)	Colonies restricted, conidiophores very short	*P. restrictum* series
		Colonies spreading, conidiophores over 60 μm long	*P. citreonigrum* series

4.2.2. Key to Biverticillata Symmetrica (*Biverticillium*)

1		Colonies producing synnemata	Section *coremigenum*
		Colonies not producing synnemata	2 Section *simplicium*
2	(1)	Growth rapid	3 Series *miniolutea*
		Growth restricted	7 Series *islandica*

4.2.2a. Series *miniolutea*

3	(2)	Conidia definitely rough	4
		Conidia smooth or nearly so	5
4	(3)	Conidia elliptical, producing red pigment	*P. purpurogenum*
		Conidia globose	*P. verruculosum*

5	(3)	Colonies very funiculose	*P. funiculosum*
		Not so	6
6	(5)	Colonies dominated by yellow-pigmented hyphae	*P. pinophilum*
		Colonies lacking in definite characters	*P. minioluteum*

4.2.2b. Series *islandica*

7	(2)	Conidia forming conical masses	*P. piceum*
		Not so	8
8	(7)	Conidia globose	9
		Conidia elliptical	10
9	(8)	Conidia smooth	*P. primulinum*
		Conidia rough	*P. aculeatum*
10	(8)	Conidia rough	11
		Conidia smooth, yellow or orange mycelium	12
11	(10)	Producing purple red pigmentation and narrow funnel-shaped heads	*P. purpurogenum*
		Yellow or orange mycelium	*P. rugulosum*
12	(10)	Compact, grayish to blue-green conidia, mycelium orange to brown	*P. islandicum*
		Grayish green, colonies velvety, mycelium yellow	*P. variabile*

4.3. Key to Common Species of *Penicillium*

4.3.1. Nonfasciculate Species

1		Species with smooth conidiophores	2
		Species with rough conidiophores	8
2	(1)	Conidia globose	3
		Conidia cylindrical	4
3	(2)	Spreading colonies, divergent columns of conidia	*P. chrysogenum*
		Restricted colonies, compact tangled heads	*P. brevicompactum*
4	(2)	Rotting citrus fruit	5
		Not so	6
5	(4)	Very restricted growth on Czapek's agar	*P. digitatum*
		Growth on Czapek's agar—gray rot	*P. italicum* (Fasciculata)
6	(4)	Gray restricted growth, spreading heads, very short conidiogenous cells, 6 μm or less	*P. griseofulvum*
		Not so	7
7	(6)	Spreading growth, silky colonies, very large conidia	*P. oxalicum*
		Less spreading, blue rot of apples, fasciculate colonies	*P. expansum*
8	(1)	Floccose colonies, white or very pale green	*P. camembertii*
		Not so	9
9	(8)	Funiculose colonies not green	*P. pallidum* (*Geosmithia*)
		Not so	10

10 (9)	Velvety colonies, spreading poor growth on Czapek's agar, black reverse	*P. roquefortii*
	Fasciculate colonies	11 Fasciculata

4.3.2. Fasciculate Species

11 (10)	Smooth conidiophores	12
	Rough conidiophores	15
12 (11)	Blue rot of apples	*P. expansum*
	Not rotting apples	13
13 (12)	Soft rot of citrus	*P. italicum*
	Not rotting citrus	14
14 (13)	Gray restricted colonies, spreading heads, conidiogenous cells short, 6 μm or less	*P. griseofulvum*
	Dense velvety to fasciculate colonies, bluish gray	*P. puberulum*
15 (11)	Conidia rough	*P. echinulatum*
	Conidia smooth	16
16 (15)	Producing pink sclerotia	*P. gladioli*
	Not so	17
17 (16)	Producing coremia with pink stalks	*P. claviforme*
	Not as above, with fasciculate colonies, smooth globose conidia	18 *P. cyclopium* group
18 (17)	Separation of these is under investigation. Distinctive characters of some possible species are:	
	Yellow-green restricted colonies	*P. verrucosum/viridicatum*
	Blue-green spreading colonies	*P. cyclopium*
	Blue-green colonies with wine-red drops on liliaceus bulbs	*P. corymbiferum*
	Spreading colonies, conidia forming crusts on malt agar	*P. crustosum*

5. APPENDIXES

5.1. Names Used in This Book for Species of *Acremonium* and Related Genera

Names	Cross reference with other names
Cephalosporium acremonium	See *A. kiliense*; *A. strictum*
A. alternatum	—
C. aphidicola	see *Verticillium lecanii*
A. butyri	*Nectria viridescens*
C. caerulens	?*Fusarium* sp.
Hymenula cerealis	*C. gramineum*
A. chrysogenum	*C. chrysogenum*
A. coenophialum	—
A. crotocinigenum	*C. crotocinigenum*

Names	Cross reference with other names
A. diospyri	*C. diospyri*
A. fungicola	*N. violacea*
A. furcatum	*C. furcatum*
A. fusidioides	*Paecilomyces fusidioides*
C. gramineum	See *Hymenula cerealis*
Phialophora gregata	*C. gregata*
A. implicatum	*A. terricola*
A. kiliense	*C. acremonium*; *C. kiliense*
Verticillium lecanii	*C. aphidicola*
Gliomastix luzulae	*A. luzulae*; *Sagrahamala luzulae*
Emericellopsis minima	*C. salmosynnemata*
C. mycophilum	Status uncertain
Monocillium nordinii	*C. nordinii*
A. recifei	*C. recifei*
C. roseum	See *A. rutilum*
A. rutilum	*C. roseum*
C. salmosynnematum	See *E. minima*
A. sordidulum	*S. sordidula*
A. strictum	*C. acremonium*; *C. kiliense*
A. terricola	See *A. implicatum*; *P. terricola*
A. zonatum	*C. zonatum*

5.2. Names Used in This Book for Species with "*Penicillium*-like" Anamorphs

Names	Cross reference with other names
Penicillium amagasakiense	See *Paecilomyces marquandii*
P. arenicola	*P. canadense*
P. atramentosum	—
P. aurantiogriseum	*P. cyclopium*; *P. verrucosum* var. *cyclopium*
P. avellaneum	*P. ingelheimense*; *Merimbla ingelheimense*; *Talaromyces avellaneus*
Eupenicillium baarnense	*P. baarnense*; *P. vanbeymae*; *P. turbatum*
E. brefeldianum	*P. brefeldianum*; *P. dodgei*; ?*P. simplicissimum*; *P. janthinellum*; *E. ehrlichii*; *E. javanicum*
P. brevicompactum	*P. stoloniferum*
P. brunneum	—
P. camembertii	*P. caseicola*; *P. candidum*
P. candidum	See *P. camembertii*
P. canescens	—
P. capsulatum	—

Names	Cross reference with other names
P. casei	?*P. roquefortii*
P. caseicola	See *P. camembertii*
P. charlsii	See *P. fellutanum*; ?*E. cinnamopurpureum*; ?*P. dierckxii*
P. chermesinum	*E. meridianum*; *P. decumbens*
P. chrysogenum	*P. notatum*; *P. flavidomarginatum*
E. cinnamopurpureum	*P. phoeniceum*; *P. cinnamopurpureum*; ?*P. dierckxii*
P. citreonigrum	*P. citreoviride*; *P. toxicarium*; *E. euglaucum*
P. citreoviride	See *P. citreonigrum*
P. citrinum	*P. steckii*
P. claviforme	—
P. clavigerum	?*P. duclauxii*
P. commune	?*P. puberulum*
P. concentricum	?*P. italicum*
P. corylophilum	—
P. corymbiferum	?*P. hirsutum*; ?*P. hordei*; *P. verrucosum* var. *corymbiferum*
E. crustaceum	?*P. gladioli*; *P. nilense*; *P. molle*
P. crustosum	?*P. verrucosum* var. *cyclopium*
P. cyaneofulvum	See *P. griseoroseum*
P. cyaneum	—
P. cyclopium	See *P. aurantiogriseum*; *P. verrucosum* var. *cyclopium*
P. daleae	—
P. dangeardii	*P. vermiculatum*; see *T. flavus*
P. decumbens	?*P. chermesinum*; *E. meridianum*
P. dierckxii	See *E. cinnamopurpureum*
P. digitatum	—
P. dimorphosporum	*E. stolkiae*; *P. velutinum*
P. diversum	—
P. dodgei	*P. brefeldianum*; see *E. brefeldianum*; *E. javanicum*
P. duclauxii	See *P. clavigerum*
P. dupontii	See *T. thermophilus*
P. echinulatum	*P. cyclopium* var. *echinulatum*
E. egypticum	*P. egypticum*; *P. nilense*; *E. crustaceum*
Geosmithia emersonii	*P. emersonii*
P. emersonii	See *Geosmithia emersonii*
P. emmonsii	*P. emmonsii*; see *T. stipitatus*
P. erythromollis	—
E. euglaucum	*P. citreonigrum*

Names	Cross reference with other names
P. expansum	?*P. glaucum*
P. fellutanum	*P. charlsii*; ?*E. cinnamopurpureum*
P. fenneliae	—
P. flavidostipitatum	—
T. flavus	*P. dangeardii*; *T. vermiculatus*; *P. vermiculatum*
P. frequentans	See *P. glabrum*
P. funiculosum	—
T. galapagensis	*P. galapagense*
P. glabrum	*P. frequentans*
P. gladioli	See *E. crustaceum*; Stolk and Samson accept
P. glaucum	See *P. expansum*
T. gossypii	*P. gossypii*
P. granulatum	—
P. griseofulvum	*P. patulum*; *P. urticae*
P. griseoroseum	*P. cyaneofulvum*
T. helicus	*P. spirillum*; *P. helicum*
P. herquei	—
E. hirayamae	*P. hirayamae*; *E. euglaucum*
P. hirsutum	?*P. corymbiferum*; *P. hordei*
P. hordei	?*P. corymbiferum*; ?*P. hirsutum*
P. humuli	—
P. implicatum	?*E. euglaucum*
P. isariiforme	—
P. islandicum	—
P. italicum	?*P. concentricum*
P. janczewskii	*P. nigricans*; *P. canescens*
P. janthinellum	?*P. simplicissimum*; ?*E. javanicum*
E. javanicum	*P. javanicum*; ?*P. janthinellum*; *P. simplicissimum*; ?*P. brefeldianum*
P. jensenii	—
P. kloeckeri	See *T. wortmanii*
E. lapidosum	*P. lapidosum*
G. lavendula	*P. lavendelum*
P. lavendulum	See *G. lavendula*
P. lehmanii	*T. spiculisporus*; *P. spiculisporum*; see *T. trachyspermus*
P. lignorum	—
Paec. lilacinus	*P. lilacinum*
P. lividum	—
T. luteus	*P. udagawawa*; *P. luteum*
P. madritii	—

Names	Cross reference with other names
P. marneffei	—
Paecilomyces marquandii	—
P. megasporum	—
P. melinii	—
E. meridianum	?*P. decumbens*; ?*P. chermesinum*
P. miczynskii	—
P. minioluteum	—
E. molle	*P. molle*; *E. crustaceum*
P. nigricans	See *P. janczewskii*
P. nilense	See *E. crustaceum*
P. notatum	See *P. chrysogenum*
P. novaezeelandii	—
P. ochraceum	See *P. olivicolor*; *P. verrucosum* var. *ochraceum*
P. ochrochloron	—
E. ochrosalmoneum	*P. ochrosalmoneum*
P. olivicolor	*P. ochraceum*; *P. verrucosum* var. *ochraceum*
E. osmophilum	*P. osmophilum*
P. oxalicum	—
P. pallidum	See *G. putterillii*
T. panasenkoi	*P. panasenkoi*
P. papuanum	See *E. parvum*
E. parvum	*P. papuanum*; *P. parvum*
P. patulum	See *P. griseofulvum*
P. paxilli	—
P. phoeniceum	See *E. cinnamopurpureum*
P. piceum	—
E. pinetorum	*P. pinetorum*
P. pinophilum	—
P. primulinum	—
P. puberulum	?*P. commune*; *P. lanosum*
P. pulvillorum	See *P. simplicissimum*
P. purpurescens	—
P. purpurogenum	*P. rubrum*
G. putterillii	*P. pallidum*
P. raistrickii	?*P. raciborskii*
P. resedanum	—
P. restrictum	?*P. citreonigrum*
E. reticulisporum	*P. reticulisporum*; ?*E. abidjanum*
P. rolfsii	?*P. ochrochloron*
P. roquefortii	?*P. casei*

Names	Cross reference with other names
P. roseopurpureum	?*E. euglaucum*; ?*P. citreonigrum*
P. rubrum	See *P. purpurogenum*
P. rugulosum	—
P. sclerotiorum	—
P. simplicissimum	?*P. janthinellum*; ?*E. javanicum*; ?*P. pulvillorum*
T. spiculisporum	See *T. trachyspermus*
P. spinulosum	—
P. steckii	See *P. citrinum*
T. stipitatus	*P. stipitatum*; *P. emmonsii*
P. stoloniferum	See *P. brevicompactum*
T. thermophilus	*P. dupontii*
P. thomii	?*P. lividum*
P. toxicarium	See *P. citreonigrum*; ?*P. lividum*
T. trachyspermus	*T. spiculisporum*; *P. lehmanii*; *P. spiculisporum*
P. turbatum	See *E. baarnense*
P. udagawae	See *T. luteus*
P. urticae	See *P. griseofulvum*
P. vanbeymae	See *E. baarnense*
P. variabile	—
P. velutinum	?*E. stolkiae*; ?*P. dimorphosporum*
T. vermiculatus	See *T. flavus*
P. verrucosum	?*P. viridicatum*
P. verrucosum var. *corymbiferum*	See *P. corymbiferum*
P. verrucosum var. *cyclopium*	See *P. aurantiogriseum*
P. verrucosum var. *melanochloron*	—
P. verrucosum var. *ochraceum*	See *P. olivicolor*
P. verruculosum	—
P. vinaceum	?*E. euglaucum*
P. viridicatum	See ?*P. verrucosum*
P. waksmanii	—
T. wortmannii	*P. wortmannii*; *P. kloeckeri*
E. zonatum	*P. zonatum*; ?*E. javanicum*

Acknowledgments. The work on *Penicillium* was carried out in conjunction with work on SERC Contract No. SO/17/84: "Systematics of microfungi of biotechnological and industrial importance." We are also grateful to Z. Lawrence of MAFF for the scanning electron micrographs and to our colleagues, especially D. L. Hawksworth, for their help and advice.

REFERENCES

Abe, S., 1956, Studies on the classification of the Penicillia, *J. Gen. Appl. Microbiol.* **2:**1–344.

Anné, J., 1985, Taxonomic implications of hybridization of *Penicillium* protoplasts, in: *Advances in Penicillium and Aspergillus Systematics* (R. A. Samson and J. I. Pitt, eds.), Plenum Press, New York and London, pp. 337–350.

Benjamin, C. R., 1955, Ascocarps of *Aspergillus* and *Penicillium*, *Mycologia* **47:**669–687.

Biourge, P., 1923, Les moisissures du groupe *Penicillium* Link, *Cellule* **33:**7–331.

Bridge, P. D., 1985, An evaluation of some physiological and biochemical methods as an aid to the characterization of species of *Penicillium* subsection *Fasciculata*, *J. Gen. Microbiol.* **131:**1887–1895.

Bridge, P. D., and Hawksworth, D. L., 1984, The APIZYM enzyme testing system as an aid to the identification of *Penicillium* isolates, *Microbiol. Sci.* **1:**232–234.

Bridge, P. D., Hawksworth, D. L., Kozakiewicz, Z., Onions, A. H. S., Paterson, R. R. M., and Sackin, M. J., 1985, An integrated approach to *Penicillium* systematics, in: *Advances in Penicillium and Aspergillus Systematics* (R. A. Samson and J. I. Pitt, eds.), Plenum Press, New York and London, pp. 281–309.

Ciegler, A., Lee, L. S., and Dunn, J. J., 1981, Production of naphthoquinone mycotoxins and taxonomy of *Penicillium viridicatum*, *Appl. Environ. Microbiol.* **42:**446–449.

Clare, B. G., Flentje, N. T., and Atkinson, M. R., 1968, Electrophoretic patterns of oxidoreductases and other proteins as criteria in fungal taxonomy, *Aust. J. Biol. Sci.* **21:**275–295.

Corda, A. C. J., 1839, *Icones Fungorum 3*, Prague.

Cowan, S. T., 1974, *Cowan and Steel's Manual for the Identification of Medical Bacteria*, 2nd ed., Cambridge University Press, Cambridge.

Dickinson, C. H., 1968, *Gliomastix* Guéguen, *Mycol. Pap.* **115:**1–24.

Dierckx, R. P., 1901, Un essai de revision du genre *Penicillium* Link, *Ann. Soc. Sci. Bruxelles* **25:**83–89.

Evans, H. C., and Samson, R. A., 1982, Entomogenous fungi from the Galápagos Islands, *Can. J. Bot.* **60:**2325–2333.

Fassatiová, O., 1977, A taxonomic study of *Penicillium* series *Expansa* Thom emend. Fassatiová, *Acta Univ. Carol. Biol.* **12:**283–335.

Frisvad, J., and Filtenborg, O., 1983, Classification of terverticillate penicillia based on profiles of mycotoxins and other secondary metabolites, *Appl. Environ. Microbiol.* **46:**1301–1310.

Gams, W., 1968, Typisierung der Gattung *Acremonium*, *Nova Hedwigia* **16:**141–145.

Gams, W., 1971, *Cephalosporium-artige Schimmelpilze (Hyphomycetes)*, Gustav Fischer Verlag, Stuttgart.

Gams, W., 1975, *Cephalosporium*-like Hyphomycetes: Some tropical species, *Trans. Br. Mycol. Soc.* **64:**389–404.

Gams, W., 1978, Connected and disconnected chains of phialoconidia and *Sagenomella* gen. nov. segregated from *Acremonium*, *Persoonia* **10:**97–112.

Gams, W., and Lacey, J., 1972, *Cephalosporium*-like Hyphomycetes: Two species of *Acremonium* from heated substrates, *Trans. Br. Mycol. Soc.* **59:**519–522.

Hawksworth, D. L., 1984, Recent changes in the international rules affecting the nomenclature of fungi, *Microbiol. Sci.* **1:**18–21.

Hawksworth, D. L., 1985, Typification and citation of the generic name *Penicillium*, in: *Advances in Penicillium and Aspergillus Systematics* (R. A. Samson and J. I. Pitt, eds.), Plenum Press, New York and London, pp. 3–8.

Kreger-van Rij, N. J. W. (ed.), 1984, *The Yeasts: A Taxonomic Study*, 3rd ed., Elsevier, Amsterdam.

Link, H. F., 1809, Observations in ordines plantarum naturales, Dissertatio 1ma, *Berl. Gesamte Naturkd.* **3:**1–42.

Minter, D. W., Kirk, P. M., and Sutton, B. C., 1982, Holoblastic phialides, *Trans. Br. Mycol. Soc.* **79:**75–93.

Minter, D. W., Kirk, P. M., and Sutton, B. C., 1983a, Thallic phialides, *Trans. Br. Mycol. Soc.* **80**:39–66.

Minter, D. W., Sutton, B. C., and Brady, B. L., 1983b, What are phialides anyway?, *Trans. Br. Mycol. Soc.* **81**:109–120.

Onions, A. H. S., Bridge, P. D., and Paterson, R. R. M., 1984, Problems and prospects for the taxonomy of *Penicillium, Microbiol. Sci.* **1**:185–189.

Paterson, R. R. M., and Hawksworth, D. L., 1985, Detection of secondary metabolites in dried cultures of *Penicillium* preserved in herbaria, *Trans. Br. Mycol. Soc.* **85**:95–100.

Pidoplichko, N. M., 1972, *Penicillia,* Polygraph, Kiev.

Pitt, J. I., 1979a, *Geosmithia* gen. nov. for *Penicillium lavendulum* and related species, *Can. J. Bot.* **57**:2021–2030.

Pitt, J. I., 1979b, *The Genus Penicillium and Its Teleomorphic States Eupenicillium and Talaromyces,* Academic Press, London.

Polonelli, L., Castagnola, M., D'Urso, C., and Morace, G., 1985, Serological approaches for identification of *Aspergillus* and *Penicillium* species, in: *Advances in Penicillium and Aspergillus Systematics* (R. A. Samson and J. I. Pitt, eds.), Plenum Press, New York and London, pp. 267–280.

Ramirez, C., 1982, *Manual and Atlas of the Penicillia,* Elsevier, Amsterdam.

Raper, K. B., and Thom, C., 1949, *A Manual of the Penicillia,* Williams and Wilkins, Baltimore.

Samson, R. A., Stolk, A. C., and Hadlock, R., 1976, Revision of the subsection Fasciculata of *Penicillium* and some allied species, *Stud. Mycol.* **11**:1–47.

Samson, R. A., Hadlock, R., and Stolk, A. C., 1977a, A taxonomic study of the *Penicillium chrysogenum* series, *Antonie van Leeuwenhoek J. Microbiol. Serol.* **43**:169–175.

Samson, R. A., Eckardt, C., and Orth, R., 1977b, A taxonomy of *Penicillium* species from fermented cheeses, *Antonie van Leeuwenhoek J. Microbiol. Serol.* **43**:341–350.

Samuels, G. J., 1973, Myxomyceticolous species of Nectria, *Mycologia* **65**:401–420.

Stolk, A. C., and Samson, R. A., 1972, The genus *Talaromyces* and related genera II, *Stud. Mycol.* **2**:1–65.

Stolk, A. C., and Samson, R. A., 1983, The ascomycete genus *Eupenicillium* and related *Penicillium* anamorphs, *Stud. Mycol.* **23**:1–149.

Stolk, A. C., and Scott, De B., 1967, Studies on the genus *Eupenicillium* Ludwig. I. Taxonomy and nomenclature of *Penicillia* in relation to their sclerotial ascocarpic states, *Persoonia* **4**:391–405.

Subramanian, C. V., 1972, Conidial chains, their nature and significance in the taxonomy of hyphomycetes, *Curr. Sci.* **41**:43–49.

Thom, C., 1906, Fungi in cheese ripening: Camembert and Roquefort, *Bull. Bur. Anim. Ind. U.S. Dept. Agric.* **82**:1–39.

Thom, C., 1910, Cultural studies of species of *Penicillium, Bull. Bur. Anim. Ind. U.S. Dept. Agric.* **118**:1–109.

Thom, C., 1930, *The Penicillia,* Williams and Wilkins, Baltimore.

Tubaki, K., 1955, Studies on the Japanese Hyphomycetes. II. Fungicolous group, *Nagaoa* **5**:15.

Udagawa, S., and Horie, Y., 1973, Some *Eupenicillium* from soils of New Guinea, *Trans. Mycol. Soc. Jpn.* **14**:370–387.

Voss, E. G., Burdet, H. M., Chaloner, W. G., Demoulin, V., Hiepko, P., McNeill, J., Meikle, R. D., Nicolson, D. H., Rollins, R. C., Silva, P. C., and Greuter, W. (eds.), 1983, International Code of Botanical Nomenclature adopted by the Thirteenth International Botanical Congress, Sydney, August 1981, *Regnum Vegetabile,* Vol. 111, Bohn, Scheltema and Holkema, Utrecht.

Vuillemin, P., 1904, Les Isaria du genre *Penicillium, Soc. Mycol. Fr. Bull. Trimest.* **20**:214–222.

Wei, R., and Liu, G., 1978, PR toxin production in different *Penicillium roquefortii* strains, *Appl. Environ. Microbiol.* **35**:797–799.

Westling, R., 1911, Uber die grünen Spezies der Gattung *Penicillium, Arkiv für Botanik.* **11**:1–156.

Morphology and Physiology of *Penicillium* and *Acremonium* 2

MAURICE O. MOSS

1. GROWTH AND MORPHOGENESIS

1.1. Growth on Solid Media: Colony Morphology

Both *Penicillium* and *Acremonium* (=*Cephalosporium*) are common and widespread in soil; they are particularly associated with senescent and dead plant material, so that the majority of floristic studies of a wide range of soil types and plant surfaces include species of these genera in the lists of fungi isolated. Although the two genera are especially associated through the production of β-lactam antibiotics by one, or at the most a few, species, they differ markedly in the visual impact that they make when growing as colonies on plates of solid media. On the one hand, mature colonies of *Penicillium* are characterized by the production of large numbers of dry conidiospores that are usually gray-green to blue-green, although strains of *P. camembertii* may remain persistently white and *P. humuli* has a distinctly brownish coloration. Colonies of *Acremonium*, on the other hand, are characterized by their pale colors—often turning pink or orange, although *A. butyri* produces pale yellow-green colonies—and the production of wet or slimy spores.

The production of dry or slimy spores is reflected in the frequency with which *Penicillium* species are isolated from the atmosphere compared with *Acremonium*. Indeed, new species of *Penicillium* are still being described from air-exposure plates (Ramirez and Martinez, 1980; Ramirez *et al.*, 1980). Because of the ease with which their spores are dispersed through the atmosphere, cultures of *Penicillium* need to be handled with particular care in the laboratory.

MAURICE O. MOSS • Department of Microbiology, University of Surrey, Guildford GU2 5XH, England.

The metabolic competence of *Penicillium* and *Acremonium* is reflected in the fact that the majority of species grow well on a relatively simple medium such as malt extract agar and many will grow on a defined medium such as Czapek's solution agar. The variation in colony morphology from one species to another on a particular medium is sufficiently diverse to provide useful characteristics for the identification of isolates. Some of the media most widely used for this purpose are listed in Table I.

An extensive formulary of fungal culture media is given by Booth (1971). J. I. Pitt (1979) describes a small range of media and growth conditions that he used to provide useful colony characteristics for the identification of *Penicillia.*

Glycerol nitrate agar (Table I) has a water activity of 0.93, and glycerol was found to be the humectant of choice for providing a usefully diagnostic medium (J. I. Pitt, 1973). Mislivec (1975) showed that the incorporation of 20 ppm of botran (2,6-dichloro-4-nitroaniline) into Czapek–Dox and malt extract agar reproducibly enhanced fascicle formation (see Section 1.3.1), making it easier to use this characteristic in identification.

Table I. Media Most Widely Used for the Cultivation of *Penicillium* and *Acremonium*

Medium	Components (g/liter)
Czapek's solution agar	$NaNO_3$, 3.0; K_2HPO_4, 1.0; $MgSO_4 \cdot 7H_2O$, 0.5; KCl, 0.5; $FeSO_4 \cdot 7H_2O$, 0.01; sucrose, 30; agar, 15
Malt extract agar	Malt extract, powdered, 20; peptone, 1; glucose, 20; agar 15
Czapek concentrate	$NaNO_3$, 300; KCl, 50; $MgSO_4 \cdot 7H_2O$, 5; $FeSO_4 \cdot 7H_2O$, 1.0
Czapek yeast autolysate agar	KH_2PO_4, 1.0; yeast autolysate, 5; sucrose, 30; agar, 15; Czapek concentrate, 10 ml
25% Glycerol nitrate agar (g/750 ml)	K_2HPO_4, 0.75; yeast autolysate, 3.7; glycerol, 250; agar, 12; Czapek concentrate, 7.5 ml; water, 750 ml
Nitrite–sucrose agar	$NaNO_2$, 3; $K_2HPO_4 \cdot 3H_2O$, 1.3; KCl, 0.5; $MgSO_4 \cdot 7H_2O$, 0.5; $FeSO_4 \cdot 7H_2O$, 0.01; sucrose, 30; agar, 20
Creatine–sucrose agar	$KH_2PO_4 \cdot 3H_2O$, 1.3; KCl, 0.5; $MgSO_4 \cdot 7H_2O$, 0.5; creatine, $1H_2O$, 3; sucrose, 30; agar, 5; adjust to pH 8 with sterile 1 N NaOH after autoclaving

Table II. Reactions of *Penicillia* on Creatine Sucrose Agar[a]

Species	Growth[b]	Acid[c]	Base[d]
P. roquefortii	++	—	NR
P. brevicompactum	—	−/w	—
P. citrinum	—	−/w	—
P. expansum	++	++	++
P. hirsutum	−/w	++	—
P. griseofulvum	—	−/w	—
P. chrysogenum	—	−/w	—

[a] From Frisvad (1981).
[b] ++, Nearly as good as a Czapek yeast autolysate agar; −/w, thin to fairly good growth; —, as on water agar.
[c] ++, Abundant acid (violet to yellow in 5–7 days); −/w, slight acid (change to yellow or red just beneath colony in 5 days); —, no color change.
[d] NR, Agar always violet; ++, production of basic metabolites after acid (color change back from yellow to violet after 8–21 days); —, no change back from acid reaction.

Frisvad (1981) has suggested the use of a range of solid media on which *Penicillia* show an increased diversity of response in the growth, morphology, and color of colony formation. Nitrite–sucrose agar and creatine–sucrose agar (Table I) were found to be particularly useful in the identification of common asymmetric *Penicillia*.

Creatine–sucrose agar provides a number of reactions including growth or no growth and acid production with or without subsequent reversion to an alkaline reaction (see Table II).

Although studies of secondary metabolism are usually carried out in liquid culture, some important investigations on the correlation of secondary metabolism with morphological differentiation have been made using colonies growing on agar-based media. Generally, the fungus is cultured on a disk of dialysis membrane placed on the agar surface, thus allowing the clean removal of biomass from the agar surface. Using such techniques, Pearce *et al.* (1981) showed that 6-methyl salicylic acid, a tetraketide precursor of patulin, is produced by *P. griseofulvum* only when an aerial mycelium is present. Bird *et al.* (1981) used the yellow-green fluorescence of brevianamides A and B under UV light to show that these metabolites not only are produced after conidiation has started, but also are present in the upper regions of the conidiophores of *P. brevicompactum.* Bird and Campbell (1982) further demonstrated that growing this organism between two layers of dialysis membrane, which precluded the differentiation of aerial mycelium, blocked the synthesis of the brevianamides, mycophenolic acid, asperphenamate, and ergosterol. However, all these metabolites were produced following the removal of the top layer of dialysis membrane and the development of aerial mycelium. Mycophenolic acid was found mainly in the medium and brevianamide and asperphenamate in the aerial hyphae, while ergosterol was shown to be present throughout the mycelium.

1.2. Growth in Liqid Media

Species of *Acremonium* and *Penicillium* grow well in submerged culture, and most physiological investigations and industrial exploitations use liquid media. Studies related to penicillin production by *P. chrysogenum* were among the first to demonstrate that a filamentous fungus could be grown in continuous culture (Pirt and Callow, 1960). Using batch cultures, Pirt and Callow obtained a value of 0.075/hr for μ_{max} of their strain, which provides the upper limit for the dilution rate possible in continuous cultures. Using both continuous and batch-fed fermenters, Ryu and Hospodka (1980) confirmed that the optimum specific production rate of penicillin requires a positive specific growth rate, and for their particular industrial strain (KA1S-12690) growing on a penicillin-producing medium, a specific growth rate of 0.015/hr was obtained. They were also able to obtain specific uptake rates for a number of nutrients under conditions for optimum penicillin production (Table III).

Colony growth on solid media, when measured as increase in colony diameter, does not always reflect the capacity of a mold for biomass formation in liquid culture. Miles and Trinci (1983) showed that the optimum temperature for colony expansion was 20°C, but for μ_{max}, the optimum temperature was 30°C. The reduction in the radial extension of colonies at temperatures above 20°C, despite an increase in exponential growth rate up to 30°C, is related to a decrease in hyphal growth-unit length. The hyphal growth unit, which is the mean length of hypha associated with each hyphal tip, is inversely related to branching frequency at any particular growth rate, and the increase in biomass will be a function of both the rate of

Table III. Specific Uptake Rates for *Penicillium chrysogenum* Growing at $\mu = 0.015$/hr[a]

Substrate	Specific uptake rate (g/cell · hr)
Hexose carbon	0.33 mmole
Oxygen	1.6 mmole
NH_3-N	2.0 mg
PO_4-P	0.6 mg
SO_4	2.8 mg

[a] From Ryu and Hospodka (1980).

hyphal-tip extension and the number of hyphal tips, which will be influenced by the frequency of branching.

Filamentous fungi growing in shaken or stirred liquid culture can exhibit a range of morphologies from dispersed mycelium forming a homogeneous suspension of high viscosity to spherical mycelial pellets forming a suspension of relatively low viscosity. The form taken has a considerable influence on industrial processes that use molds such as *P. chrysogenum* and *A. strictum*; some of the factors that influence pellet formation have been reviewed by Whittaker and Long (1973). The kinetics of mycelial growth have been analyzed by Righelato (1978). A phenomenon that influences the performance of cultures in liquid media in laboratory fermenters and shake flasks is the attachment to the walls of vessels. Ugalde and Pitt (1983a) have shown that the growth rates of cultures of *P. aurantiogriseum* grown in siliconized flasks, in which wall growth was reduced, were twice those in control flasks.

Although the full expression of differentiation and subsequent dispersal of *Acremonium* and *Penicillium* is usually associated with growth on solid substrates, sporulation of at least some species in both genera can occur in liquid culture. Indeed, *A. strictum* (= *Cephalosporium acremonium*) produces both arthrospores and conidiospores in submerged culture. Nash and Huber (1971) showed that the period of growth when hyphae were differentiating into arthrospores coincided with the period of maximum biosynthesis of β-lactam antibiotics. Although the two phenomena are not obligately linked, it is possible to separate arthrospores from hyphae by gradient centrifugation, and such preparations enriched in arthrospores have increased antibiotic-producing capability compared with hyphae.

The formation of conidia by *P. chrysogenum* growing in submerged culture is induced by high concentrations of calcium ions. The phenomenon has been characterized by D. Pitt and Poole (1981), who also review the earlier literature on calcium-induced sporulation in both this species and *P. griseofulvum*. They showed that in aerated culture vessels, increased sporulation occurred with an increase in calcium-ion concentration up to 5 mM without any limitation in the availability of major nutrients. The growth of *P. chrysogenum* is essentially aerobic, but vegetative growth can occur at relatively low oxygen concentration, whereas sporulation is considerably reduced. Working with *P. chrysogenum*, D. Pitt *et al.* (1983) demonstrated that the addition of calcium ions to a liquid medium within 24 hr of inoculation not only induced sporulation, but also reduced the levels of glucose oxidase and gluconate accumulation from glucose and sucrose. The presence of calcium ions in the medium was associated with increased lateral branching as a prelude to the differentiation of hyphal tips into conidiogenous cells and the production of conidia (Ugalde and Pitt, 1983b). This process of differentiation in shake flasks appeared to be fairly synchronous and took about 7 hr for completion. Using the radioisotope ^{45}Ca, Ugalde and Pitt (1984) demonstrated that calcium is bound principally to membrane components of the cytoplasm such as the plasma membrane and mitochondria. For these studies, density-gradient centrifugation in silica–sol gradients was used to fractionate organelles liberated from mycelium homogenized with acid-washed glass beads in a Braun MSK cell homogenizer. In an attempt to overcome problems of interpretation because of the possible redistribution of calcium ions during homogenization, the authors also studied the distribution of ^{45}Ca by autoradiography of glutaraldehyde–paraformaldehyde-fixed material. This process itself is not without technical difficulties, but the authors concluded that the results from both techniques provided evidence for the association of calcium with the plasma membrane, the tonoplast, and the mitochondrial membrane.

Penicillium simplicissimum also responds to the presence of calcium in submerged culture in stirred fermenters by sporing as well as producing the tremorgenic secondary metabolite verruculogen (Day *et al.*, 1980). Neither of these processes occurs in the absence of calcium. However, this association between sporulation and verruculogen production is not universal among verruculogen-synthesizing species of *Penicillium. Penicillium raistrickii* produces high yields of this metabolite in stirred fermenters during the active growth phase, but does not spore under these conditions (Mantle and Wertheim, 1982).

Cell-Wall Composition

The structure, plasticity, and development of the hyphal wall must be intimately involved in the morphology, growth, and development of the hypha. Rizza and Kornfeld (1969) showed that the wall material isolated from hyphae and conidia of *P. chrysogenum* had different compositions (Table IV).

There is a dramatic increase in the level of protein and polypeptide in

Table IV. Comparison of the Composition of Walls from Hyphae and Conidia of *P. chrysogenum*[a]

Component	Conidial walls	Hyphal walls
Glucose (mg/100 mg wall)	26.3	33.9
Galactose (mg/100 mg wall)	19.4	7.1
Mannose (mg/100 mg wall)	2.6	2.7
Amino acids (μg N/mg wall)	14.6	2.0
Glucosamine (μg N/mg wall)	9.0	14.1

[a] From Rizza and Kornfeld (1969).

conidial walls, and the detailed amino acid composition is also different from that of mycelium-wall material, the amino acid composition of hyphal walls being richer in the acidic amino acids aspartate and glutamate.

Unger and Hayes (1975) analyzed the walls of the mycotoxigenic species *P. purpurogenum* (=*P. rubrum*) and collected the results of earlier analyses of the walls of *P. roquefortii*, *P. digitatum*, *P. chrysogenum*, and *P. italicum*. It is not always possible to be certain whether the differences observed are a reflection of species or age and the physiological state of the mycelium. Gomez-Miranda *et al.* (1984) have demonstrated the influence of the age of the culture on polysaccharide composition of *P. pinophilum* (=*P. allahabadense*). A detailed study of the changes in wall composition during the germination of conidia of *P. chrysogenum* by Martin *et al.* (1973a) essentially confirms the results of Rizza and Kornfeld (1969), although there are quantitative differences that must reflect the difficulty of these analyses. Martin *et al.* (1973b) correlated the chemical changes observed with ultrastructural changes that occur during the germination of conidia. They distinguished four layers in the cell walls of conidia; the outermost layer is released during spore swelling, and the inner layers extend to form the wall of the germ tubes.

The use of tritium- or ^{14}C-labeled substrates (Gander and Fang, 1976) and of ^{13}C nuclear magnetic resonance (Matsunaga *et al.*, 1981) has provided means of probing the fine structure of the walls of species of *Penicillium*, although there seem to be no comparable studies on *Acremonium*.

A number of species of *Penicillium* secrete extracellular polymers of varying complexity. *Penicillium charlesii* (=*P. fellutanum*) secretes a polygalactomannan glycopeptide, the peptide moiety of which is rich in serine, threonine, and ethanolamine (Gander *et al.*, 1974). In common with some species of *Aspergillus*, a number of *Penicillium* species produce the glucan nigeran, which can be extracted from hyphal walls with hot water and which has been studied as a potential phylogenetic marker (Bobbitt and Nordin, 1978, 1980). Acidic polysaccharides containing malonic acid have been reported from a number of *Penicillia*, including *P. erythromellis* (Ruperez *et al.*, 1983), *P. luteum* (=*P. udagawae*) (Anderson *et al.*, 1939), and *P. islandicum* (Baddiley *et al.*, 1953).

1.3. Asexual Reproduction

The evolution of an increasing utilization of terrestrial environments by the filamentous fungi has required the coevolution of mechanisms for dispersal that are independent of motility and the aquatic environment. The terrestrial fungi depend on agencies such as animals, the seeds of higher plants, and the atmosphere for the dispersal of specialized propagules that may be produced as a result of a sexual process or asexually. *Penicillium* and

Acremonium produce small single-spored conidia that have been isolated from the atmosphere (Gregory, 1973). Indeed, *Penicillium* frequently dominates the air spora inside buildings, and contamination from this source is often followed by the biodeterioration of industrial commodities and the spoilage of foods.

Asexual reproduction is a specialized aspect of normal vegetative growth in which mitotic division of nuclei is followed by delimitation, and eventually secession, of a portion of the hypha as a conidium. Specialization and differentiation of the hyphae from which the conidium separates may range from almost none at all to the production of highly sophisticated structures. Such a range is well illustrated by a comparison of the two genera *Penicillium* and *Acremonium.*

1.3.1. Structure and Development of the Conidiophore

Whereas *Penicillium* is characterized by the production of a well-defined aerial conidiophore, from which a more or less complex arrangement of branches, metulae, and phialides arises, *Acremonium* produces individual phialides directly from hyphae growing on or in the substrate (Fig. 1). The relationship of the conidiophores and substrate in *Penicillium* provides a characteristic texture to colonies growing on plates that was used by Raper and Thom (1949) to separate a number of colony types. Thus, if the conidiophores arise from hyphae that are growing on or just below the surface of the medium, they produce a uniform "velvety" texture (Fig. 2a). In many species, although the conidiophores arise from the surface of the agar plate, they are aggregated into clumps, producing a mealy, or "fasciculate," texture (Fig. 2b). A number of species produce conidiophores from a mass of aerial hyphae, producing a cotton-woolly, or "lanose," appearance (Fig.2c). The fourth of Raper and Thom's colony types is produced as a result of the aggregation of aerial hyphae into ropy bundles from which the conidiophores grow and is referred to as "funiculose" (Fig. 2d). The conidiophore of *Penicillium* is not such a clearly differentiated structure as that in the related genus *Aspergillus,* although the surface is sometimes ornamented with a roughened texture that distinguishes it from the mycelium from which it has developed.

The genus *Penicillium* is a rich source of secondary metabolites, some of which have important biological activities such as mammalian toxicity or antibiotic properties. The morphological differentiation of these organisms is frequently associated with the biochemical differentiation that leads to the production of these secondary metabolites. Nover and Luckner (1974) clearly demonstrated that cyclopenin and related metabolites are closely associated with the differentiation of aerial structures and the production of

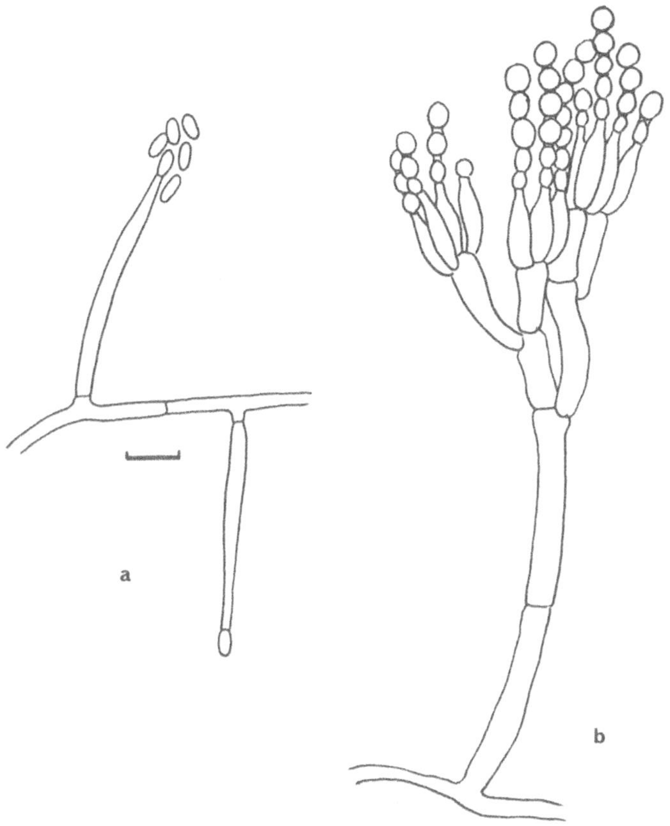

Figure 1. Conidiogenous structures of *Acremonium* (a) and *Penicillium* (b). Scale bar: 10 μm.

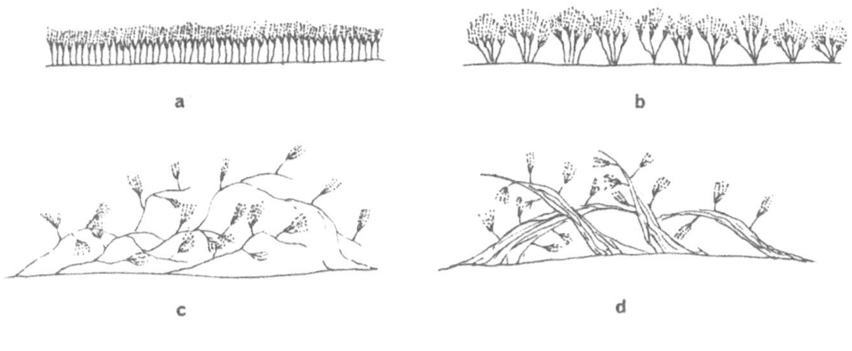

Figure 2. Relationship between colony texture and the production of conidiophores in *Penicillium.* (a) Velvety; (b) Fasciculate; (c) Lanose; (d) Funiculose.

spores by *P. aurantiogriseum.* Indeed, cyclopenase, a key enzyme in the biosynthesis of these metabolites, is known to be present only in the spores.

A number of such associations between morphological and biochemical differentiation have been demonstrated in *Penicillium.* Thus, Pearce *et al.* (1981) have shown that 6-methyl salicylic acid, a precursor of patulin, is produced in solid-state cultures of *P. patulum* only when an aerial mycelium has developed. Bird *et al.* (1981) found that the brevianamides were synthesized by *P. brevicompactum* only after conidiation had started, and this resulted in a yellow-green fluorescence to the penicillus head and the upper portion of the conidiophores when viewed under UV light.

The conidiophores of *Penicillium* are usually hyaline and colorless, although *P. flavio-stipitatum*, a newly described species isolated from sand dunes, has yellow conidiophores (Ramirez and Gonzalez, 1984).

1.3.2. Phialide and Conidium

Although the conidiogenous cell of both *Penicillium* and *Acremonium* is described as a phialide, implying that the conidia are produced endogenously from the open tip of a clearly defined structure, the outer wall of which makes no contribution to the spore wall, it is not necessarily the case that the detailed mechanism of spore formation is identical in both genera. In a detailed analysis of the nature of the phialide, Minter *et al.* (1983b) concluded that the term is imprecise and proposed a terminology for each stage in a developmental sequence (Minter *et al.*, 1982, 1983a).

By analogy with *Aspergillus* and *Fusarium*, the first conidium is probably produced holoblastically. Hanlin (1976) carried out a detailed transmission-electron-microscope study of the development of conidia in *Aspergillus clavatus*, and Minter *et al.* (1983a) based their description of the ontogeny, delimitation, and maturation of the conidia of this species on Hanlin's study. The first conidium is the result of wall-building activity at the apex of the conidiogenous cell, which also produces an additional inner wall both in the conidium and at the top of the conidiogenous cell. After delimitation of this first spore, the wall-building apex is no longer functional, but a wall-building ring comes into operation just below the septum that delimits the first conidium. Subsequent inner-wall formation in this region leads to fracture of the outer wall between the first conidium and the developing phialide with the production of a fine collarette. The alternation of activity of the wall-building ring and delimitation by the production of septa leads to a chain of spores that are still attached to each other by wall material.

Zachariah and Fitz-James (1967) studied the structure of the conidiogenous cell of *P. claviforme* by a combination of light and electron microscopy. They showed that the single nucleus always divided at the apex of the conidiogenous cell and the distal daughter nucleus immediately

Table V. Sizes of Conidiospores of a Selection of *Acremonium* and *Penicillium* spp.

Species	Length × width (μm)
Acremonium kiliense	2.5 – 6(9.5) × 0.8 – 2
A. recifei	4 – 6 × 1.2 – 2
A. zonatum	4 – 7 × 2 – 2.5
A. strictum	3 – 6 × 1 – 2
Penicillium chrysogenum	2.5 – 4.0 × 2.2 – 3.5
P. digitatum	6 – 8(15) × 2.5 – 5(6)
P. expansum	3 – 3.5 × 2.5 – 3
P. megasporium	6 – 7 × 6 – 7
P. purpurogenum	3 – 3.5 × 2.5 – 3

entered the developing conidium. They also demonstrated that the apparent thickening of the wall in the neck region of the conidiogenous cell was due to the formation of spore wall material on the inner surface of the wall at the top of the conidiogenous cell, so that the development of the conidium was endogenous. Fletcher (1971) also confirmed that the conidiogenous cell of *Penicillium* is uninucleate. Using *P. claviforme*, *P. clavigerum*, and *P. corymbiferum*, he demonstrated that the conidial chain was covered with a thin, electron-opaque surface layer, so that the chains remain intact. The remarkably stable chains of conidia, which may contain several hundred spores, are a feature of species of *Penicillium* and related genera.

The conidia of both *Penicillium* and *Acremonium* are small (Table V), single-celled, and probably uninucleate. The fine structure of the relatively large spinose spores of *P. megasporium* was studied by Sassen *et al.* (1967), who showed that the wall consisted of several layers and was covered by ordered arrays of rodlets that were themselves composed of a linear array of particles approximately 5 nm in diameter (Fig. 3). A layer of rodlets over the surface of conidiophore and the conidium is widespread among the Hyphomycetes, and the studies of the orientation during different stages in conidium development have been useful in furthering understanding of conidiogenesis (Cole and Samson, 1979). In the case of *Trichophyton mentagrophytes*, this rodlet layer has been identified as a protein–glucomannan complex with a very regular, possibly crystalline, structure (Hashimoto *et al.*, 1976).

Sassen *et al.* (1967) also showed that the plasma membrane of the conidia of *P. megasporium* had a complex structure and was also covered with particles as well as having a number of linear invaginations about 0.2–1.0 μm long and 30–40 nm wide. These invaginations seemed to be filled with the material of the innermost layer of the spore wall.

The pattern of rodlets on the outer surface of the spore has been found to show a degree of species specificity (Fig. 4). Hess *et al.* (1968) studied the

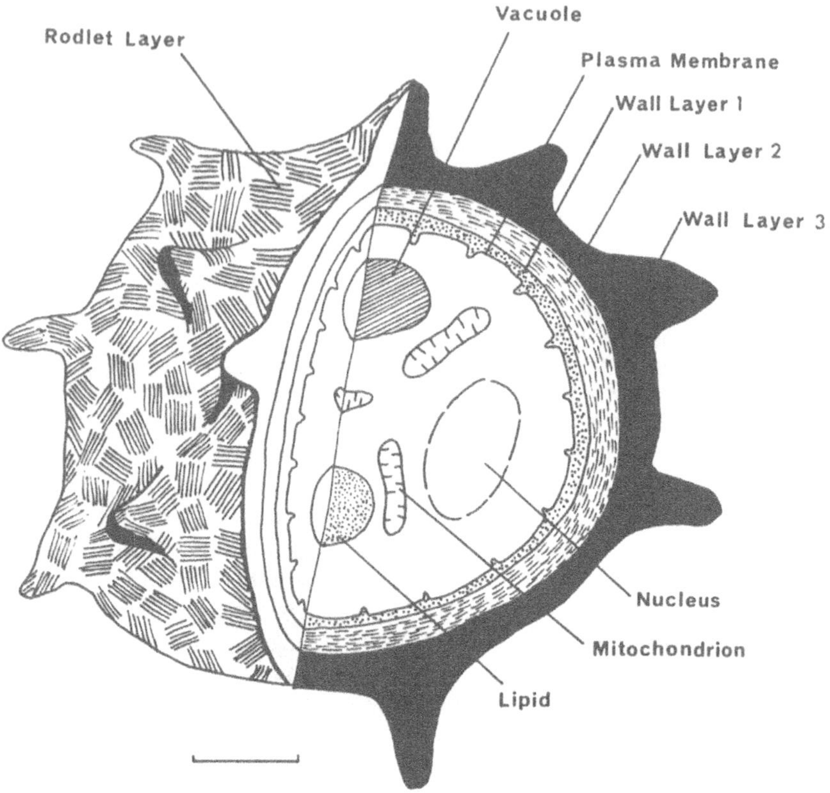

Figure 3. Structure of conidium of *Penicillium megasporium*. Scale bar: 1 μm. Based on Sassen *et al.* (1967).

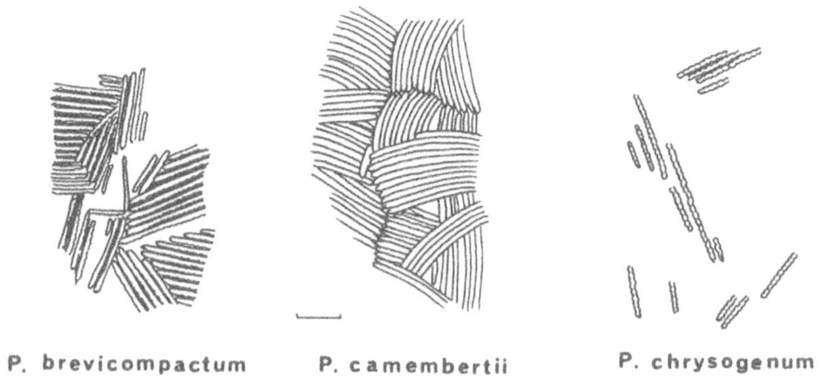

Figure 4. Variation in surface rodlet patterns on the conidia of *Penicillia*. Scale bar: 100 nm. Based on Hess *et al.* (1968).

spores of ten species of *Penicillium* and recognized distinct patterns in each. As well as the fine structure of rodlets, many species of *Penicillium* produce spores with a distinctively sculptured surface that may be seen with the light microscope, but is most easily demonstrated in the scanning electron microscope (Martinez *et al.*, 1982; Ramirez, 1982). The surfaces may vary from completely smooth (*P. lavendulum* = *Geosmithia lavendula*) to very delicately roughened (*P. funiculosum*) to warty (*P. crustosum*) to strongly echinulate (*P. melinii*). Some species of *Penicillium* are characterized by the presence of ridges over the surfaces that may be arranged longitudinally (*P. restrictum*) or laterally (*P. daleae*) (Fig. 5).

A feature unusual in *Penicillium* is the production of two distinct morphological types of conidia by *P. dimorphosporum*, first described by Swart (1970) from mangrove swamp soils. The first conidia produced by young colonies are ovoid, smooth, and slightly greenish when viewed by transmitted light. Conidia produced by older colonies are globose, coarsely roughened, and brownish by transmitted light. That both spore types belong to the same species was confirmed by studying cultures from single spores. *Acremonium fusidioides*, strains of which produce the antibacterial agent fusidic acid, also has two types of conidia, one of which is slightly pigmented, fusiform with truncate ends, and the other globose, hyaline, and slightly warty (Domsch *et al.*, 1980).

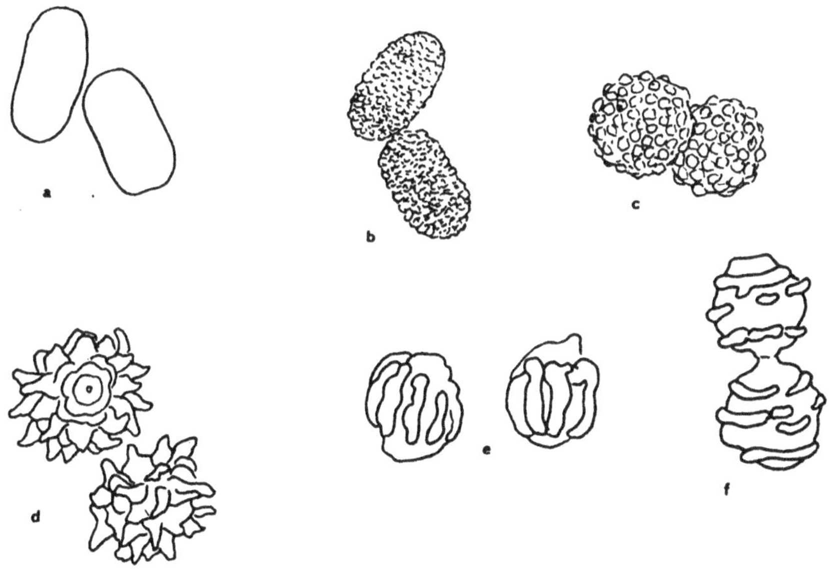

Figure 5. Surface ornamentation in the conidia of *Penicillia*. (a) *Geosmithia lavendula* (= *P. lavendulum*); (b) *P. funiculosum*; (c) *P. crustosum*; (d) *P. melinii*; (e) *P. restrictum*; (f) *P. dahliae*. Based on Martinez *et al.* (1982).

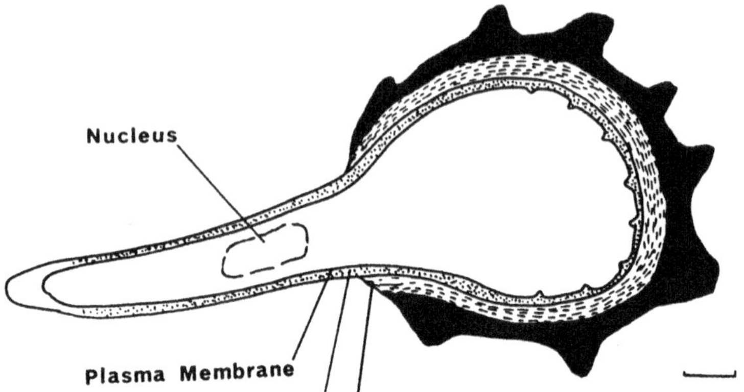

Figure 6. Germination of the conidium of *Penicillium megasporium*. Scale bar: 1 μm. Simplified from Remsen *et al.* (1967).

The germination of *P. megasporium* conidia was studied by Remsen *et al.* (1967), who showed that the wall of the germ tube was continuous with the innermost layer of the spore wall. The outer layers of the wall split to allow growth of the developing germ tube (Fig. 6). This confirmed the observations of Hawker (1966) on *P. frequentans*.

Microcycle conidiation, described by Smith *et al.* (1977) as the recapitulation of conidiation following spore germination without an intervening phase of mycelial growth, has been studied in *P. griseofulvum* (Sekiguchi *et al.*, 1975a–c). When incubated at 37°C in a medium favoring growth, spores grew in size from 2.8–3.2 to 6–7.5 μm. This growth was associated with a loss of refractility and a disturbance of the characteristic pattern of surface rodlets. Transfer of these enlarged cells to a nitrogen-poor medium at 35°C resulted in synchronous germination, limited outgrowth, and a roughly synchronous morphogenesis leading to the production of conidiospores. These studies demonstrated nuclear migration into the emerging germ tube and further confirmed that the new cell wall of the germ tube was continuous with the innermost layer of the spore wall. The formation of a septum, which had a pore filled with particulate material, occurred at the neck of the germ tube. A further distinction between the structure of the germ tube and that of the original spore was the absence of the furrows in the plasma membrane that are characteristic of that of the spore. The new crop of conidiospores that developed at the end of this microcycle process had the typical furrows in the membrane and the rodlet pattern on the outer surface.

Zeidler and Margalith (1973) showed that the presence of glutamic acid as the sole nitrogen source stimulated microcycle conidiation in *P. digitatum*.

1.3.3. Coremium Formation

The proliferation of conidiophores from a single region of fertile hyphae results in a bundle of closely packed penicilli referred to as a fascicle, and the presence of such fascicles over the surface of a colony gives a granular texture. When a large enough number of conidiophores become sufficiently closely associated, they may grow to form a macroscopic structure—the coremium (Fig. 7)—that often has the appearance of a miniature matchstick with a determinate structure. The best-known species of *Penicillium* associated with coremium formation is *P. claviforme*, in which a mass of greenish-gray spores is borne on a white to pink stalk that may be 2–4 or even 10 mm long. A similar-looking structure, which may be referred to as a "synnema" (plural "synnemata"), arises from the aggregation of fertile hyphae to form a massive aerial structure from the surface of which conidiophores may grow. Unlike the coremium, the synnema is essentially an indeterminate structure that, in *P. isariiforme*, may be several centimeters long.

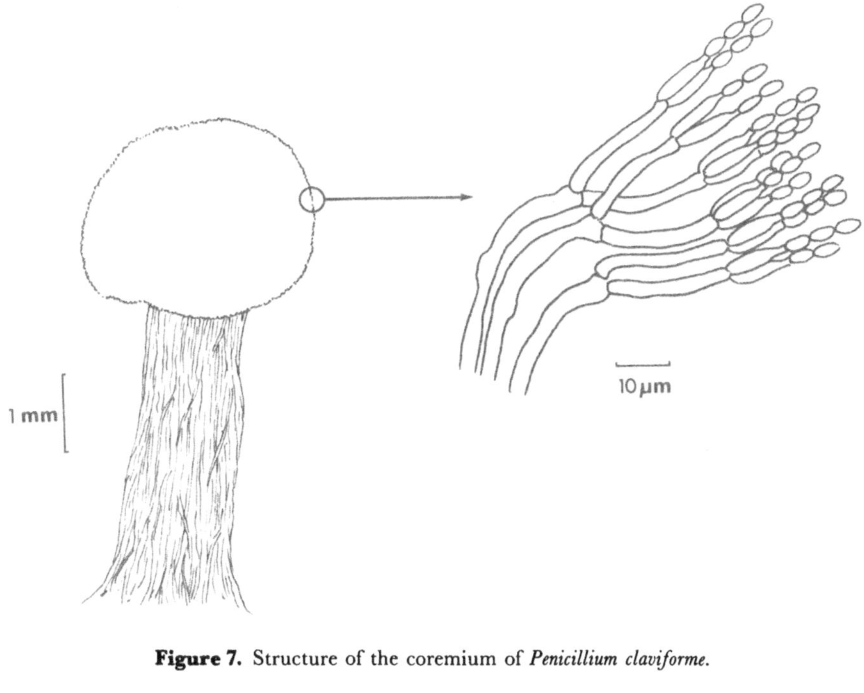

Figure 7. Structure of the coremium of *Penicillium claviforme*.

Because of the usefulness of reproducible fascicle formation in the identification of *Penicillia*, factors that influence this phenomenon have been studied. The presence of botran (2,6-dichloro-4-nitroaniline) at a concentration of 20 ppm enhances fascicle formation in *P. cyclopium* (=*P. aurantiogriseum*), *P. urticae* (=*P. griseofulvum*), and *P. viridicatum* (Mislivec, 1975). It was incorporated into one of the differential media used by Frisvad (1981).

The nutritional factors that influence coremium formation have been reviewed by Watkinson (1979). Mycelia need to achieve a level of maturity to support the formation of coremia, which may develop in a synchronous flush resulting in the formation of periodic rings of coremia in plate cultures. A model was developed to explain periodic coremium formation on the supposition that the uptake of effective nutrients by the hyphal tips results in accumulation to a critical level. Subsequent growth of coremia consumes this material, so that the initiation of further coremia is inhibited until the hyphal tips have taken up enough nutrient from freshly exploited medium. This hypothesis is supported by studies on the accumulation and fate of ^{14}C-labeled amino acids into the mycelium of *P. claviforme* (Watkinson, 1981). L-Glutamate is especially favorable in showing a concentration-dependent influence on coremium formation. This same amino acid accelerates sporulation of *P. digitatum* by 10–12 hr (Zeidler and Margalith, 1973).

Many factors influence coremium formation, as would be expected of such a complex structure. It is useful to consider the formation of a coremium as the result of three stages in development: primordium formation, elongation and formation of the stalk, and sporulation (Watkinson, 1975). Light has an important influence on the early stages (see Section 3.4), and high concentrations of carbon dioxide in the gas phase inhibit the later stages (Graafmans, 1973). Some known metabolic inhibitors such as 2-deoxyglucose completely inhibit primordium development at concentrations that still allow growth of undifferentiated hyphae (Watkinson, 1977).

The separation of coremium development into three phases based on morphology corresponds to three distinct phases in nutritional activity. Thus, primordium development corresponds to a period of active nutrient uptake and storage; the elongation of the stalk and the early phases of sporulation do not require exogenous nutrients even if they are available, whereas the continuing production of spores on the fully developed coremium again requires an exogenous supply of nutrients. As Watkinson (1975) pointed out, the association of the most actively differentiating stage of coremium formation with a restriction to the use of endogenous nutrients is a good example of the general observation that nutritional isolation is often a prerequisite for differentiation (Wright, 1973).

1.4. Sexual Reproduction

The anamorph genera *Penicillium* and *Acremonium* are each associated with several teleomorph genera in the Ascomycetes (see Chapter 1). *Penicillium*, as presently understood (J. I. Pitt, 1979), is clearly a more homogeneous group, some species being anamorphs of either *Eupenicillium* or *Talaromyces*, both of which are placed in the Eurotiales by many authorities (Hawksworth *et al.*, 1983). *Eupenicillium* is characterized by the production of almost spherical asci, singly or in short chains, in cleistothecia, the walls of which may be white to buff, yellow to brown, pink, or gray. Ascospores are single-celled, hyaline to yellow, with smooth to spinose walls, and may or may not have equatorial ridges (Udagawa and Horie, 1973b). *Eupenicillium* is associated only with *Penicillium* anamorph states, and it is usually the anamorph that is most apparent in culture. Cleistothecia will often require several weeks and sometimes months to mature.

Species of *Talaromyces*, on the other hand, have asci in gymnothecia that are structures of intermediate size with ill-defined walls of loosely woven hyphae. The asci and ascospores have a range of structure similar to that of *Eupenicillium* (Stolk and Samson, 1972), but *Talaromyces* is associated with anamorphic states in *Penicillium*, *Paecilomyces*, *Geosmithia*, and *Merimbla*.

The heterogeneity of the form genus *Acremonium* is indicated by its association with teleomorph genera classified in different orders of Ascomycetes. They include *Nectria*, *Neocosmospora*, and *Neohenningsia* in the Hypocreales and *Emericellopsis* in the Eurotiales (see Cannon and Hawkesworth, 1984; Samuels, 1976; Tubaki, 1973; Von Arx, 1981).

2. NUTRITIONAL REQUIREMENTS FOR GROWTH

The genus *Penicillium* has a remarkable biosynthetic ability, as illustrated by the capability of the majority of species to grow on the simple chemically defined medium most widely used for their study, namely, Czapek–Dox. *Penicillium digitatum*, the gray-green mold of citrus fruits, is probably one of the most commonly encountered species that does not grow at all well on this medium. A number of species of *Acremonium* have rather specialized relationships to other living organisms (see Section 4), but even those isolated from soil produce colonies of 1–2 cm only after 10 days' incubation on 2% malt extract agar at 20°C (Domsch *et al.*, 1980).

2.1. Carbon Metabolism

Species of *Penicillium* can usually grow on one of a wide range of monosaccharides, disaccharides, and sugar alcohols as the sole source of carbon; indeed, trehalose (α-D-glucopyranosyl α-D-glucopyranoside) and polyols such as mannitol are frequently important storage compounds in both conidia and vegetative mycelium (Ballio *et al.*, 1964; Thevelein, 1984). Species of both *Penicillium* and *Acremonium* can degrade polysaccharides. In a study of soil fungi, Domsch and Gams (1969) identified *Cephalosporium roseum* (= *Acremonium rutilum*) as degrading pectin and carboxymethyl cellulose and *C. nordinii*, *C. furcatum* (= *Acremonium furcatum*), and *C. acremonium* (= *A. strictum*) as degrading xylans. *Acremonium butyri* has been reported to degrade pectin (Tubaki, 1958) and *A. kiliense* to degrade starch (Mangallam *et al.*, 1967). There are several species of cellulolytic *Penicillia*, including *P. funiculosum*, *P. janthinellum*, and *P. chrysogenum*. *Penicillium janthinellum* produces an extracellular cellulase (endo-1,4-β-glucanase), which is induced by cellobiose, as well as a constitutive cellobiase (1,4-β-glucosidase). The latter is also primarily extracellular, but a significant activity is strongly associated with the cell wall (Rapp *et al.*, 1981). The thermophilic species *Talaromyces emersonii* can grow on starch as the only source of carbon (Oso, 1979), and *P. waksmanii* is able to grow on an artificial seawater medium with agar and carrageenan as sole carbon source (Dewey *et al.*, 1983).

The degradation of pectin is widespread among the species of *Penicillium* that are associated with the spoilage of fresh fruits and other plant commodities (see Section 4.2). In the case of *P. oxalicum*, isolated from rotting yams, production of oxalic acid is thought to act synergistically with the production of endopolygalacturonase in the process of infection and rotting of yams (Ikotun, 1984).

Eupenicillium zonatum can grow at the expense of paraffin-based crude oil, utilizing hydrocarbons with chain lengths of C_8–C_{20} (Hodges and Perry, 1973), and although they may not be primary biodeteriogens, species such as *P. coryophilum* and *P. frequentans* have frequently been isolated from stored diesel fuel (Hettige and Sheridan, 1984).

Several species of *Penicillium* are known to produce lipase, enabling them to grow at the expense of fats; the lipases of several species are important in cheese ripening (see Chapter 7). One of the most intensively studied lipases of this species is the glycoprotein known as phospholipase B (EC 3.1.1.5). This enzyme was isolated from mycelium during studies of the hydrolysis of lecithin by *P. chrysogenum* (Dawson, 1958). Initially, it was considered that the enzyme could hydrolyze lecithin only in the presence of activating lipids, but it was subsequently shown that deacylation of purified egg lecithin could occur in the absence of such activators after ultrasonic dispersion of the substrate (Beare and Kates, 1967). The enzyme shows a degree of specificity in the order in which it cleaves acyl groups from a phospholipid, removing the 2-acyl group first, with subsequent removal of the 1-acyl group to yield a glyceryl phosphoryl base and free fatty acids (Nishijima and Nojima, 1977). The native enzyme, which contains 30%

carbohydrate made up of *N*-acetyl glucosamine, mannose, and glucose linked via asparagine, has been purified and modified during studies of its mode of action (Kawasaki *et al.*, 1975; Okumura *et al.*, 1981). It has been shown that exposure of the carbohydrate component by protease treatment leads to a reduction in activity, but also makes it possible to remove the carbohydrates with an endoglycosidase. This leads to an increase in lipolytic activity, presumably because the removal of the exposed lipophobic groups once more allows the enzyme to interact with the lipid.

Fatty acids are important constituents of the lipids of fungi, and they are synthesized *de novo* from acetyl coenzyme A. The fatty acids of *Penicillium* show a pattern similar to those of other fungi, palmitic, stearic, oleic, and linoleic being the major acids in *P. cyaneum* (Koman *et al.*, 1969) and *P. pulvillorum* (Nakajima and Tanebaum, 1968). The possibility that such differences as do occur among species of *Penicillium* may be useful taxonomic features has been explored by Dart *et al.* (1976). Unsaturated fatty acids are important components of fungal lipids, and they are usually formed by desaturation of the corresponding saturated acid (Bennett and Quackenbush, 1969). In *P. chrysogenum*, the triene acid, linolenic acid, is formed by two alternative pathways: by elongation of shorter trienoic acids and by direct desaturation of the C_{18} acids (Richards and Quackenbush, 1974).

A reflection of the biochemical capability of species of *Penicillium* is the ability of different species to break down a very wide range of organic molecules, including the antifungal agent sorbic acid. A number of species can grow on up to 8400 ppm sorbic acid, converting it to the volatile hydrocarbon 1,3-pentadiene (Finol *et al.*, 1982). The most resistant isolates seemed to be strains of *P. puberulum*, which could completely degrade 12,000 ppm potassium sorbate (equivalent to 8400 ppm sorbic acid) in 12 days when grown at 21 °C. Other active isolates include *P. roquefortii* and an atypical strain of *P. aurantiogriseum*. Loss of viability of conidia exposed to sorbic acid is correlated with the complete loss of detectable ATP, and although there is an initial reduction in ATP in resistant strains in the presence of sorbic acid, levels soon build up again as a result of continuing biochemical activity in the conidia (Liewen and Marth, 1985).

2.2. Nitrogen Metabolism

The majority of species of *Penicillium* grow with nitrate as the sole source of nitrogen, although growth is usually more rapid with the addition of peptone. At least two uncharacterized isolates of *Penicillium* have been reported to produce nitrite and nitrate from reduced forms of nitrogen (Eylar and Schmidt, 1959). Amino acids are assimilated, and there appear to be several amino acid permease systems of varying specificity (Benko *et al.*, 1969). There is a sulfur-regulated permease for L-methionine, a relatively specific constitutive transport system for the basic amino acids L-lysine and L-arginine, a transport system specific for the acidic amino acids L-glutamate and L-aspartate, and a nonspecific amino acid permease in *P. chrysogenum* (Hunter and Segel, 1971). Weak acids, such as sorbate, benzoate, and propionate, at or below their pK_a, inhibit the transport of amino acids and cause a decrease in the levels of ATP (Hunter and Segel, 1973).

Acremonium species utilize a range of complex nitrogenous compounds, and a number of species have proteolytic activity (Pisano *et al.*, 1963; Oleniacz and Pisano, 1968). An alkaline protease (Yagi *et al.*, 1972) and a blood-coagulating protease (Satoh *et al.*, 1977) have both been isolated, purified, and characterized from *Acremonium.* In a study of the conditions that favor the production of protease by *Acremonium*, Kimura and Tsuchiya (1982) showed that the presence of peptone in a complex medium inhibited production, and it was only after growth had completely ceased and peptone had been completely utilized from the medium that protease production occurred. They showed that protease could be produced in a growth-associated manner on a semisynthetic medium containing (g/liter): glucose, 10; $NaNO_3$, 10; NaCl, 2.0; KH_2PO_4, 1.0; $MgCl_2 \cdot 6H_2O$, 0.5; yeast extract, 1.0.

2.3. Other Nutrient Requirements

Elements other than carbon and nitrogen can be mobilized from inorganic sources, and there is no requirement for complex growth factors or vitamins by the majority of species of *Penicillium* and probably of *Acremonium.* There are usually active transport processes for nutrients such as sulfate, although the specificity may not be complete; thus, in the case of *P. chrysogenum*, molybdate is transported into the cell by the sulfate carrier system (Tweedie and Segel, 1970). Metal ions such as iron, which are required for growth, can be scavenged from very low concentrations in the environment. The production of specific chelating agents, such as triacetyl fusarinine by an unidentified species of *Penicillium*, may be a factor in the capability of scavenging low concentrations of metal ions (Moore and Emery, 1976). This particular compound is a trimer of a hydroxamic acid derivative of ornithine and 3-methyl-5-hydroxypent-2-enoic acid, stabilized by acetylation (Fig. 8). The complete stereochemistry of the iron complex has been elucidated by X-ray crystallography (Hossain *et al.*, 1980).

The possibility that fungal mycelium, including that of *Penicillium*, may have a high affinity for heavy metals has been studied in the context of the recovery of metals such as uranium from low-grade ores (Galum *et al.*, 1983).

Although copper is also an essential element for optimum growth, it is usually toxic to molds, especially at low pH. Some species of *Penicillium* are remarkably tolerant of high concentrations of copper. *Penicillium spinulosum*

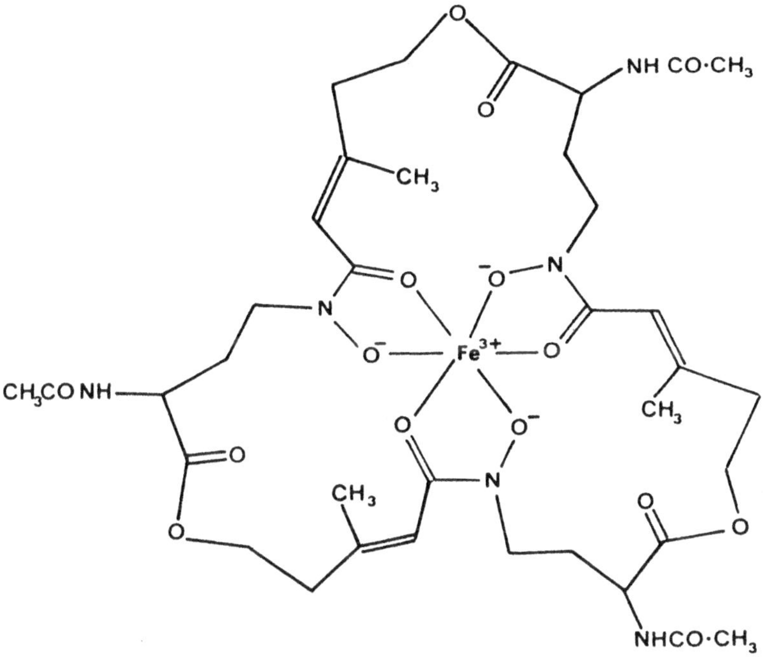

Figure 8. Structure of *N*, *N′*, *N″*-triacetyl fusarinine–iron complex.

(Murphy and Levy, 1983) and *P. ochrochloron* (Stokes and Lindsay, 1979) are both able to produce oxalic acid, with the result that a large amount of copper may be rendered harmless as the relatively insoluble copper oxalate. However, in the case of *P. ochrochloron*, which can grow in the presence of 5000 ppm copper in solutions at pH 2.0, this cannot be the only mechanism of copper tolerance.

3. INFLUENCE OF EXTRINSIC FACTORS ON GROWTH AND SURVIVAL

3.1. Temperature

The majority of species of *Penicillium* and *Acremonium* grow optimally between 20 and 30°C, 25°C being used for routine diagnostic cultures of *Penicillium* and 20°C for *Acremonium*. The ability to grow at 37°C is sufficiently restricted as to be considered by J. I. Pitt (1979) to be a useful diagnostic test. In his extensive study of *Penicillium*, Pitt found that no species classified in the subgenus *Biverticillium* or in the genus *Talaromyces* could produce colonies of 3-mm diameter or more on Czapek's yeast autolysate

Table VI. Species of *Penicillium* Able to Grow at Low (5°C) and High (37°C) Temperatures within 7 Days[a]

Low temperature	High temperature
E. baarnense	*E. abidjanum*
E. crustaceum	*E. brefeldianum*
E. egyptiacum	*E. catenatum*
E. pinetorum	*E. ehrlichii*
P. arenicola	*E. erubescens*
P. aurantiogriseum	*E. fractum*
P. brevicompactum	*E. hirayamae*
P. camembertii	*E. javanicum*
P. chrysogenum	*E. levitum*
P. crustosum	*E. ludwigii*
P. echinulatum	*E. ochrosalmoneum*
P. expansum	*E. parvum*
P. glabrum	*E. rubidurum*
P. granulatum	*E. zonatum*
P. griseofulvum	*P. capsulatum*
P. hirsutum	*P. chermesianum*
P. italicum	*P. diversum*
P. lividum	*P. funiculosum*
P. olivicolor	*P. janthinellum*
P. puberulum	*P. oxalicum*
P. purpurescens	*P. piceum*
P. raistrickii	*P. pinophilum*
P. resticulosum	*P. purpurogenum*
P. roquefortii	*P. rolfsii*
P. spinulosum	*P. simplicissimum*
	P. verruculosum
P. thomii	*T. flavus*
P. turbatum	*T. galapagensis*
P. verrucosum	*T. gossypii*
P. viridicatum	*T. helicus*
	T. panasenkoi
	T. stipitatus
	T. thermophilus
	T. trachyspermus

[a] From Pitt (1979).

agar at 5°C in 7 days, nor could any species in the subgenus *Penicillium* produce colonies up to 30 mm on this medium at 37°C in 7 days. Table VI summarizes the results of these studies.

Species of *Penicillium* are frequently associated with the spoilage of foods stored at low temperatures, *P. roquefortii*, *P. aurantiogriseum*, and *P. crustosum* being isolated from moldy refrigerated foods with particular frequency.

Among the five isolates of *Eupenicillium* from soils of New Guinea, a single species (*E. papuanum* = *E. parvum*) grew better at 37 than at 25°C (Fig. 9) (Udagawa and Horie, 1973a).

Oso (1979) described the isolation of *Talaromyces emersonii* from stacks of oil palm kernels in Nigeria. This species, which is an active amylase producer, grows optimally at 40°C, but the only truly thermophilic species is *P. dupontii* (= *Talaromyces thermophilus*) with an optimum temperature of 50°C (Cooney and Emerson, 1964). This species, which was first isolated from manure and self-heating damp hay, shows no growth at 25°C and very little growth at 30°C.

The heat resistance of the vegetative and asexual spores is not widely recorded, but would be expected to be comparable to that of other mesophilic Hyphomycetes. Fergus (1982) reports values of survival times of 10 min at 50°C for *A. butyri* and 10 min at 60°C for *P. oxalicum*. Ascospores, on the other hand, are usually more resistant to exposure at elevated temperatures, and Scott (1968) recommends heating soil samples with sterile distilled water (2 g to 16 ml) at 60°C for 30 min for the isolation of *Eupenicillium*.

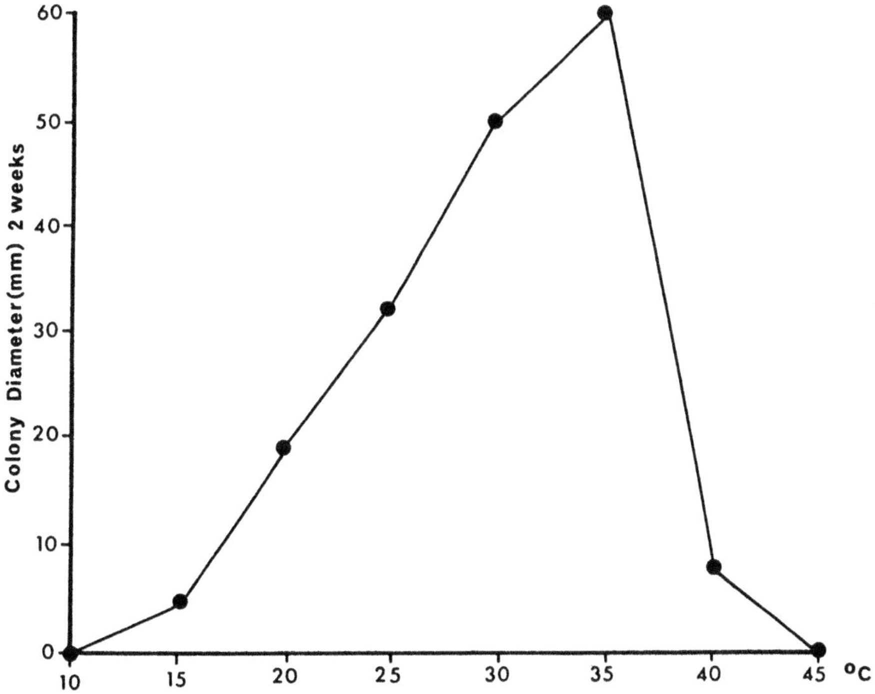

Figure 9. Influence of temperature on growth of *Eupenicillium parvum*. From Udagawa and Horie (1973a).

3.2. Water Activity

Although there is very little information available, members of the genus *Acremonium* would be expected to be mesophilic to hydrophilic with respect to water activity, as are species of the related genus *Fusarium*. Magan and Lacey (1984a) found that a water activity (a_w) of 0.89 was the minimum at which the growth of *Fusaria* could occur, even at their optimum growth temperature. Optimum a_w values are in the range 0.99–0.995, except in the case of *F. poae* growing at 30°C, its optimum a_w then being as low as 0.98.

By contrast, the genus *Penicillium* contains a large number of species that can grow at water activities below 0.85 (Table VII). Indeed, the growth response of *Penicillia* on a medium containing 25% glycerol ($a_w = 0.93$) has been a useful diagnostic tool in the identification of members of this genus by J. I. Pitt (1973), who found a frequently strong correlation between penicillus type and the extent of growth on this medium. *Eupenicillium osmophilum* has been reported as capable of growth and asexual reproduction on a medium containing 70% sucrose (Stolk and Veenbaas-Rijks, 1974), but such a medium still has a water activity higher than 0.85.

Many authors have described the interaction that occurs between tem-

Table VII. Minimum a_w for the Germination of Spores of the More Xerophilic Species of *Penicillium*[a]

Minimum a_w	Species
0.84	*P. purpurogenum*
0.83	*P. islandicum*
0.83	*P. oxalicum*
0.82	*P. expansum*
0.82	*P. restrictum*
0.82	*P. vanbeymae*
0.81	*P. aurantiogriseum*
0.81	*P. glabrum*
0.80	*P. citrinum*
0.80	*P. corylophilum*
0.80	*P. spinulosum*
0.80	*P. viridicatum*
0.78	*P. brevicompactum*
0.78	*P. chrysogenum*
0.78	*P. fellutanum*
0.78	*P. implicatum*
0.78	*P. jancewskii*
0.78	*P. pheoniceum*

[a] From Pitt (1979).

Table VIII. Minimum a_w for Different Stages of Development of *Penicillium* Species at 25°C[a]

Species	Germination	Linear growth	Asexual sporulation
P. brevicompactum	0.80	0.82	0.85
P. aurantiogriseum	0.80	0.83	0.85
P. hirsutum	0.80	0.83	0.86
P. piceum	0.79	0.85	0.89
P. roquefortii	0.83	0.83	0.83

[a] From Magan and Lacey (1984b).

perature and water activity and influences the growth of filamentous fungi. This is particularly well illustrated for *Fusarium* in the paper by Magan and Lacey (1984a). Mislivec and Tuite (1970) studied the temperature and water-activity requirements of species of *Penicillium* isolated from maize kernels either directly from the field or from storage. Of the total of 14 species of *Penicillium* isolated, those isolated from the field could grow within the temperature range 8–35°C, with optima close to 30°C. This group, exemplified by *P. oxalicum* and *P. funiculosum*, could not grow at water activities less than 0.86. In contrast, the species isolated from maize in storage, exemplified by *P. aurantiogriseum*, *P. brevicompactum*, and *P. viridicatum*, could grow at temperatures as low as −2°C, with optima of 23°C, and were also able to grow at water activities as low as 0.81.

The different stages of fungal development usually differ in the critical values of these environmental parameters. In a study of the effect of temperature and pH on the water-activity requirements of molds isolated from freshly harvested and stored wheat, Magan and Lacey (1984b) found that the minimum water activity for germination of the spores of a number of *Penicillia* at 25°C ranged from 0.79 to 0.83, whereas for sporulation, the range was 0.83–0.89 (Table VIII).

3.3. Hydrogen-Ion Concentration

There is little information on the influence of pH on the growth of *Acremonium*. Read and Seviour (1984) used glucose-limited continuous culture to study the influence of pH on the maximum specific growth rate and morphology of *A. diospyri*. For this species, the maximum specific growth rate was at pH 6.0, and growth was less at pH 3.0 than at pH 9.0 (Table IX). Read and Seviour observed very significant changes in morphology in cultures growing at $D = 0.028$/hr, from thick, short, sporulating hyphal fragments at pH 3.0 to branched mycelial clumps at pH 6.0 and long, thin, unbranched filaments at pH 9.0.

Most species of *Penicillium* grow well over a pH range of 3.0–8.0, and

Table IX. Effect of pH on the Maximum Growth Rate (μ_{max}) of *Acremonium diospyri* in Continuous Culture[a]

pH	μ_{max}
3.0	0.060/hr
4.5	0.055/hr
6.0	0.120/hr
7.5	0.100/hr
9.0	0.082/hr

[a] From Read and Seviour (1984).

P. lignorum seems to be unusual in requiring acidic conditions for growth. Isolated from dead wood, this species will not grow on ordinary Czapek agar (pH 7.1), but grows rapidly on the same medium adjusted to pH 3.6 (Stolk, 1969).

3.4. Light

Light may have a profound influence on the growth and development of fungi, but in the case of *Acremonium*, the most obvious effects of light are on pigmentation. Codner and Platt (1959) surveyed a large number of *Cephalosporia* and divided them into three groups on the basis of their reaction to light. Group 1 (25 isolates) produced no pink or orange pigments when grown in the light or dark; group 2 (48 isolates) produced pink or orange pigments in the light but not in the dark; group 3 (2 isolates), although they produced pigments in the dark, showed enhanced pigmentation when grown in the light. In a more detailed study of *A. diospyri*, Seviour and Codner (1973) showed that this dimorphic species produced carotenoids only when growing in the filamentous form after exposure to light. The single-celled yeastlike form, on the other hand, was able to produce pigments even when grown in the dark. The effect of light on *A. diospyri* could be mimicked by the presence of organomercurial compounds such as *p*-hydroxymercuribenzoate that are considered to react with sulfydryl groups (Parn and Seviour, 1974). Under these conditions β- and γ-carotenes, lycopene, torulene, and neurosporoxanthin are produced. This photomimetic effect of organomercurials is restricted to *A. diospyri* and *A. strictum* and is not shown by other species that respond to light (Seviour and Read, 1983).

The influence of light on the development of coremia and synnemata by some species of *Penicillium* has already been discussed (see Section 1.3.3).

The number and distribution of coremium initials of *P. claviforme* are influenced by light (Carlile *et al.*, 1961), although in this species a phototropic response occurs only during the early stage of the elongation of the coremium stalk. Light is essential both for initiation and for continued growth of the synnemata of *P. isariiforme*, as well as for subsequent sporulation (Carlile *et al.*, 1962). Graafmans (1974) showed that light influenced the levels of citric acid in the medium supporting growth of *P. isariiforme*.

4. ECOLOGY AND SIGNIFICANCE

4.1. Interactions with Plants

Although members of the genus *Penicillium* are known primarily as saprophytes growing on a wide range of organic substrates, including dead plant material, a few species have evolved a degree of specificity for living tissue. Thus, *P. expansum* is especially associated with apples, although it can be readily isolated from soil and organic material of plant origin. It forms a soft brown rot of apples, but may also attack other fruit such as cherries and grapes. The rot develops rapidly throughout the fruit, and the mold eventually forms pustules of conidiophores on the surface from which the characteristic blue-green conidia develop. It is the growth of this species in apples that is responsible for the occurrence of patulin in samples of unfermented apple juice produced from poor-quality fruit.

Penicillium digitatum and *P. italicum* are, respectively, the green and blue molds of citrus fruits. Both usually initiate infection through wounds in the skin, although they can invade healthy fruit from contact with a heavy inoculum. They may cause serious losses in citrus-growing regions, and control has inevitably led to widespread use of fungicides and the subsequent emergence of fungicide resistance. The fourth species of *Penicillium* that can be a nuisance is *P. gladioli*, first isolated from gladiolus corms, but also causing a rot of the bulbs and corms of other plant species.

Two of the species of *Acremonium* that are plant pathogens have been considered as potential biological control agents for specific weed species. *Acremonium diospyri* is pathogenic to persimmon (*Diospyros virginiana*), causing a wilt disease, and has been considered as a control agent against weed persimmon (Wilson, 1965). *Acremonium zonatum* causes a leaf spot disease of coffee and a wide range of other plant species. It is widespread throughout tropical countries of Africa, Asia, and the Americas and is known to cause a zonal leaf spot of water hyacinth (*Eichhorrnia crassipes*), for which it has also been considered as a potential biological control agent (Rintz, 1973).

Syngonium podophyllum, a decorative climbing plant of the family Araceae from tropical America, which is widely cultivated as an indoor pot plant, is susceptible to a leaf spot disease caused by *A. crotocinigenum* (Uchida and Aragaki, 1982). Although *A. kiliense* is better known as an animal pathogen (see Section 4.2), it has been found to be pathogenic to the green alga *Cladophora glomerata* (Bott and Rogenmuser, 1980).

Acremonium coenophialum is known as an endophyte of tall fescue grass (*Festuca arundinacea*) and is considered to be related to the anamorph of *Epichloe typhina* (Morgan-Jones and Gams, 1982). The presence of this endophyte is associated with fescue toxicosis of cattle, and methods have been devised for the rapid histochemical detection of the fungus in the grass. It can be detected in seed by staining with aniline blue after nitric acid treatment (Clark *et al.*, 1983). The organism can be cultured but grows very slowly, forming a colony of about 25 mm after 8 weeks on a medium containing (g/liter): yeast extract, 2; KH_2PO_4, 5; $MgSO_4$, 0.5 g; glucose, 5; agar, 20.

An endophyte of perennial ryegrass (*Lolium perenne*), which may be related to that in tall fescue, is considered to be the causative agent of ryegrass staggers of sheep in New Zealand. This disease is caused by the presence of complex tremorgenic compounds known as lolitrems that are present in the grass when it is infected by the endophyte (Gallagher *et al.*, 1981). Although the structure of the major neurotoxin, lolitrem B (Fig. 10), has been published (Gallagher *et al.*, 1984), and it shows a remarkable similarity to a number of tremorgens of fungal origin, the source of the lolitrems is still uncertain. They may be genuine mold metabolites produced and secreted by the endophyte, or they may be plant metabolites comparable to phytoalexins and produced by the plant in response to the presence of the endophyte. If they are mold metabolites, then they are of particular interest, not only because of their mammalian toxicity, but also because of their chemical relationship with metabolites of *Penicillium* such as penetrem and paxiline, on one hand, and metabolites of *Claviceps* such as paspalinine, on the other (see Chapter 7).

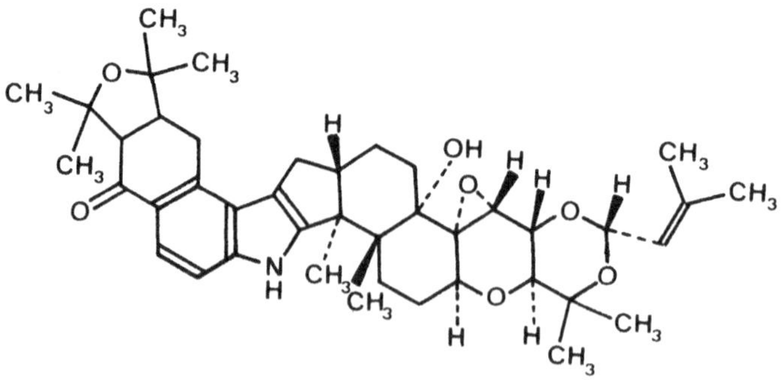

Figure 10. Structure of lolitrem B.

4.2. Interactions with Animals

Although several species of *Penicillium* are very important in animal and human health through the production of mycotoxins (see Chapter 7), only *P. marneffei* is associated with mycotic disease (Segretain, 1959). *Acremonium kiliense* and *A. recifei* are involved in mycetomas of man. *Acremonium kiliense* is a common soil fungus that can cause madura foot, skin infections, or infections of the toenails in man. *Acremonium recifei* has also been isolated from a mycetoma of the foot in man, especially following injury during agricultural work, but it is also widespread on a number of plant hosts. A number of entomogenous species of *Cordyceps* and *Torrubella* have anamorphs in *Acremonium* (Roberts and Humber, 1981).

4.3. Interactions with Fungi

Penicillium purpurogenum can parasitize *Aspergillus niger* and has occasionally been a nuisance when the latter is used industrially for the production of citric and gluconic acids. A number of species of *Acremonium* have been shown to be facultative necrotrophs of other molds (Rudakov, 1978). *Acremonium sordidulum* was isolated as a mycoparasite of *Colletotrichum dematium* (Singh *et al.*, 1978), and *A. fungicola* has been isolated from the myxomycete *Fuligo septica* (Samuels, 1973).

REFERENCES

Anderson, C. G., Haworth, W. N., Raistrick, H., and Stacey, M., 1939, Polysaccharides synthesized by microorganisms. IV. The molecular constitution of luteose, *Biochem. J.* **33:**272–279.

Baddiley, J., Buchanan, J. G., and Thain, E. M., 1953, The polysaccharide of *Penicillium islandicum* Sopp., *J. Chem. Soc.* **1953:**1944–1946.

Ballio, A., Divitorio, V., and Russi, S., 1964, The isolation of trehalose and polyols from the conidia of *Penicillium chrysogenum* Thom, *Arch. Biochem. Biophys.* **107:**177–183.

Beare, J. L., and Kates, M., 1967, Properties of the phospholipase B from *Penicillium notatum*, *Can. J. Biochem.* **45:**101–113.

Benko, P. V., Wood, T. C., and Segel, I. H., 1969, Multiplicity and regulation of amino acid transport in *Penicillium chrysogenum*, *Arch. Biochem. Biophys.* **129:**498–508.

Bennett, A. S., and Quackenbush, F. W., 1969, Synthesis of unsaturated fatty acids by *Penicillium chrysogenum*, *Arch. Biochem. Biophys.* **130:**567–572.

Bird, B. A., and Campbell, I. M., 1982, Disposition of mycophenolic acid, brevianamide A, asperphenamate and ergosterol in solid cultures of *Penicillium brevicompactum*, *Appl. Environ. Microbiol.* **43:**345–348.

Bird, B. A., Remaley, A. T., and Campbell, I. M., 1981, Brevianamides A and B are formed only after conidiation has begun in solid cultures of *Penicillium brevicompactum*, *Appl. Environ. Microbiol.* **42:**521–525.

Bobbitt, T. F., and Nordin, J. H., 1978, Hyphal nigeran as a potential phylogenetic marker for *Aspergillus* and *Penicillium* species, *Mycologia* **70:**1201–1211.

Bobbitt, T. F., and Nordin, J. H., 1980, A survey of *Aspergillus* and *Penicillium* species producing an exocellular nigeran–protein complex, *Mycologia* **72:**637–640.

Booth, C., 1971, Fungal culture media, in: *Methods in Microbiology* (C. Booth, ed.), Academic Press, New York, pp. 49–111.

Bott, T. L., and Rogenmuser, K., 1980, Fungal pathogen of *Cladophora glomerata* (Chlorophyta), *Appl. Environ. Microbiol.* **40:**977–980.

Cannon, P. F., and Hawkesworth, D. L., 1984, A revision of the genus *Neocosmospora* (Hypocreales), *Trans. Br. Mycol. Soc.* **82:**673–688.

Carlile, M. J., Lewis, B. G., Mordue, E. M., and Northover, J., 1961, The development of coremia. I. *Penicillium claviforme, Trans. Br. Mycol. Soc.* **44:**129–133.

Carlile, M. J., Dickens, S. W., Mordue, E. M., and Schipper, M. A. A., 1962, The development of coremia. II. *Penicillium isariiforme, Trans. Br. Mycol. Soc.* **45:**457–461.

Clark, E. M., White, J. F., and Patterson, R. M., 1983, Improved histochemical techniques for the detection of *Acremonium coenophialum* in tall fescue (*Festuca arundinacea*) and methods of *in vitro* culture of the fungus, *J. Microbiol. Methods* **1:**149–156.

Codner, R. C., and Platt, B. C., 1959, Light induced production of carotenoid pigments by cephalosporia, *Nature (London)* **184:**741–742.

Cole, G. T., and Samson, R. A., 1979, *Patterns of Development in Conidial Fungi*, Pitman, London.

Cooney, D. G., and Emerson, R., 1964, *Thermophilic Fungi*, Freeman, San Francisco.

Dart, R. K., Stretton, R. J., and Lee, J. D., 1976, Relationships of *Penicillium* species based on their long chain fatty acids, *Trans. Br. Mycol. Soc.* **66:**525–529.

Dawson, R. M. C., 1958, Studies on the hydrolysis of lecithin by a *Penicillium notatum* phospholipase B preparation, *Biochem. J.* **70:**559–570.

Day, J. B., Mantle, P. G., and Shaw, B. I., 1980, Production of verruculogen by *Penicillium estinogenum* in stirred fermenters, *J. Gen. Microbiol.* **117:**405–410.

Dewey, F. M., Donnelly, K. A., and Foster, D., 1983, *Penicillium waksmanii* isolated from a red seaweed *Eucheuma striatum, Trans. Br. Mycol. Soc.* **81:**433–434.

Domsch, K. H., and Gams, W., 1969, Variability and potential of a soil fungus population to decompose pectin, xylan and carboxymethyl cellulose, *Soil Biol. Biochem.* **1:**29–36.

Domsch, K. H., Gams, W., and Anderson, T.-H., 1980, *Compendium of Soil Fungi*, Academic Press, London.

Eylar, D. R., and Schmidt, E. L., 1959, A survey of heterotrophic micro-organisms from soil for ability to form nitrite and nitrate, *J. Gen. Microbiol.* **20:**473–481.

Fergus, C. L., 1982, The heat resistance of some mesophilic fungi isolated from mushroom compost, *Mycologia* **74:**149–152.

Finol, M. L., Marth, E. H., and Lindsay, R. C., 1982, Depletion of sorbate from different media during growth of *Penicillium* species, *J. Food Protect.* **45:**398–404.

Fletcher, J., 1971, Conidium ontogeny in *Penicillium, J. Gen. Microbiol.* **67:**207–214.

Frisvad, J. C., 1981, Physiological criteria and mycotoxin production as aids in identification of common asymmetric penicillia, *Appl. Environ. Microbiol.* **41:**568–579.

Gallagher, R. T., White, E. P., and Mortimer, P. H., 1981, Ryegrass staggers: Isolation of potent neurotoxins lolitrem A and lolitrem B from staggers-producing pastures, *N. Z. Vet. J.* **29:**189–190.

Gallagher, R. T., Hawkes, A. D., Steyn, P. S., and Vleggaar, R., 1984, Tremorgenic neurotoxins from perennial ryegrass causing ryegrass staggers disorder of livestock: Structure elucidation of lolitrem B, *J. Chem. Soc. Chem. Commun.* **1984:**614–616.

Galum, M., Keller, P., Feldstein, H., Galum, E., Siegel, S., and Siegel, B., 1983, Recovery of uranium VI from solution using fungi. 2. Release from uranium loaded *Penicillium* biomass, *Water Air Soil Pollut.* **20:**277–286.

Gander, J. E., and Fang, F., 1976, The occurrence of ethanolamine and galactofuranosyl residues attached to *Penicillium charlesii* cell wall saccharides, *Biochem. Biophys. Res. Commun.* **71:**719–725.

Gander, J. E., Jentoft, N. H., Drewes, L. R., and Rick, P. D., 1974, The 5-*O*-*β*-d-galactofuranosyl containing exocellular glycopeptide of *Penicillium charlesii*: Characterization of the phosphogalactomannan, *J. Biol. Chem.* **249:**2063–2072.

Gomez-Miranda, B., Guerrero, C., and Leal, J. A., 1984, Effect of culture age on cell wall polysaccharides of *Penicillium allahabadense*, *Exp. Mycol.* **8:**298–303.

Graafmans, W. D. J., 1973, The influence of carbon dioxide on morphogenesis in *Penicillium isariiforme*, *Arch. Mikrobiol.* **91:**67–76.

Graafmans, W. D. J., 1974, Metabolism of *Penicillium isariiforme* on exposure to light, with special reference to citric acid synthesis, *J. Gen. Microbiol.* **82:**247–252.

Gregory, P. H., 1973, *The Microbiology of the Atmosphere*, 2nd ed., Leonard Hill, Aylesbury.

Hanlin, R. T., 1976, Phialide and conidium development in *Aspergillus clavatus*, *Am. J. Bot.* **63:**144–155.

Hashimoto, T., Wu-Yuan, C. D., and Blumenthal, H. J., 1976, Isolation and characterization of the rodlet layer of *Trichophyton mentagrophytes* microconidial walls, *J. Bacteriol.* **127:**1543–1549.

Hawker, L. E., 1966, Germination, morphology and anatomical changes in the fungus spore, in: *The Fungus Spore* (M. F. Madelin, ed.), Butterworth, London, pp. 151–161.

Hawkesworth, D. L., Sutton, B. C., and Ainsworth, G. E., 1983, *Ainsworth and Bisby's Dictionary of the Fungi*, C.M.I., Kew.

Hess, W. M., Sassen, M. M. A., and Remsen, C. C., 1968, Surface characteristics of *Penicillium* conidia, *Mycologia* **60:**290–303.

Hettige, G., and Sheridan, J. E., 1984, Mycoflora of stored diesel fuel in New Zealand, *Int. Biodeterior. Bull.* **20:**225–227.

Hodges, C. S., and Perry, J. J., 1973, A new species of *Eupenicillium* from soil, *Mycologia* **65:**697–702.

Hossain, M. B., Eng-Wilmot, D. L., Loghry, R. A., and Van der Helm, D., 1980, Circular dichroism, crystal structure, and absolute configuration of the siderophore ferric *N*, *N*′, *N*″-triacetylfusarinine $FeC_{39}H_{57}N_6O_{15}$, *J. Am. Chem. Soc.* **102:**5766–5773.

Hunter, D. R., and Segel, I. H., 1971, Acidic and basic amino acid transport systems in *P. chrysogenum*, *Arch. Biochem. Biophys.* **144:**168–183.

Hunter, D. R., and Segel, I. H., 1973, Effect of weak acids on amino acid transport by *Penicillium chrysogenum*: Evidence of a proton or charge gradient as the driving force, *J. Bacteriol.* **113:**1184–1192.

Ikotun, T., 1984, Production of oxalic acid by *Penicillium oxalicum* in culture and in infected yam tissue and interaction with macerating enzyme, *Mycopathologia* **88:**9–14.

Kawasaki, N., Sugatani, J., and Saito, K., 1975, Studies on a phospholipase B from *Penicillium notatum*: Purification, properties and mode of action, *J. Biochem.* **77:**1233–1244.

Kimura, T., and Tsuchiya, K., 1982, Characteristics of protease production by *Cephalosporium* sp., *Appl. Environ. Microbiol.* **43:**654–658.

Koman, V., Betina, V., and Baruth, Z., 1969, Fatty acid lipid and cyanein production by *Penicillium cyaneum*, *Arch. Microbiol.* **65:**172.

Liewen, M. B., and Marth, E. H., 1985, Viability and ATP content of conidia of sorbic acid-sensitive and resistant strains of *Penicillium roquefortii* after exposure to sorbic acid, *Appl. Microbiol. Biotechnol.* **21:**113–117.

Magan, N., and Lacey, J., 1984a, Water relations of some *Fusarium* species from infected wheat ears and grain, *Trans. Br. Mycol. Soc.* **83:**281–285.

Magan, N., and Lacey, J., 1984b, Effect of temperature and pH on water relations of field and storage fungi, *Trans. Br. Mycol. Soc.* **82:**71–81.

Mangallam, S., Menon, M. R., Sukapure, R. S., and Gopalkrishnan, K. S., 1967, Amylase production by some *Cephalosporium* species, *Hindustan Antibiot. Bull.* **10:**194–199.

Mantle, P. G., and Wertheim, J. S., 1982, Production of verruculogen during growth of *Penicillium raistrickii*, *Trans. Br. Mycol. Soc.* **79:**348–349.

Martin, J. F., Nicolas, G., and Villanueva, J. R., 1973a, Chemical changes in the cell wall of conidia of *Penicillium notatum* during germination, *Can. J. Microbiol.* **19:**789–796.

Martin, J. F., Uruburu, F., and Villanueva, J. R., 1973b, Ultrastructural changes in the conidia of *Penicillium notatum* during germination, *Can. J. Microbiol.* **19:**797–801.

Martinez, A. T., Calvo, M. A., and Ramirez, C., 1982, Scanning electron microscopy of *Penicillium* conidia, *Antonie van Leeuwenhoek J. Microbiol. Serol.* **48:**245–255.

Matsunaga, T., Okubo, A., Fukami, M., Yamazaki, S., and Toda, S., 1981, Identification of β-galactofuranosyl residues and their rapid internal motion in the *Penicillium ochro-chloron* cell wall probed by ^{13}C NMR, *Biochem. Biophys. Res. Commun.* **102:**525–530.

Miles, E. A., and Trinci, A. P. J., 1983, Effect of pH and temperature on morphology of batch and chemostat cultures of *Penicillium chrysogenum*, *Trans. Br. Mycol. Soc.* **81:**193–200.

Minter, D. W., Kirk, P. M., and Sutton, B. C., 1982, Holoblastic phialides, *Trans. Br. Mycol. Soc.* **79:**75–93.

Minter, D. W., Kirk, P. M., and Sutton, B. C., 1983a, Thallic phialides, *Trans. Br. Mycol. Soc.* **80:**39–66.

Minter, D. W., Sutton, B. C., and Brady, B. L., 1983b, What are phialides anyway?, *Trans. Br. Mycol. Soc.* **81:**109–120.

Mislivec, P. B., 1975, The effect of botran on fascicle production by species of *Penicillium*, *Mycologia* **67:**194–198.

Mislivec, P. B., and Tuite, J., 1970, Temperature and relative humidity requirements of species of *Penicillium* isolated from yellow dent corn kernels, *Mycologia* **62:**75–88.

Moore, R. E., and Emery, T. F., 1976, *N*-Acetylfusarinines: Isolation, characterization and properties, *Biochemistry* **15:**2719–2723.

Morgan-Jones, G., and Gams, W., 1982, Notes on Hyphomycetes. XLI. An endophyte of *Festuca arundinacea* and the anamorph of *Epichloe typhina*: New taxa in one of two new sections of *Acremonium*, *Mycotaxon.* **15:**311–318.

Murphy, R. J., and Levy, J. F., 1983, Production of copper oxalate by some copper tolerant fungi, *Trans. Br. Mycol. Soc.* **81:**165–168.

Nakajima, S., and Tanebaum, S. W., 1968, The fatty acids of *Penicillium pulvillorum*, *Arch. Biochem. Biophys.* **127:**150–156.

Nash, C. H., and Huber, F. M., 1971, Antibiotic synthesis and morphological differentiation of *Cephalosporium acremonium*, *Appl. Microbiol.* **22:**6–10.

Nishijima, M., and Nojima, S., 1977, Positional specificity of phospholipase B of *Penicillium notatum*, *J. Biochem.* **81:**533–537.

Nover, L., and Luckner, M., 1974, Expression of secondary metabolism as part of the differentiation processes during idiophase development of *Penicillium cyclopium* Westling, *Biochem. Physiol. Pflanz.* **166:**293–305.

Okumura, T., Sugatani, J., and Saito, K., 1981, Role of the carbohydrate moiety of phospholipase B from *Penicillium notatum* in enzyme activity, *Arch. Biochem. Biophys.* **211:**419–429.

Oleniacz, W. S., and Pisano, M. A., 1968, Proteinase production by a species of *Cephalosporium*, *Appl. Microbiol.* **16:**90–96.

Oso, A., 1979, Mycelial growth and amylase production by *Talaromyces emersonii*, *Mycologia* **71:**520–536.

Parn, P., and Seviour, R. J., 1974, Pigments induced by organomercurial compounds in *Cephalosporium diospyri*, *J. Gen. Microbiol.* **85:**229–236.

Pearce, J. N., Bartman, C. D., Doerfler, D. L., and Campbell, I. M., 1981, 6-Methylsalicylic acid production in solid cultures of *Penicillium patulum* occurs only when an aerial mycelium is present, *Appl. Environ. Microbiol.* **41:**1407–1412.

Pirt, S. J., and Callow, D. S., 1960, Studies of the growth of *Penicillium chrysogenum* in continuous flow culture with reference to penicillin production, *J. Appl. Bacteriol.* **23:**87–98.

Pisano, M. A., Oleniacz, W. S., Mason, R. T., Fleischman, A. I., Vaccaro, S. E., and Catalano, G. R., 1963, Enzyme production by species of *Cephalosporium*, *Appl. Microbiol.* **11:**111–115.

Pitt, D., and Poole, P. C., 1981, Calcium-induced conidiation in *Penicillium notatum* in submerged culture, *Trans. Br. Mycol. Soc.* **76:**219–230.

Pitt, D., Mosley, M. J., and Barnes, J. C., 1983, Glucose oxidase activity and gluconate production during calcium-induced conidiation of *Penicillium notatum* in submerged culture, *Trans. Br. Mycol. Soc.* **81:**21–27.

Pitt, J. I., 1973, An appraisal of identification methods for *Penicillium* species: Novel taxonomic criteria based on temperature and water relations, *Mycologia* **55:**1135–1157.

Pitt, J. I., 1979, *The Genus Penicillium and Its Teleomorphic States Eupenicillium and Talaromyces*, Academic Press, New York.

Ramirez, C., 1982, *Manual and Atlas of the Penicillia*, Elsevier, Amsterdam.

Ramirez, C., and Gonzalez, C. C., 1984, *Penicillium flavido-stipitatum* sp. nov., *Mycopathologia* **88:**3–7.

Ramirez, C., and Martinez, A. T., 1980, Some species of *Penicillium* recovered from the atmosphere in Madrid and from other substrata, *Mycopathologia* **72:**181–191.

Ramirez, C., Martinez, A. T., and Berenguer, J., 1980, Four new species of *Penicillium* isolated from the air, *Mycopathologia* **72:**27–34.

Raper, K. B., and Thom, C., 1949, *A Manual of the Penicillia*, Williams and Wilkins, Baltimore.

Rapp, P., Grote, E., and Wagner, F., 1981, Formation and location of 1,4-β-gluconases and 1,4-β-glucosidases from *Penicillium janthinellum*, *Appl. Environ. Microbiol.* **41:**857–866.

Read, M. A., and Seviour, R. J., 1984, Effect of pH on maximum specific growth rate (μ_{max}) of *Acremonium diospyri*, *Trans. Br. Mycol. Soc.* **82:**159–161.

Remsen, C. C., Hess, W. M., and Sassen, M. A. A., 1967, Fine structure of germinating *Penicillium megasporium* conidia, *Protoplasma* **64:**439–451.

Richards, R. L., and Quackenbush, F. W., 1974, Alternate pathways of linolenic acid biosynthesis in growing cultures of *Penicillium chrysogenum*, *Arch. Biochem. Biophys.* **165:**780–786.

Righelato, R. C., 1978, The kinetics of mycelial growth, in: *Fungal Walls and Hyphal Growth* (J. H. Burnett and A. P. J. Trinci, eds.), Cambridge University Press, Cambridge, pp. 385–401.

Rintz, R. E., 1973, A zonal leaf spot of water hyacinth caused by *Cephalosporium zonatum*, *Hyacinth Control J.* **11:**41–44.

Rizza, V., and Kornfeld, J. M., 1969, Components of conidial and hyphal walls of *Penicillium chrysogenum*, *J. Gen. Microbiol.* **58:**307–315.

Roberts, D. W., and Humber, R. A., 1981, Entomogenous fungi, in: *Biology of Conidial Fungi*, Vol. 2 (G. T. Cole and B. Kendrick, eds.), Academic Press, New York, pp. 201–236.

Rudakov, O. L., 1978, Physiological groups in mycophilic fungi, *Mycologia* **70:**150–159.

Ruperez, P., Gomez-Miranda, B., and Leal, J. A., 1983, Extracellular β-malanoglucan from *Penicillium erythromellis*, *Trans. Br. Mycol. Soc.* **80:**313–318.

Ryu, D. D. Y., and Hospodka, J., 1980, Quantitative physiology of *Penicillium chrysogenum* in penicillin fermentation, *Biotechnol. Bioeng.* **22:**289–298.

Samuels, G. J., 1973, The myxomyceticolous species of *Nectria*, *Mycologia* **65:**401–420.

Samuels, G. J., 1976, Perfect states of *Acremonium*: The genera *Nectria*, *Actiniopsis*, *Ijuhya*, *Neohenningsia*, *Ophiodictyon* and *Peristomialis*, *N. Z. J. Bot.* **14:**231–260.

Sassen, M. M. A., Remsen, C. C., and Hess, W. M., 1967, Fine structure of *Penicillium megasporum* conidiospores, *Protoplasma* **64:**75–88.

Satoh, T., Beppu, T., and Arima, K., 1977, Purification and properties of blood-coagulating protease from *Cephalosporium* sp., *Agric. Biol. Chem.* **41**:293–298.

Scott, D. B., 1968, *The genus Eupenicillium Ludwig*, C.S.I.R., Pretoria.

Segretain, G., 1959, Description d'une nouvelle espèce de *Penicillium*: *Penicillium marneffei* n. sp., *Bull. Trimest. Soc. Mycol. Fr.* **75**:412–416.

Sekiguchi, J., Gaucher, G. M., and Costerton, J. W., 1975a, Microcycle conidiation in *Penicillium urticae*: An ultrastructural investigation of spherical spore growth, *Can. J. Microbiol.* **21**:2048–2058.

Sekiguchi, J., Gaucher, G. M., and Costerton, J. W., 1975b, Microcycle conidiation in *Penicillium urticae*: An ultrastructural investigation of conidial germination and outgrowth, *Can. J. Microbiol.* **21**:2059–2068.

Sekiguchi, J., Gaucher, G. M., and Costerton, J. W., 1975c, Microcycle conidiation in *Penicillium urticae*: An ultrastructural investigation of conidiogenesis, *Can. J. Microbiol.* **21**:2069–2083.

Seviour, R. J., and Codner, R. C., 1973, Effect of light on carotenoid and riboflavin production by the fungus *Cephalosporium diospyri*, *J. Gen. Microbiol.* **77**:403–415.

Seviour, R. J., and Read, M. A., 1983, Organomercurials as photomimetic compounds in carotenoid production by species of *Acremonium* and *Cephalosporium*, *Trans. Br. Mycol. Soc.* **81**:163–165.

Singh, U. P., Vishwakarma, S. N., and Basuchaudhuy, K. C., 1978, *Acremonium sordidulum* mycoparasitic on *Colletotrichum dematium* f. *truncata* in India, *Mycologia* **70**:453–455.

Smith, J. E., Anderson, J. G., Deans, S. G., and Davis, B., 1977, Asexual development in *Aspergillus*, in: *Genetics and Physiology of Aspergillus* (J. E. Smith and J. A. Pateman, eds.), Academic Press, London, pp. 23–58.

Stokes, P. M., and Lindsay, J. E., 1979, Copper tolerance and accumulation in *Penicillium ochrochloron* isolated from copper-plating solution, *Mycologia* **71**:796–806.

Stolk, A. C., 1969, Four new species of *Penicillium*, *Antonie van Leeuwenhoek J. Microbiol. Serol.* **35**:261–274.

Stolk, A. C., and Samson, R. A., 1972, The genus *Talaromyces*, *Stud. Mycol.* **2**:1–67.

Stolk, A. C., and Veenbaas-Rijks, J. W., 1974, *Eupenicillium osmophilum* sp. n., *Antonie van Leeuwenhoek J. Microbiol. Serol.* **40**:1–5.

Swart, H. J., 1970, *Penicillium dimorphosporum* sp. nov., *Trans. Br. Mycol. Soc.* **55**:310–313.

Thevelein, J. M., 1984, Regulation of trehalose mobilisation in fungi, *Microbiol. Rev.* **48**:42–59.

Tubaki, K., 1958, Studies on the Japanese hyphomycetes. 5. Leaf and stem group with a discussion of the classification of hyphomycetes and their perfect stages, *J. Hattori Bot. Lab.* **21**:142–244.

Tubaki, K., 1973, Aquatic sediment as a habitat of *Emericellopsis* with a description of an undescribed species of *Cephalosporium*, *Mycologia* **65**:938–941.

Tweedie, J. W., and Segel, I. H., 1970, Specificity of transport processes for sulphur, selenium and molybdenum anions by filamentous fungi, *Biochem. Biophys. Acta* **196**:95–106.

Uchida, J. Y., and Aragaki, M., 1982, *Acremonium* leaf spot of *Syngonium podophyllum* cultivar Green Gold: Nomenclature of the causal organism and chemical control, *Plant Dis.* **66**:421–423.

Udagawa, S., and Horie, Y., 1973a, Some *Eupenicillium* from soils of New Guinea, *Trans. Jpn. Mycol. Soc.* **14**:370–387.

Udagawa, S., and Horie, Y., 1973b, Surface ornamentation of ascospores in *Eupenicillium* species, *Antonie van Leeuwenhoek J. Microbiol. Serol.* **39**:313–319.

Ugalde, U. O., and Pitt, D., 1983a, Silicone coating to prevent acretion on glass walls by *Penicillium cyclopium* grown in shaken flask culture, *Trans. Br. Mycol. Soc.* **81**:412–415.

Ugalde, U., and Pitt, D., 1983b, Morphology and calcium-induced conidiation of *Penicillium cyclopium* in submerged culture, *Trans. Br. Mycol. Soc.* **80**:319–325.

Ugalde, U. O., and Pitt, D., 1984, Subcellular sites of calcium accumulation and relationships with conidiation in *Penicillium cyclopium*, *Trans. Br. Mycol. Soc.* **83**:547–555.

Unger, P. D., and Hayes, A. W., 1975, Chemical composition of the hyphal wall of a toxigenic fungus, *Penicillium rubrum* Stoll, *J. Gen. Microbiol.* **91**:201–206.

Von Arx, J. A., 1981, *The Genera of Fungi Sporulating in Pure Culture*, J. Cramer, FL-9490, Vaduz.

Watkinson, S. C., 1975, Regulation of coremium morphogenesis in *Penicillium claviforme*, *J. Gen. Microbiol.* **87**:292–300.

Watkinson, S. C., 1977, Effect of amino acids on coremium development in *Penicillium claviforme*, *J. Gen. Microbiol.* **101**:269–275.

Watkinson, S. C., 1979, Growth of rhizomorphs, mycelial strands, coremia and sclerotia, in: *Fungal Walls and Hyphal Growth* (J. H. Burnett and A. P. J. Trinci, eds.), Cambridge University Press, Cambridge, pp. 93–113.

Watkinson, S., 1981, Accumulation of amino acids during development of coremia in *Penicillium claviforme*, *Trans. Br. Mycol. Soc.* **76**:231–236.

Whittaker, A., and Long, P. A., 1973, Fungal pelleting, *Proc. Biochem.* **8**:27–31.

Wilson, C. L., 1965, Consideration of the use of persimmon wilt as a silvercide for weed persimmon, *Plant Dis. Rep.* **49**:780–791.

Wright, B. E., 1973, *Critical Variables in Differentiation*, Prentice-Hall, Englewood Cliffs, New Jersey.

Yagi, J., Yano, T., Kubochi, Y., Hattori, S., Ohashi, M., Sakai, H., Jomon, K., and Ajisaka, M., 1972, Studies on the alkaline proteinase from *Cephalosporium*. 1. Purification of the enzyme, *J. Ferment. Technol.* **50**:592–599.

Zachariah, K., and Fitz-James, P. C., 1967, The structure of phialides in *Penicillium claviforme*, *Can. J. Microbiol.* **13**:249–256.

Zeidler, G., and Margalith, P., 1973, Modification of the sporulation cycle in *Penicillium digitatum*, *Can. J. Microbiol.* **19**:481–483.

Genetics of the *Penicillia* 3

G. SAUNDERS and G. HOLT

1. INTRODUCTION

The genus *Penicillium* contains species that lack a demonstrable sexual or perfect stage as well as those with one. Many species of *Penicillium* have actual and potential applications in biotechnology (see Chapter 1), but much interest has focused on the production of secondary metabolites, especially antibiotics. For example, the production of penicillin by the imperfect species *P. chrysogenum* has been intensively developed as an industrial process for the last 40 years (Queener and Schwartz, 1979) (see also Chapter 5). However, within this time, no clear understanding of the genetic basis for the regulation of penicillin production has emerged. While this status undoubtedly reflects the absence of classic genetic studies via the sexual cycle, another significant contributing factor is the complexity of secondary-metabolite synthesis in general.

From the point of view of strain development, there are now a range of genetic tools available. These include traditional approaches to mutation induction as well as protoplast fusion, an efficient means for bringing about genetic recombination. However, at present, the barrier to analysis of these genetic changes remains the limitations imposed by parasexual analysis. It seems possible that this block may be removed with the advent of recombinant DNA techniques.

In this chapter, we will outline the means for bringing about genetic change within the *Penicillia*, drawing heavily on our experience and that of others with *P. chrysogenum*. Emphasis will be placed on the practical aspects of inducing mutation and bringing about recombination. In addition, our work on introducing recombinant DNA technology into the genetic study of

G. SAUNDERS • School of Biotechnology, Faculty of Engineering and Science, The Polytechnic of Central London, London W1M 8JS, England. G. HOLT • Biological Laboratory, University of Kent at Canterbury, Kent CT2 TNJ, England.

this industrially important fungus will be described. Finally, genetic studies in other species of *Penicillium* will be reviewed.

2. MUTATION INDUCTION

As might be expected, a wide range of chemical and physical agents have been used to induce mutation in *P. chrysogenum*. In addition to the traditionally employed mutagen, far-UV light ($\approx$254 nm), others, including X rays and chemical alkylating agents, have been used. Details of the mutagenic treatment conditions for *P. chrysogenum* with a range of physical and chemical agents can be found in P. J. M. Normansell *et al.* (1979). For many years now, it has been realized that many factors, both environmental and genetic (Saunders and Holt, 1982), can affect overall mutant yield and also mutant type isolated. Mutagen dose, treatment conditions, and the physiological state of the cells or spores can make a significant difference in the end result (Holt and Saunders, 1986). However, from studies with a wide range of organisms, it has been shown that the single most important cellular factor that affects mutation induction is DNA repair (Kimball, 1978), and this rule has been shown to apply equally to *P. chrysogenum* (Saunders and Holt, 1982). However, before we discuss the role of DNA repair in this organism, it is necessary to describe how material should be prepared for mutagenesis and how the effectiveness of a mutagen treatment is assessed.

2.1. Preparation of Material for Mutagenesis

It is of course desirable to carry out a mutagenic treatment on haploid material whenever possible, particularly if the desired mutant is likely to be recessive in nature. The majority of conidia of wild-type strains of *P. chrysogenum* can be shown to be haploid using an appropriate nuclear stain and light microscopy (Morris *et al.*, 1982). However, the ratio of multinucleate to uninucleate spores can vary from strain to strain, particularly in industrially improved isolates (S. D. Rogers and G. Holt, unpublished observations).

Spore suspensions can be prepared in either water or 0.01 % (vol./vol.) Tween 80 in distilled water. Clumping of spores can occur and is most effectively dealt with by agitating the spore suspension in a universal bottle containing sterile glass beads (3 mm in diameter). Finally, the resulting spore suspension can be filtered through a sintered glass filter (porosity 1) to remove any mycelial debris. This suspension can then be directly irradiated, or if chemical treatment is being used, the spores can be pelleted and resuspended in the appropriate buffer (Rogers and Holt, 1984a).

The best way to assess the level of mutants obtained from a particular mutagenic treatment is to score an easily detectable phenotype such as drug resistance or reversion of an auxotrophic marker. For example, in our laboratory, resistance to 5-fluorouracil (5-FU) or the polyene antibiotic candidine is routinely employed to estimate induced mutation frequency (Rogers and Holt, 1984a, b). Thus, by treating a population of spores ($\approx 5 \times 10^7$ ml) with a mutagen, the number of resistant mutants induced can be estimated by selecting on plates that contain the antimetabolite. By reference to the number of mutants obtained by a particular mutagenic dose and the surviving fraction of spores at that dose, mutation frequency per surviving spore can be estimated (Auerbach, 1976). Once a suitable system for scoring mutants in this way is obtained, optimum conditions for mutagenesis can be investigated and the relative efficiencies of different mutagens compared. In addition, one can seek conditions, either genetic or environmental, that lead to an elevation in the yield of mutants obtained. It should be remembered that a mutagenic test system optimized for, say, auxotrophic reversion or drug resistance, may not be readily translated into the best system for selecting mutations in genes for secondary metabolism, but it can often provide valuable information. For example, such a system can be used to demonstrate whether an organism or strain is actually mutated at all under the conditions being employed.

2.2. Factors That Affect Mutant Yield

One of the most common factors that affect the isolation of mutants is the necessity, in many cases, of an expression time (Auerbach, 1976; Lemontt and Lair, 1982). This represents the time required for an initial damage induced in DNA to be converted to a heritable mutation finally in the form of a recognizably altered phenotype. This time varies among organisms, strains, and often the type of mutation being scored. An expression period can also help reduce any problems associated with the use of multinucleate cells (Rowlands and Normansell, 1983; Rowlands, 1984a). In many cases, it is sufficient to obtain expression by allowing growth to take place directly on the agar plates onto which a mutagenized spore suspension is spread. Usually, the need for an expression time and its duration must be determined empirically for the mutation and organism concerned. For example, with strains derived from *P. chrysogenum* NRRL 1951, after far-UV treatment, an expression time of 6 hr is routinely allowed on nonselective agar before selection for 5-FU- or candidine-resistant mutants is made. Technically, this can be achieved in one of two ways, either by spreading the mutagenized spore suspension onto a cellophane disk that can be transferred from a nonselective plate to a selective one or by overpouring the selective agent onto an initially nonselective medium (Saunders and Holt, 1982; Rogers and Holt, 1984a). Expression times vary with strains, and for the high-penicillin-producing strain of *P. chrysogenum* P2 (Queener and Schwartz, 1979), no apparent expression time is required to obtain maximum yield of 5-FU-resistant mutants.

2.3. DNA Repair

In this connection, a large number of repair-deficient strains of *P. chrysogenum* have been isolated and shown to be altered in their respective kinetics of induction of mutations by far-UV light (Rogers and Holt, 1984a). The 28 strains isolated by the latter authors on the basis of an increased sensitivity to far-UV light or the chemical mutagen 4-nitroquinoline-1-oxide (4-NQO) can be broadly divided into two groups on the basis of whether or not they are mutated by UV light. A simple DNA-repair assay has been devised and used to demonstrate physically a repair impairment in two strains. One strain, HP508, was shown to be hypermutable—i.e., it showed an increased overall induced mutation frequency—and illustrates well the type of mutation that could be of significant use in industrial strain-development programs. Wild-type strains of *P. chrysogenum* have been shown to produce extracellular deoxyribonuclease (DNase) and ribonuclease (RNase). Certain of the repair-deficient isolates are either DNase- or RNase-deficient (Rogers and Holt, 1984a). More recently, mutants sensitive to the base analogue 2-aminopurine have been isolated (S. D. Rogers and G. Holt, unpublished observations). The types and characteristics of repair-deficient strains of *P. chrysogenum* are summarized in Table I.

Table I. Types of *P. chrysogenum* Mutants Altered in Their Response to a Range of Mutagenic Agents

Type	Mutagen sensitivity[a]	Phenotypic characteristics[a]
1	Sensitive to UV and 4-NQO	Decreased UV and 4-NQO mutability. Sensitive to a range of other mutagens including EMS, MMS, and X rays. Some members do not produce extracellular DNase or RNase.
2	Sensitive to UV and 4-NQO	Increased UV and 4-NQO mutability. Some members show this effect only at low doses. DNA-repair assay used to demonstrate inability in a representative of this group to remove SI-nuclease-sensitive sites from its DNA.
3	Sensitive to 2-AP	Strain demonstrates an increased spontaneous mutation frequency. DNA shown by chromatographic analysis to be undermethylated.

[a] (UV) Ultraviolet light; (4-NQO) 4-nitroquinoline-1-oxide; (EMS) ethylmethanesulfonate; (MMS) methylmethanesulfonate; (2-AP) 2-aminopurine.

2.4. Isolation of Specific Mutant Types

The type of mutant required will often be dictated by the type of genetic investigation taking place. For mapping studies, a wide range of variability is desirable, to permit assignment of diverse genes to linkage groups and ultimately their ordering on that linkage group relative to one another (see Section 3). Of particular use in this respect are spore-color mutations (including brown, yellow, white, and pale green in *P. chrysogenum*). Their use becomes particularly apparent in the isolation and subsequent breakdown of diploids. Similarly, resistance markers are of considerable value, and a large number have been isolated including resistance to 5-FU, candidine, amphotericin B, iodoacetate, methyl-1-(butylcarbamyl-1)-2-benzimidazole carbamate (benlate), and 2-deoxyglucose. Both color mutations and resistance mutations can be isolated with relative ease. Other types of mutations, e.g., particularly auxotrophic mutants and those impaired in penicillin production, are harder to isolate, requiring somewhat lengthier screening procedures, although certain auxotrophs can be directly isolated by selecting for resistance to antimetabolites such as selenate and chlorate (Birkett and Rowlands, 1981). Where screening is necessary, potential auxotrophs are usually identified by replicating master plates of (≈ 50) mutagenized survivors to a minimal medium and scoring for colonies that fail to show growth relative to a control replicate on complete medium. Various enrichment procedures can be employed to increase the proportion of auxotrophic mutants among survivors. A period of incubation in or on minimal medium to allow germination of prototrophs is usually followed by a treatment designed to discriminate between germinated and nongerminated spores (Fincham *et al.*, 1979). In *P. chrysogenum*, Macdonald (1963) has used the polyene antibiotic nystatin to successfully enrich for auxotrophs.

Mutants impaired in the production of penicillin have greatly assisted the study of the biosynthetic pathway of this antibiotic (see Section 3). Isolation of such Npe strains (P. J. M. Normansell *et al.*, 1979) can be carried out in a number of ways, all of which are rather laborious. Usually, bioactivity assays of mutagenized colonies are performed directly on agar plates by overpouring with agar seeded with the test organism or by growing colonies on agar plugs and transferring them at a later time to the surfaces of seeded assay plates. Alternatively, mutagenized survivors can be subcultured into liquid fermentation medium and assayed for penicillin production. Unless the conditions are carefully defined and controlled, a poor correlation is often observed between production on surface culture and liquid culture. In industry, where tens of thousands of mutagenized survivors of *P. chrysogenum* are screened for improved penicillin titer, almost on a weekly basis, miniaturized or automated screens or both have been successfully applied (Queener and Schwartz, 1979; Rowlands, 1984b).

3. PARASEXUAL CYCLE AND PROTOPLAST FUSION

Although *P. chrysogenum* lacks a sexual cycle, it has proven possible to carry out genetic mapping studies employing the alternate parasexual route (Pontecorvo *et al.*, 1953). The parasexual cycle features recombination occurring during mitosis, as opposed to meiosis, and usually proceeds in three distinct stages (see Fig. 1). First, a heterokaryon is formed (in practice, this involves complementation of easily scored phenotypes, e.g., color mutations and auxotrophic requirements). The second stage involves the isolation of stable diploids derived by nuclear fusion in the heterokaryon. Finally, segregants are derived, of either diploid or haploid nature, and formed by nondisjunction or genetic crossover during mitotic division of the heterozygous diploid. Thus, the overall effects of the parasexual cycle are similar to those of a sexual cycle, with the biggest difference occurring in the frequency of recombination events. In *P. chrysogenum*, the frequencies of diploid formation and subsequent segregation have been estimated to be 1 in 10^6 and 1 in 10^2, respectively (cited in Ball, 1983). Thus, in parasexual genetics, it is necessary to use selective techniques to isolate both diploids and subsequent segregation products. Standard procedures used to achieve these successive steps in the parasexual process are outlined below.

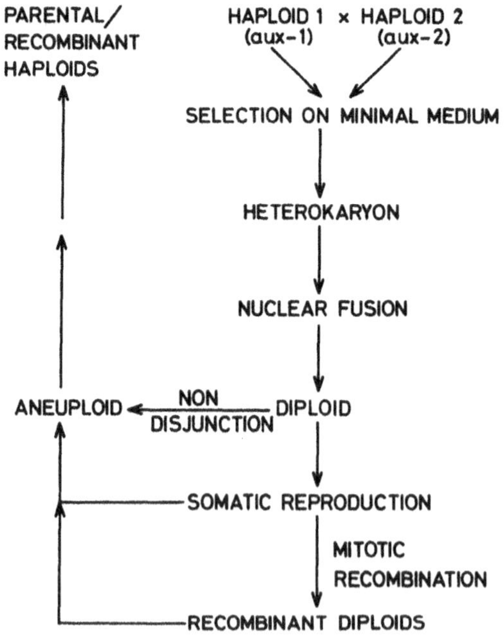

Figure 1. Schematic representation of the standard parasexual cycle.

3.1. Isolation of Heterokaryons and Diploids

Prior to the advent of protoplast-fusion techniques, heterokaryons were usually isolated by plating spores of two different auxotrophically marked strains as a mixture on complete medium followed by transfer to appropriate unsupplemented minimal medium. If complementary spore-color mutations were used, well-mixed heterokaryotic growth could be recognized. Putative diploid sectors, arising by rare fusion of nuclei within the heterokaryon, could be recognized as fast-growing sectors arising from the heterokaryotic growth. Once again, if complementary spore-color mutations were used, diploids would possess the wild-type green spore coloration. Subsequently, the diploid status was further confirmed by measuring the size of spores from the putative diploid using a microscope with an eyepiece micrometer, the diploid having a diameter of 1.3 times that of the haploid. Today, heterokaryons and diploids are more commonly obtained by fusion of protoplasts of the two strains participating in the cross. Protoplast formation from *P. chrysogenum* was first reported by Fawcett *et al.* (1973), who used the protoplasts obtained for the generation of cell-free enzymatic systems. Subsequently, the use of protoplasts to bring about recombination in this organism employing polyethylene glycol (PEG) and calcium chloride was described (Anné and Peberdy, 1975, 1976). Since this time, the use of protoplast fusion to bring about industrial development of penicillin fermentations has become widespread (Elander, 1981). In our laboratories, protoplasts from *P. chrysogenum* are formed by incubation of mycelial pellets in an enzyme lytic mix, typically composed of *Helix pomatia* juice 1% (vol./vol.) and 5 mg/ml Novo cellulase, in 0.6 M potassium chloride. In practice, 0.2–0.5 g wet weight of 36-hr-old mycelium grown in Aspergillus Complete Medium (ACM) (Ditchburn *et al.*, 1974) at 26°C on a reciprocal shaker (200 rev/min) is incubated, with gentle agitation, at 26°C for 2–4 hr. During this time, cell-wall digestion occurs and protoplasts are released into the osmotically stabilized medium. Subsequently, mycelial debris is removed by passage of the suspension through a porosity 1 sintered glass filter. Protoplasts are then pelleted by centrifugation and washed three times with 0.6 M KCl. Typically, 1 ml containing 5×10^7 to 1×10^8 viable protoplasts is obtained. Mixtures of protoplasts from two different strains are then fused using a 30% (wt./vol.) PEG solution (molecular weight 4000 or 6000) supplemented with 10 mM calcium chloride at pH 8 (e.g., to 1 ml of a mixed protoplast preparation is added 9 ml PEG solution). The viscous protoplast–PEG mixture is then mixed gently for 5–10 min before being diluted in 0.6 M KCl and plated onto osmotically buffered regeneration agar (ACM supplemented with 0.6 M KCl). After 7–10 days' growth at 26°C, heterokaryons can be seen as irregularly growing colonies giving rise to both parental colors.

3.2. Analysis of Segregants from the Diploid

After diploid formation, segregation may occur at one or both of the two subsequent steps in the parasexual cycle; mitotic crossover during division of the diploid may occur, in addition to segregation of whole chromosomes from an initial aneuploid state to give hyperhaploids or hyperdiploids. For *P. chrysogenum*, 5-FU (Ball and Azevedo, 1976) increases the frequency of mitotic crossover, while *p*-fluorophenylalanine and benlate are commonly used to enhance the frequency of aneuploidy (Edwards *et al.*, 1975). Whichever chemical or combination of chemicals is employed, the resulting segregants may be analyzed in the normal way both to assign genes to particular linkage groups and to estimate distances between genes (Macdonald and Holt, 1976). The stability of strains of *P. chrysogenum* (haploid or diploid) can vary enormously, a property that is clearly advantageous in certain situations (i.e., parasexual analysis) but undesirable when continued stability is required [e.g., in strain improvement (Ball, 1982)]. The reasons for such variable stability remain unclear, although the existence of transposonlike DNA elements in *P. chrysogenum* could go some way to explaining the observations made (Calos and Miller, 1980). However, for the purposes of parasexual analysis, both mitotic crossover and chromosome loss are promoted by chemical agents incorporated into growth medium.

The use of the parasexual cycle for investigations of this type can be illustrated with reference to work on the genetic analysis of penicillin production by *P. chrysogenum*. Initially, 78 non-penicillin-producing isolates of *P. chrysogenum* were obtained in this laboratory after mutagenesis and screening of survivors for lack of biological activity against *Bacillus subtilis*. Subsequently, complementation analysis of these mutants by crossing, in all possible pairwise combinations, to obtain diploids allowed the division of the large number of mutants into five complementation groups. This could be taken as evidence that at least five genes (designated *npe* V, W, X, Y, and Z) directly govern the biosynthetic pathway of pencillin. Biochemical analysis has shown members of complementation groups X, Y, and Z to be blocked before the synthesis of the tripeptide intermediate of penicillin biosynthesis (P. J. M. Normansell *et al.*, 1979). Members of groups V and W, however, do synthesize tripeptide and are therefore most likely to be blocked at ring closure or in the acyltransferase terminal reactions. Formation of diploids between Npe strains and others genetically marked on separate chromosomes allowed assignment of *npe* genes to their respective linkage groups. Three genes, *npe* W, Y, and Z, were found to be on linkage group I, while *npe* V was assigned to linkage group II and data indicate that *npe* X is not on linkage group I, II, III, or V.

The most detailed haploidization map of *P. chrysogenum* NRRL 1951 so far obtained shows six chromosomes. The majority of spore-color markers and three genes directly involved in penicillin biosynthesis were found to be on the same linkage group together with a gene that confers resistance to 5-FU (C. M. Blake and G. Holt, unpublished observations). Mitotic crossover has also been used to assign positions of *npe* alleles on linkage group I. Available data indicate that *npe* W, Y, and Z are all on the same arm of this chromosome, suggesting the possible occurrence of a gene cluster. Ordering of these genes relative to others on the same chromosome indicates *npe* W and Y to be linked to *flu* A (5-FU resistance) and genes involved in spore color. Other groups have presented haploidization maps of high-titer derivatives of *P. chrysogenum*, and these results tend to confirm the clustering of spore-color markers to one linkage group (Sermonti, 1956; Macdonald *et al.*, 1965; Ball, 1971).

4. RECOMBINANT DNA TECHNIQUES IN *PENICILLIUM*

The advantages afforded to the geneticist by an approach involving recombinant DNA techniques have been thoroughly discussed before (see, for example, Timmis, 1981; Elander, 1982). Gene-cloning systems available at present for the filamentous fungi lag far behind those available for prokaryotes, but several systems are now available for genetically well-characterized species such as *Neurospora crassa* (Stohl and Lambowitz, 1983; Grant *et al.*, 1984) and *Aspergillus nidulans* (Ballance *et al.*, 1983).

4.1. Extraction, Purification, and Base Analysis of Total DNA

The isolation of DNA from filamentous fungi is typically harder to achieve than is similar isolation from most bacteria. This difficulty stems primarily from the structural complexity and strength of the cell wall, which necessitates somewhat more vigorous cell-breakage techniques. Enzymatic hydrolysis of the cell wall can be used, but usually requires more than one enzyme. Typically, yields of DNA from *P. chrysogenum* are low when protoplasts or osmotically fragile mycelia are used as the source. Good yields of DNA can be obtained after cell breakage by following the procedure outlined in Saunders and Holt (1982). Cell breakage is best achieved by grinding lyophilized mycelium. Mycelium from 1–2 liters of liquid ACM (40-hr growth in ACM at 26°C, 200 rpm) is filtered, using either Whatman filter paper (22.0 cm in diameter) or sterile netting, blotted dry, and frozen in plastic petri dishes at −70°C for 1 hr. Subsequently, the petri dishes are placed in a Chemlab freeze drier for 48–72 hr, kept cold at −20°C and under vacuum (< 1 m torr). When the dishes are dry, sterile glass helices are mixed with the mycelium and the whole is ground to a fine powder with a cold pestle and mortar.

The final DNA pellet is dissolved in a convenient volume of TE buffer (10 mM Tris, 1 mM EDTA, pH 8). This crude DNA solution can be subsequently purified either by centrifugation in a cesium chloride gradient or by chromatography on a Sepharose 4B or Sephacryl-1000 column. For purification on CsCl, 1.05 g CsCl is added to each 1 ml of DNA solution with centrifugation at 35,000 rpm in an MSE Europa 55 centrifuge for 60–72 hr (either a 10 × 10 ml or an 8 × 25 ml angle rotor may be used, depending on the amount of DNA to be purified). After centrifugation, the total DNA (constituting both nuclear and mitochondrial fractions) can be visualized, removed, and finally purified according to standard methods (Cummings *et al.*, 1979).

Sepharose 4B and Sephacryl-1000 are gel-filtration chromatography matrices that will separate molecules primarily on the basis of molecular weight. Thus, using a Sepharose 4B column, it is possible to separate high-molecular-weight DNA from low-molecular-weight RNA and protein. In the case of derivatives of *P. chrysogenum* NRRL 1951, a prior RNase treatment is essential, to digest to low-molecular-weight pieces the double-stranded viral RNA (I. D. Normansell and Holt, 1978) that is present. Without this precaution, separation of DNA from viral RNA will not occur; indeed, the same problem can arise on CsCl gradients. The use of Sepharose 4B has certain attractive advantages, particularly with species or strains that do not contain viral RNA. Indeed, the majority of wild-type isolates of *P. chrysogenum* do not contain RNA viruses (P. Ford, G. Saunders, and G. Holt, unpublished observations), and therefore crude DNA extracts can be loaded onto a Sepharose 4B column without protease or RNase treatment. Pure DNA, in the solvent of choice, can be obtained in as little as 8 hr. For details of flow rate through the column and column dimensions, see Saunders *et al.* (1984).

Using high-performance liquid chromatographic techniques, it has been possible to quantitate the overall base composition of *P. chrysogenum* total DNA as being 51.4% G + C (Rogers *et al.*, 1986) (see Table II). This figure

Table II. Quantitation of the Base Content of *P. chrysogenum* DNA by High-Performance Liquid Chromatography

Nucleotide base	Mole %
Adenine	24.8
Cytosine	25.6
Guanine	25.8
Thymine	23.6
N^6-Methyl-adenine	0.1

correlates well with that obtained from buoyant density measurements (Saunders *et al.*, 1984). In addition, N^6-methyl-adenine has been detected in the DNA from this organism and is present at a level, relative to other bases, of 0.1 mole%. There have been few reports on the occurrence of methylated DNA in fungi. Those that have been published, however, indicate no universal pattern of methylation. The necessity for this methylation remains unclear; however, there are indications from our studies that in *P. chrysogenum*, methylation is at least associated with a "mismatch" type of repair system possibly analogous to those established in prokaryotes (Glickman and Radman, 1980). Its involvement in other cellular processes, e.g., restriction-modification systems, cannot, however, be ruled out at present.

4.2. Fractionation of DNA

Total DNA from a filamentous fungus can be considered to be composed of three types. Mitochondrial DNA comprises one type, and DNA from the nucleus consists of highly repeated sequences, usually coding for ribosomal RNA, in addition to the bulk heterogeneous DNA of single or low-copy-number genes. Mitochondrial and nuclear DNA from strains of *P. chrysogenum* can be separated in one of two ways using either density-gradient centrifugation, in combination with a dye specifically able to bind preferentially to A + T-rich DNA, or ion-exchange chromatography on RPC-5 Analog or NaCs-37. Mitochondrial DNA obtained from wild-type isolates of *P. chrysogenum* is shown in Figure 2.

In the first case, DNA extracted in the manner described in Section 4.1 is treated with protease and ribonuclease A. CsCl is then added to the DNA solution (1.05 g CsCl/ml liquid), and 0.5 mg bisbenzimide dye is added. Usually, the dye is allowed to dissolve in the DNA solution for 4–18 hr. After centrifugation at 35,000 rpm (MSE Europa 55M Centrifuge) at 18°C for 60–72 hr, the DNA bands in the tube are visualized in the same manner as described for CsCl–ethidium bromide gradients. Two distinct bands are usually observed (fluorescing purple), the upper (and fainter) band consisting of mitochondrial DNA, the lower, which in fact consists of two very close bands, comprising the nuclear DNA. The upper of these poorly resolved bands is enriched for repeated DNA sequences presumably analogous to the ribosomal DNA isolated from other filamentous fungi in the same way.

An alternative approach involving ion-exchange chromatography on RPC-5 Analog or NaCs-37 resins can be used; like the use of Sepharose 4B, this method offers certain advantages over density-gradient centrifugation. DNA is dissolved in TES buffer (10 mM Tris · HCl, 10 mM EDTA, 0.5 M NaCl, pH 7.2) and bound to an RPC-5 Analog column

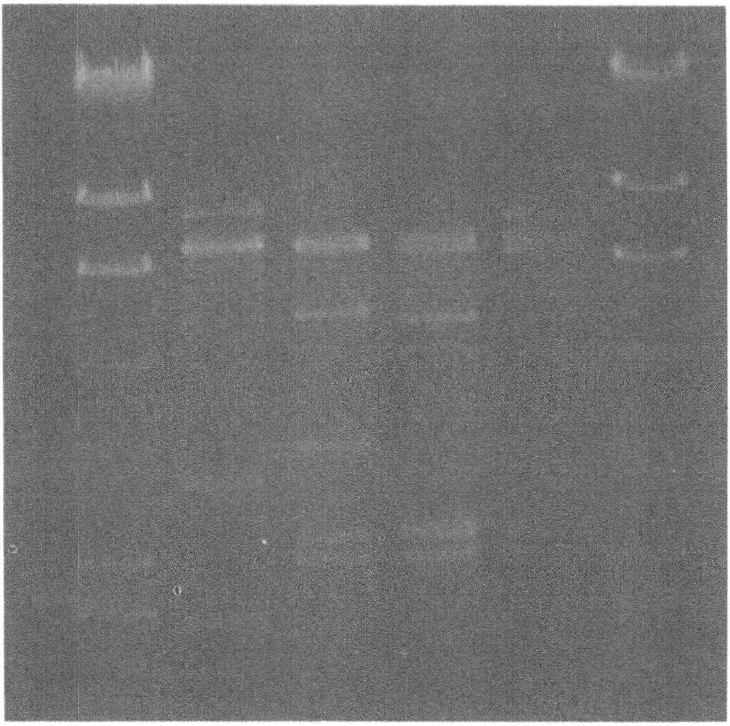

Figure 2. Variation in the mitochondrial DNA among *P. chrysogenum* NRRL 1951 and three other wild-type isolates (domestic codes PS1, PS2, and PS3). *Right to left*: (lanes 1 and 6) *Hin-d*III-digested; (2) NRRL 1951; (3) PS1; (4) PS2; (5) PS3. DNA in lanes 2–5 was digested with *Eco*RI.

(10 cm × 1 cm) equilibrated in the same buffer. A salt gradient is then applied to the column (Saunders *et al.*, 1984), and fractions are collected from the end of the column. The column itself is a Biorad Econo column connected to a peristaltic pump, which in turn is controlled by an LKB Ultragrad gradient maker. The bottom of the column is connected to a fraction collector.

Using a flow rate of 0.15 ml/min over 24 hr, 7.5 g RPC-5 Analog, and up to 1 mg DNA, complete fractionation of DNA into three types is observed. The first (type I) and second (type II) types to elute comprise the nuclear DNA highly enriched for the ribosomal DNA repeat unit. The third and last type to elute is mitochondrial DNA. In addition to yielding a more complete DNA fractionation, again, the use of chromatography of this sort does not require the use of an ultracentrifuge, and pure nucleic acid is obtained in a simple aqueous buffer. The presence of the high concentrations of NaCl (as high as 2.0 M) necessary to remove mitochondrial DNA from the column usually necessitates the removal of NaCl prior to enzymatic treatment. This can be done either by dialysis or by dilution prior to ethanol precipitation. Alternatively, sodium acetate may be used to elute nucleic acids, in which case concentration of DNA can take place without prior dialysis or dilution. The whole process can be much simplified if the aim is simply to isolate mitochondrial DNA, e.g., by using a stepwise salt gradient rather than one generated by a gradient maker. In this procedure, total DNA is bound to RPC-5 Analog or NaCs-37 in a universal bottle by resuspension of pelleted resin in the DNA solution. Successive washes and pelleting with 0.5, 0.75, and 2.0 M TES results in successive supernatants containing RNA, total nuclear DNA, and mitochondrial DNA. Using DNA purified and fractionated by a combination of Sepharose 4B and RPC-5 Analog, a mitochondrial restriction-enzyme map for *P. chrysogenum* NRRL 1951 has been constructed, and both ribosomal and mitochondrial DNA have been cloned in *E. coli* plasmid vectors (Smith and Holt, 1984). Most of the mitochondrial genome has also now been cloned in our laboratory in a yeast–*Escherichia coli* shuttle vector designed to permit the isolation of autonomously replicating sequences (*ars*) in yeast (P. Ford, G. Saunders, and G. Holt, unpublished observations).

4.3. A General Strategy for Vector Construction and the Generation of Gene Libraries

The construction of cloning vectors for fungi has assumed a distinct character of its own (see Fig. 3). As a consequence of the almost complete lack of plasmids and few usable dominantly acting resistance markers, vectors for filamentous fungi are based, with some exceptions, on *E. coli* replicons, typically pBR322 or pBR328. Identification of transformants is usually on the basis of complementation, by the vector, of a biosynthetic deficiency of the recipient strain.

The first step in this strategy involves the construction of a gene library of the fungus in a plasmid such as pBR322. The recombinant plasmids so obtained can subsequently be used in transformation experiments with auxotrophically marked strains of *E. coli* (e.g., *leuB*, *trpC*). Selection of transformants is carried out on a minimal medium lacking the appropriate amino

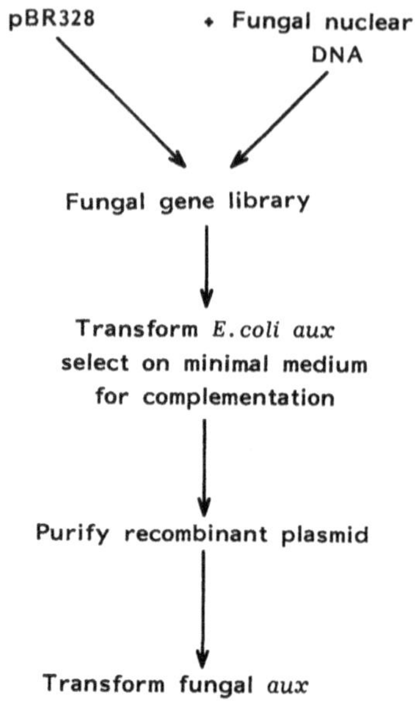

Figure 3. General strategy for the construction of fungal cloning vectors.

acid supplement. It is hoped that any colonies that grow will contain an insert of fungal DNA capable of complementing the mutant gene in the *E. coli* strain. Although there are barriers to the expression of eukaryotic DNA in *E. coli*, many steps in amino acid metabolism are shared in common by prokaryotes and eukaryotes. The concept of this form of complementation has been successfully proven for yeast, *Neurospora crassa*, *Aspergillus nidulans*, *Acremonium chrysogenum*, and *P. chrysogenum*. Once such a recombinant plasmid is obtained, it can be used in an attempt to transform similar biochemically deficient strains of the fungus. Such vectors, usually acting in an integrative manner, can subsequently be made to convert to the autonomous mode of replication (see Grant *et al.*, 1984).

The total genome size of *P. chrysogenum* has been estimated to be 11.5×10^9 daltons (Gupalo *et al.*, 1978). In constructing any gene library, the number of separate clones required to represent the entire genome obviously depends on the average insert size, which, as might be expected, gives an inverse relationship between these two parameters. Possibly the method of choice in the case of fungi, as a consequence of their larger genome size, is the use of a cosmid vector able to accommodate 45-kilobase (kb) pieces of DNA. However, cosmid cloning has some hidden pitfalls in that there is often a high incidence of nonrecombinant cosmids obtained. To overcome this problem, the use of a cosmid that requires an insert to allow positive selection of the vector is desirable and greatly assists the rapidity of construction of libraries of large pieces of DNA (Maniatis *et al.*, 1982).

By far the easiest initial route to library–vector construction, and the one most commonly used, employs either pBR322 or pBR328 into which size-fractionated partial digests of fungal DNA can be readily inserted. DNA fragments of size between 10 and 20 kb can be satisfactorily accommodated by pBR328. As little as 10 μg *P. chrysogenum* and 5 μg vector DNA can easily yield a library with a high probability of having all the genome represented. In our laboratory, using such quantities of DNA, gene libraries have been constructed of both *Bam*HI- and *Hind*III-generated NRRL 1951 fragments (5–10 and 10–20 kb pieces) using *E. coli* HB101 as recipient. Transformant colonies obtained can then be pooled by scraping them off the agar surfaces into L-broth [supplemented 50% (vol./vol.) with glycerol] to generate a clone bank. These cells can be stored frozen at −20°C for several years. Subsequently, aliquots of the clone bank can be spread onto selective minimal medium to score for complementation of the auxotrophic markers carried by the initially transformed strain (in the case of HB101 *pro*A, *leu*B). Alternatively, preparations of plasmid DNA from the pooled cells can be prepared and used to transform other *E. coli* auxotrophic derivatives in an effort to obtain complementation of other biosynthetic deficiencies.

5. GENETIC STUDIES IN OTHER SPECIES OF *PENICILLIUM*

In-depth genetic studies of other species of *Penicillium* are rarely found. Indeed, in those species that have a sexual cycle, meiotic or Mendelian analysis has not been reported, in contrast to the elegant systems established in fungi such as *Neurospora crassa* and *Aspergillus nidulans*. However, parasexual analysis of phytopathogenic species such as *P. digitatum*, *P. expansum*, and *P. italicum* has been undertaken and has generated linkage-group information (Stromnaes *et al.*, 1964; Fjeld and Stromnaes, 1966). Also, in a genetic study of alkaloid production by species of *Penicillium*, the biosynthesis of both cyclopenin and cyclopenol was found to be under nuclear control (Mohamed *et al.*, 1984). Heterokaryons were formed between auxotrophic derivatives of *P. cyclopium* (synthesizing cyclopenol and cyclopenin) and *P. viridicatum* (producing only traces of the alkaloids) and were found to produce both alkaloids in amounts similar to those produced by the *P. cyclopium* parent. This result indicates that alkaloid formation in this species is not subject to control by cytoplasmic genetic material. There have been a few reports of the use of mutagenesis to improve the production of other metabolites, e.g., mycophenolic acid by *P. stoloniferum* (Queener *et al.*, 1982) and gluconic acid by *P. funiculosum* (Mandal and Chatterjee, 1985). In the former investigation, use was made of the fact that mycophenolic acid has a precursor that is also common to the synthesis of ergosterol, a major component of fungal membranes. It was reasoned that selection for mutants resistant to polyene antibiotics would involve the selection of strains altered in sterol biosynthesis and hence also mycophenolic acid synthesis, and such selection was clearly demonstrated. Recently, mutagenesis has also been used to isolate hypercellulolytic mutants of *P. funiculosum* (Joglekar and Karanth, 1984).

An expanding volume of work involving interspecific recombination within the *Penicillia* via protoplast fusion has accumulated in the past 10 years. For example, Anné (1983) lists several species that have been used in interspecific crosses with *P. chrysogenum*, including *P. citrinum*, *P. roqueforti*, *P. patulum*, *P. stoloniferum*, and *P. cyclopium*. In one particular case, interspecific hybrids between *P. baarnense* and *P. chrysogenum* have been formed, and one aim was to obtain recombinants able to synthesize penicillin but possessing the ability to recombine sexually, a character of *P. baarnense* (Mellon *et al.*, 1983). Although recombinants were obtained at high frequency, no information was available at the time of publication as to the success of combining the two characters of sexuality and penicillin production. Indeed, recombination among a wide range of different species of *Penicillium* has now been reported (for a summary, see Anné, 1983). In certain instances, there have been indications of alterations of genetic regulation in recombinants when compared to the original parents. For example, it was reported by Anné (1977) that interspecific hybrids of *P. chrysogenum* and *P. roqueforti* produced a red pigment that neither parent could produce alone. Using isoenzyme analysis, it has been demonstrated how interspecific crosses between *Penicillium* species can be used to demonstrate the potential of altering genetic regulation by interspecific hybridizations, leading possibly to the generation of novel and useful metabolites (Mellon *et al.*, 1983).

As this brief review of investigation in the *Penicillia* demonstrates, there is much to be done to extend genetic studies, including molecular biology, to the many species of *Penicillium* important in fields such as enzyme production, plant pathogenesis, and cheese making, and, of course, in secondary metabolism.

REFERENCES

Anné, J., 1977, Somatic hybridisation between *Penicillium* species after induced fusion of their protoplasts, *Agricultura* **25:**1–117.

Anné, J., 1983, Protoplasts of filamentous fungi in genetics and metabolite production, in *Protoplasts 1983: Lecture Proceedings* (I. Potrykus, C. T. Harms, A. Hinner, R. Hutter, P. J. King, and R. D. Shillito, eds.), Birkhauser Verlag, Basel, Stuttgart, and New York, pp. 167–178.

Anne, J., and Peberdy, J. F., 1975, Conditions for induced fusion of fungal protoplasts in polyethylene glycol solutions, *Arch. Microbiol.* **105:**201–205.

Anné, J., and Peberdy, J. F., 1976, Induced fusion of fungal protoplasts following treatment with polyethylene glycol, *J. Gen. Microbiol.* **92:**413–417.

Auerbach, C., 1976, *Mutation Research: Problems, Results, Perspectives*, Chapman Hall, London.

Ball, C. 1971, Haploidisation analysis in *Penicillium chrysogenum, J. Gen. Microbiol.* **66:**63–69.

Ball, C., 1982, Genetic approaches to overproduction of beta-lactam antibiotics in eukaryotes, in: *Overproduction of Microbial Products* (V. Krumphanzl, B. Sikyta, and Z. Vanek, eds.), Academic Press, New York and London, pp. 515–532.

Ball, C., 1983, Protoplast fusion in commercially important microorganisms, in: *Bioactive Microbial Products: Development and Production* (L. J. Nisbet and D. J. Winstanley, eds.), Academic Press, New York and London, pp. 19–32.

Ball, C., and Azevedo, J. L., 1976, Genetic instability in parasexual fungi, in: *Proceedings of the Second International Symposium on the Genetics of Industrial Microorganisms* (K. D. Macdonald, ed.), Academic Press, New York and London, pp. 243–251.

Ballance, D. J., Buxton, F. P., and Turner, G., 1983, Transformation of *Aspergillus nidulans* by the orotidine-5′-phosphate decarboxylase gene of *Neurospora crassa, Biochem. Biophys. Res. Commun.* **112:**284–289.

Birkett, J. A., and Rowlands, R. T., 1981, Chlorate resistance and nitrate assimilation in industrial strains of *Penicillium chrysogenum, J. Gen. Microbiol.* **123:**281–285.

Calos, M. P., and Miller, J. H., 1980, Transposable elements, *Cell* **20:**579–595.

Cummings, D. J., Belcour, L., and Grandeschamp, C., 1979, Mitochondrial DNA from *Podospora anserina, Mol. Gen. Genet.* **171:**229–238.

Ditchburn, P., Giddings, B., and Macdonald, F. D., 1974, Rapid screening for the isolation of mutants of *Aspergillus nidulans* with increased penicillin yield, *J. Appl. Bacteriol.* **37:**515–523.

Edwards, G. F. St. L., Normansell, I. D., and Holt, G., 1975, Benlate induced haploidisation in diploid strains of *Aspergillus nidulans* and *Penicillium chrysogenum, Aspergillus Newslett.* **12:**15.

Elander, R. P., 1981, Strain improvement programmes in antibiotic producing microorganisms: Present and future strategies, in *Advances in Biotechnology*, Vol. 1 (M. Moo-Young, C. W. Robinson, and C. Vezina, eds.), Pergamon Press, Oxford, pp. 3–8.

Elander, R. P., 1982, Traditional versus current approaches to the genetic improvement of microbial strains, *Overproduction of Microbial Products* (V. Krumphanzl, B. Sikyta, and Z. Vanek, eds.), Academic Press, New York and London, pp. 353–370.

Fawcett, P. A., Loder, P. B., Duncan, M. J., Beesley, T. J., and Abraham, E. P., 1973, Formation and properties of protoplasts from antibiotic producing strains of *Penicillium chrysogenum* and *Cephalosporium acremonium, J. Gen. Microbiol.* **79:**293–309.

Fincham, J. R. S., Day, P. R., and Radford, A., 1979, *Botanical Monographs*, Vol. 4, *Fungal Genetics*, Blackwell, Oxford.

Fjeld, A., and Stromnaes, O., 1966, The parasexual cycle and linkage groups in *Penicillium expansum, Hereditas* **54:**389–403.

Glickman, B., and Radman, M., 1980, *E. coli* mutator mutants deficient in methylation instructed DNA mismatch correction, *Proc. Natl. Acad. Sci. U.S.A.* **77:**1063–1066.

Grant, D. M., Lambowitz, A. M., Rambosek, J. A., and Kinsey, J. A., 1984, Transformation of *Neurospora crassa* with recombinant plasmids containing the cloned glutamate dehydrogenase (*am*) gene: Evidence for autonomous replication of the transformation plasmid, *Mol. Cell. Biol.* **4:**2401–2051.

Gupalo, I. D., Surkov, V. V., Tikhonenko, T. I., Ergorov, A. A., and Parjenov, N. N., 1978, DNA from *Penicillium chrysogenum, Vestn. Mosk. Univ. Ser.* **16:**62–66.

Holt, G., and Saunders, G., 1986, Genetic modification of industrial microorganisms, in *Comprehensive Biotechnology*, Vol. 1 (M. Moo-Young, ed.), Pergamon Press, Oxford, pp. 51–76.

Joglekar, A. V., and Karanth, N. G., 1984, Studies on cellulase production by a mutant: *Penicillium funiculosum* UV-49, *Biotechnol. Bioeng.* **26:**1079–1084.

Kimball, R. F., 1978, The relation of repair phenomena to mutation induction in bacteria, *Mutat. Res.* **55:**85–120.

Lemontt, J. F., and Lair, S. V., 1982, Plate assay for chemical and radiation induced mutagenesis of *an*1 in yeast as a function of post treatment DNA replication: The effect of *rad*6-1, *Mutat. Res.* **93:**339–352.

Macdonald, K. D., 1963, The selection of auxotrophs of *P. chrysogenum* with nystatin, *Genet. Res.* **11:**327–330.

Macdonald, K. D., and Holt, G., 1976, Genetics of biosynthesis and overproduction of penicillin, *Sci. Prog. (Oxford)* **63:**547–573.

Macdonald, K. D., Hutchinson, J. M., and Gillet, W. A., 1965, Heterozygous diploids of *Penicillium chrysogenum* and their segregation pattern, *Genetica* **36**:378–397.

Mandal, S. K., and Chatterjee, S. P., 1985, Improved production of calcium gluconate by mutants of *Penicillium funiculosum, Curr. Sci.* **54:**149–150.

Maniatis, T., Fritsch, E. F., and Sambrook, J., 1982, *Molecular Cloning—A Laboratory Manual,* Cold Spring Harbor Laboratory, Cold Spring Harbor, New York.

Mellon, F. M., Peberdy, J. F., and Macdonald, K. D., 1983, Hybridisation of *Penicillium chrysogenum* and *Penicillium baarnense* by protoplast fusion: Genetic and biochemical analysis, in: *Protoplasts 1983: Lecture Proceedings* (I. Potrykus, C. T. Harms, A. Hinner, R. Hutter, P. J. King, and R. D. Shillito, eds.), Birkhauser Verlag, Basel, Stuttgart, and New York, pp. 310–311.

Mohamed, Z. S., Todorova-Dragonova, R. W., and Luckner, M., 1984, Nuclear inheritance of the biosynthesis of cyclopenin and cyclopenol in *P. cyclopium, Z. Allg. Mikrobiol.* **24:**615–618.

Morris, N. R., Kirsch, D. R., and Oakley, B. R., 1982, Molecular and genetic methods for studying mitosis and spindle proteins in *Aspergillus nidulans, Methods Cell Biol.* **25:**107–130.

Normansell, I. D., and Holt, G., 1978, Viruses in strains of *Penicillium chrysogenum* impaired in penicillin biosynthesis, *Mycovirus Newslett.* **6:**15.

Normansell, P. J. M., Normansell, I. D., and Holt, G., 1979, Genetic and biochemical studies of mutants of *Penicillium chrysogenum* impaired in penicillin biosynthesis, *J. Gen. Microbiol.* **112:**113–126.

Pontecorvo, G., Roper, J. A., Hemmons, L. M., Macdonald, K. D., and Bufton, A. W. J., 1953, The genetics of *Aspergillus nidulans, Adv. Genet.* **5**:141–238.

Queener, S., and Schwartz, R., 1979, Penicillins: Biosynthetic and semi-synthetic, *Economic Microbiology,* Vol. 3 (A. H. Rose, ed.), Academic Press, New York and London, pp. 35–122.

Queener, S. W., Wilkerson, S. G., and Nash, C. H., 1982, Sterol content and titre of mycophenolic acid in polyene antibiotic resistant mutants of *Penicillium stoloniferum,* in: *Overproduction of Microbial Products* (V. Krumphanzl, B. Sikyta, and Z. Vanek, eds.), Academic Press, New York and London, pp. 535–548.

Rogers, S. D., and Holt, G., 1984a, DNA repair and mutagenesis in *Penicillium chrysogenum, Trans. Biochem. Soc.* **12:**646–647.

Rogers, S. D., Rogers, M. E., Saunders, G., and Holt, G., 1986, Isolation of mutants sensitive to 2-aminopurine and alkylating agents and evidence for the role of DNA methylation in *Penicillium chrysogenum, Curr. Genet.* **10:**557–560.

Rogers, S. D., and Holt, G., 1984b, DNA repair deficient mutants from *Penicillium chrysogenum, J. Appl. Microbiol. Biotechnol.* **20:**251–255.

Rowlands, R. T., 1984a, Industrial strain improvement: Mutagenesis and random screening procedures, *Enzyme Microbiol. Technol.* **6:**3–10.

Rowlands, R. T., 1984b, Industrial strain improvement: Rational screens and genetic recombination techniques, *Enzyme Microbiol. Technol.* **6:**290–300.

Rowlands, R. T., and Normansell, I. D., 1983, Current strategies in industrial selection, in

Bioactive Microbial Products (L. J. Nisbett and D. J. Winstanley, eds.), Academic Press, New York and London, pp. 1–18.

Saunders, G., and Holt, G., 1982, Far ultra-violet light sensitive mutants of *Streptomyces clavuligerus*, *J. Gen. Microbiol.* **128**:381–385.

Saunders, G., Allsop, A. E., and Holt, G., 1980, Modern developments in mutagenesis, *J. Chem. Technol. Biotechnol.* **32**:354–364.

Saunders, G., Rogers, M. E., Adlard, M. W., and Holt, G., 1984, Chromatographic resolution of nucleic acids extracted from *Penicillium chrysogenum*, *Mol. Gen. Genet.* **194**:343–346.

Sermonti, G., 1956, Complementary genes which affect penicillin yields, *J. Gen. Microbiol.* **15**:599–608.

Smith, T. M., and Holt, G., 1984, Cloning of DNA from *P. chrysogenum*, *Trans. Biochem. Soc.* **12**:645–646.

Stohl, L. L., and Lambowitz, A. M., 1983, Construction of a shuttle vector for the filamentous fungus *Neurospora crassa*, *Proc. Natl. Acad. Sci. U.S.A.* **80**:1058–1062.

Stromnaes, O., Garba, E. D., and Beraha, L., 1964, Genetics of phytopathogenic fungi. IX. Heterokaryosis and the parasexual cycle in *Penicillium italicum* and *Penicillium digitatum*, *Can. J. Bot.* **42**:423–427.

Timmis, K. N., 1981, Gene manipulation *in vitro*, in: *Genetics as a Tool in Microbiology* (S. W. Glover and D. A. Hopwood, eds.), Cambridge University Press, Cambridge, pp. 49–109.

Genetics of *Acremonium* 4

J. F. PEBERDY

1. INTRODUCTION

In comparison with several other filamentous fungi, such as *Aspergillus nidulans* and *Neurospora crassa*, our understanding of the genetics of species of *Acremonium* is still in its infancy. Much of the work that has been done has centered on *Acremonium strictum* (*Cephalosporium acremonium*), which is not surprising in view of its commercial importance. (For the purposes of uniformity in this text the revised name *A. strictum* will be used; however, it is recognized by the author that the fungus is still generally known, especially in industrial circles, as *C. acremonium* and that it is most likely that this situation will continue.) More restricted studies have been carried out on *Cephalosporium mycophilum* (Tuveson and Coy, 1961) and on a representative perfect stage of *Acremonium*, *Emericellopsis salmonsynnemata* (Fantini, 1962).

Since *A. strictum* is an asexual fungus, genetic manipulations must be based on parasexual events and processes. The significant developments in genetic manipulation that have been made with this fungus during the past decade arise from the application of protoplast-fusion techniques that are now important tools for microbial geneticists (Peberdy, 1980; Ferenczy, 1981). With the use of these methods, genetic recombination has been readily demonstrated in *A. strictum*, in contrast to earlier attempts involving conventional approaches in which difficulty in the recovery of heterokaryons and diploids was frequently encountered. It is probably the case that these difficulties relate to the hyphal organization of uninucleate segments (Fig. 1). Hyphal anastomoses can be readily found after strains are mated; however, binucleate segments are rarely seen. It is likely that the presence of a second nucleus could disturb the mechanism that controls the proportional relationship between genome and cytoplasmic mass. *Acremonium* lacks most of

J. F. PEBERDY • Department of Botany, University of Nottingham, Nottingham NG7 2RD, England.

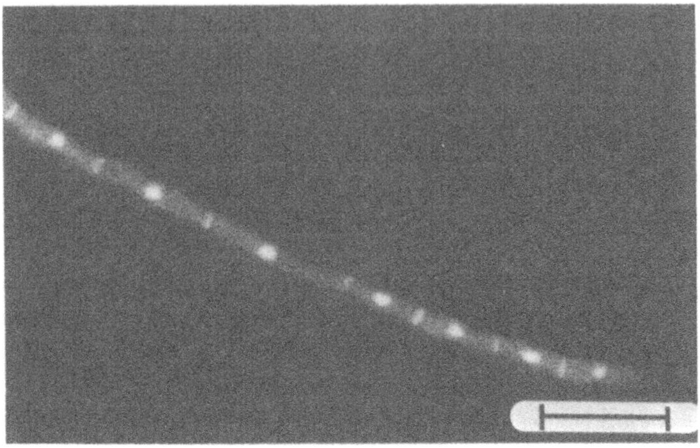

Figure 1. Hyphae of *Acremonium strictum* stained with Chromomycin and Tinopal BOPT showing nuclei and septa. Scale bar: 10 μm. From Hamlyn (1982).

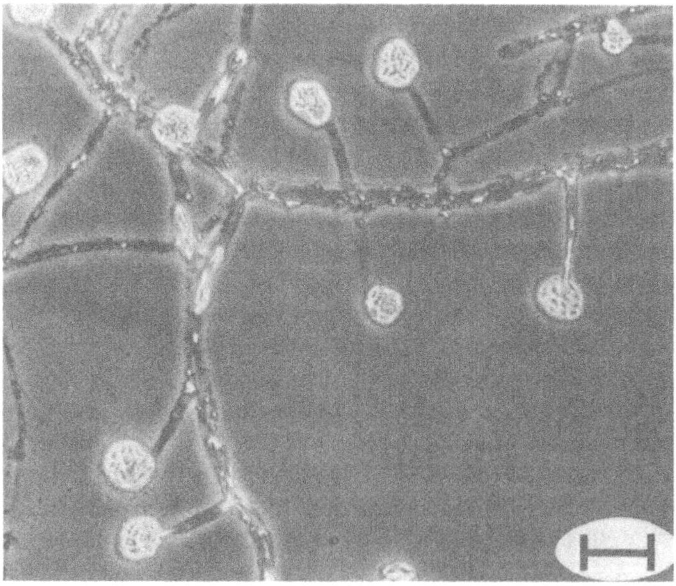

Figure 2. Conidiophores of *Acremonium strictum* in slime heads. Scale bar: 20 μm. From Hamlyn (1982).

the structures and biological features of the *Aspergilli* and the *Penicillia* that have been effectively exploited by fungal geneticists. Although *Acremonium* produces uninucleate conidia, these conidia are formed in heads covered in slime (Fig. 2) and not in regular chains. The isolation of conidia from a single head is consequently difficult, if not impossible. The spores lack pigmentation, and spore-color mutations are therefore not available. Another drawback for genetic studies is the relatively slow growth rate of the fungus. The unknown relationship between the imperfect stage and the perfect stage, *Emericellopsis*, with respect to laboratory manipulation necessitates the application of parasexual methods.

2. MUTANT ISOLATION

There are no published reports on the effects of different mutagens on *A. strictum*; however, it is likely that most, if not all, of the well-known mutagenic agents have been used in industrial strain-improvement programs. With one exception in which both UV irradiation and nitrosoguanidine were used (Lemke and Nash, 1972), all the papers that

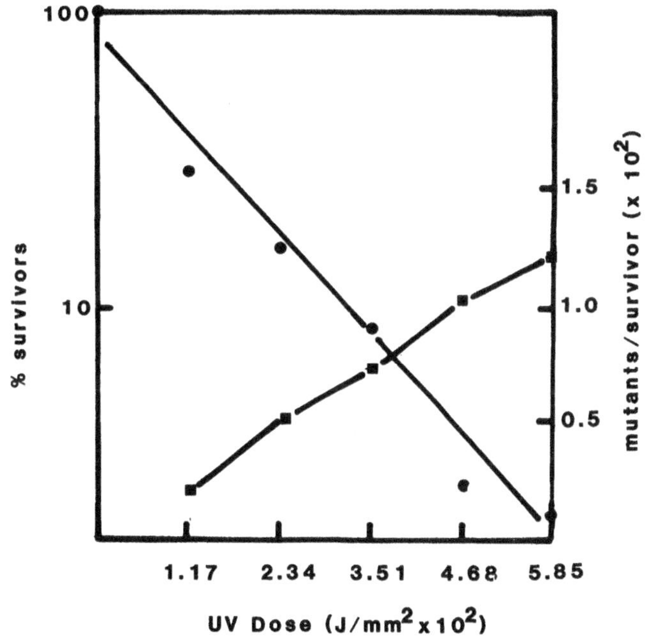

Figure 3. Lethal and mutagenic effects of UV irradiation on *Acremonium strictum*. (●) Survival; (■) number of auxotrophs. From Hamlyn (1982).

describe the procedure refer to the use of the former agent (Elander, 1975; Queener and Ellis, 1975; Queener *et al.*, 1974; Felix *et al.*, 1981; Matsumura *et al.*, 1980). Generally, conidia are used for mutagenesis, and following UV irradiation, they display typical survival characteristics with increasing frequency of mutant formation with increased exposure to the mutagen (Fig. 3).

3. PARASEXUALITY IN *ACREMONIUM STRICTUM* (*CEPHALOSPORIUM ACREMONIUM*)

3.1. Conventional Approaches to Parasexual Crosses

Since the early 1970s, several attempts to establish parasexual crossing by natural or conventional methods using auxotrophic mutants, as developed for *Aspergillus nidulans* and other fungi (Caten, 1981), have been reported. Presumptive heterokaryons were described by Nuesch *et al.* (1973), Nash *et al.* (1974), Kanzaki and Fujisawa (1976), Elander *et al.* (1976), and Hamlyn and Ball (1979). The recovery of diploids was also described by the Ciba-Geigy group (Nuesch *et al.*, 1973). The single diploid isolated by these workers was characterized by its stability, thick hyphae, and larger nuclei. When the diploid was grown in the presence of *p*-fluorophenylalanine, mitotic segregation was observed, and a restricted range of segregants was recovered, demonstrating free assortment of markers. Hamlyn and Ball (1979) attempted some 40 crosses following a conventional procedure; however, in no case was a diploid recovered. Stable prototrophic progeny were recovered from 10 crosses and unstable prototrophs from 2; 1 of the latter produced various segregants spontaneously. Exposure of the stable prototrophs to recombinogens had no effect. In the light of more recent observations, we can conclude that the stable and unstable progeny were probably haploid recombinants and aneuploids, respectively.

3.2. Protoplast-Fusion Crosses

Protoplast fusion has proved to be an effective procedure for establishing crosses in fungi with no known natural mechanism for somatic or sexual hybridization and in fastidious species in which mechanisms exist but are not readily manipulated in the laboratory. Anné and Peberdy (1976) first showed that protoplast-fusion crosses were possible in *A. strictum*, and Hamlyn and Ball (1979) developed this expertise to demonstrate genetic recombination. The essential features of a protoplast system are (1) a requirement for producing protoplasts in sufficient numbers, (2) a procedure for inducing fusion, and (3) the ability to culture the protoplasts, to induce wall regeneration and reversion to the normal cell form. The cultural step also embodies a selection procedure whereby parental protoplasts do not survive.

Protoplasts can be prepared from *A. strictum* quite readily utilizing either the *Cytophaga* lytic enzyme L1 (Fawcett *et al.*, 1973) or a cellulase preparation (Cellulose CP, from *Penicillium funiculosum*; Supplied by John and E. Sturge, Ltd., Selby, N. Yorks, UK), as described by Hamlyn *et al.*, 1981. Mycelium from 48-hr cultures, pretreated with dithiothrietol, exposed to either of these enzymes in the presence of 0.7M sodium chloride will yield $2–3 \times 10^8$ protoplasts/50 mg fresh mycelium (Hamlyn, 1982). The protoplasts undergo regeneration on general complex, e.g., Sabouraud agar, and defined media, e.g., Czapek–Dox agar, supplemented with an osmotic stabilizer. The regeneration of protoplasts does vary from strain to strain (Table I); however, the underlying cause of this variation is unknown.

The rationale generally adopted for protoplast-fusion crosses in fungi involves the use of auxotrophic mutants, allowing the selection of fusion products, normally heterokaryons, as a consequence of nutritional complementation. Anné and Peberdy (1976) used this approach with *A. strictum* and concluded that the progeny recovered were heterokaryons. However, subsequent work (Hamlyn and Ball, 1979; Hamlyn, 1982) would suggest that these products were not heterokaryons but prototrophic haploid recombinants. The clue to this view was the observation of Hamlyn and Ball (1979) in which they showed that similar fusion products were very stable and furthermore that similarly stable progeny of specific phenotypes could be recovered when a range of selective media, designed to support only non-parental types, were used.

Table I. Frequency of Protoplast Regeneration in Various Auxotrophic Strains of *Acremonium strictum*[a]

Strain	Protoplast reversion (%)
ane-5	51
azu-1, arg-2	42
arg-6	7
leu-2	27
cys-1	50
met-1	36
azu-1, arg-2, met-1	41
ane-1, ino-1	48
lys-1, arg-6, pyt-2	8
red-1, lys-1, pyt-1	64

[a] From Hamlyn (1982).

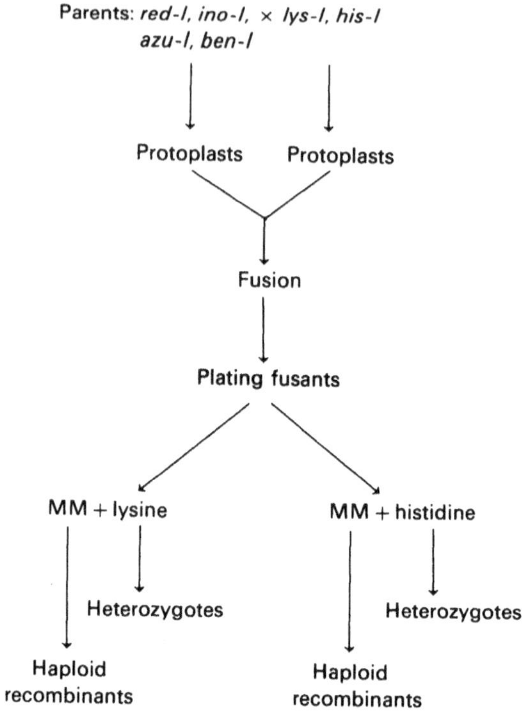

Figure 4. Protocol for protoplast fusion crosses in *Acremonium strictum*.

These observations led to the adoption of a particular experimental protocol for *A. strictum* (Fig. 4) that has similarities to selection schemes used in streptomycete genetics and that may also have application to other fungi. Ideally, strains with two or more selective markers are used, and they may also carry other neutral markers, e.g., resistance to growth inhibitors and morphological characters. Resistance markers may also be used in selection, with only a slight technical modification to the procedure being necessary (Perez-Martinez and Peberdy, 1987). An inevitable feature of the selection system based on auxotrophic markers is the loss of specific classes of non-parental segregants on certain media (Table II); however, experience has proved that this does not cause excessive distortion in the genetic analysis.

3.2.1. Progeny from Fusion Products

In the initial stages in the establishment of the protoplast-fusion system in *A. strictum*, emphasis was placed on the analysis of haploid recombinants or selectants (Hamlyn and Ball, 1979; Hamlyn, 1982). These were recovered from fusion plates and confirmed for purity by streaking before phenotypic

Table II. Genotypes of Segregants Recoverable, on Selective Media, from a Hypothetical Three-Factor Cross: *lys* × *arg*, *his*[a]

Genotypes of progeny			Arginine[b]		Histidine[b]
lys	+	+	—	(Parentals)	—
	arg	*his*	—		—
+	+	+	*		*
lys	*arg*	*his*	—		—
lys	+	*his*	—		—
+	*arg*	+	*		—
lys	*arg*	+	—		—
+	+	*his*	—		*

[a] From Hamlyn (1982).
[b] (*). Recoverable in the absence of linkage and differential viability; (—) medium not suitable for recovery.

analysis was carried out. Small, slower-growing colonies were also reported; more recently, however, these colonies have been incorporated into the overall scheme for analysis (Birkett and Hamlyn, 1985; Perez-Martinez, 1984). When transferred to nonselective medium, these colonies prove to be unstable, yielding sectors that may or may not be of the same phenotype. These slow-growing colonies have been designated heterozygotes and probably range in ploidy from diploid to aneuploid. Depending on the strains used in a cross, fusion plates may have a background growth of one parent or both parents. This growth results from one strain having a "leaky" mutation or from the two strains cross-feeding each other. Hamlyn (1982) found that this problem could be reduced by using glycerol as carbon source and by plating the fused protoplasts onto cellophane laid over the regeneration medium. After 4 days, by which time the developing colonies are osmotically stable, the sheets are transferred to nonstabilized medium.

3.2.2. Genetic Analysis Using Haploid Recombinants (Selectants) and Heterozygotes

In this section, data from a typical cross will be used to demonstrate the approach to genetic analysis and to highlight some of the factors that have to be taken into account in carrying out such an analysis. Details of the protocol involved are given in Fig. 4, and the data are given in Tables III and IV. The selectant progeny are checked for purity and stability before their characteristics are determined.

Table III. Pairwise Genetic Analysis of Haploid Segregants Recovered from Selective Media Used in the Cross *red-1, ino-1, azu-1, ben-1* × *lys-1, his-1*[a]

Selective media			Genetic analysis		
			red-1	*lys-1*	*azu-1*
			+ −	+ −	+ −
mm + lysine	*ben-1*	+	31 (34)	(45) 20	23 (42)
		−	(23) 36	40 (19)	(24) 35
	azu-1	+	17 (30)	(10) 37	
		−	(37) 40	75 (2)	
	lys-1	+	(42) 43		
		−	12 (27)		
			red-1	*his-1*	*azu-1*
			+ −	+ −	+ −
mm + histidine	*ben-1*	+	(29) 6	(6) 29	4 (31)
		−	39 (3)	6 (36)	2 40
	azu-1	+	5 (1)	(1) 5	
		−	(63) 8	1 (60)	
	his-1	+	(3) 9		
		−	65 (0)		

[a] Abbreviations of genetic markers: (*lys, his, ino, arg*) requirement for lysine, histidine, inositol, and arginine, respectively; (*azu, ben*) resistance to azuracil and benomyl, respectively; (*red*) red colony pigmentation.

Table IV. Pairwise Genetic Analysis of Segregants from Two Heterozygotes Recovered from the Cross Described in Table III[a]

		red-1	*ino-1*	*lys-1*
		+ −	+ −	+ −
his-1	−	(110) 29	(4) 35	27 (12)
	+	38 (1)	36 (3)	(31) 8
lys-1	+	(37) 21	(30) 28	
	−	11 (9)	10 (10)	
ino-1	+	35 (5)		
	−	(13) 25		

[a] See Table III for details of markers.

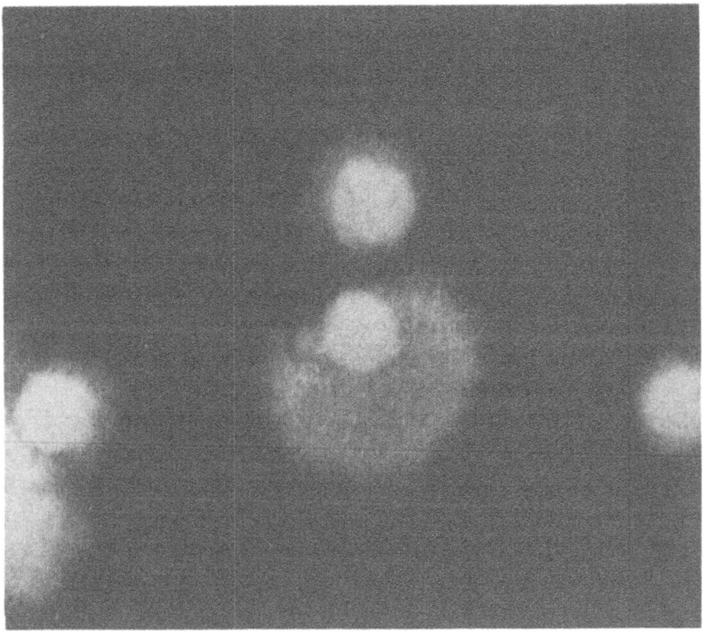

Figure 5. Heterozygote of *Acremonium strictum* undergoing segregation on complete medium.

Heterozygotes are identified on fusion plates as small, abnormal, slow-growing colonies that on transfer to a complete medium segregate normal haploid progeny. In practice, the small colonies are recovered from fusion plates and macerated, and the hyphal fragments are plated onto complete medium (Fig. 5). A single sector from each colony is sampled and purified in the same manner as the selectants, before phenotypic analysis. In view of the instability of the heterozygotes, it is most probable that some segregation event(s) may have occurred prior to isolation. For this reason, several colonies derived from the purification of the heterozygote are analyzed.

The progeny are characterized with respect to the selective medium on which they were recovered. A key factor in the analysis is the fitness of particular mutant genes, which in many instances is lower than that of their wild-type equivalents. Where this occurs, strains with the mutant allele may be selectively lost during the culture of the fusion products. Fitness of particular genes may be a factor in determining protoplast regeneration, as discussed previously. Analysis of a cross is also affected by linkage of markers: Where a marker is linked to another that is selected against, neither will be recovered in the progeny of the cross. Normally, linkage is identified where the frequency of recovery of particular nonparental phenotypes is low, against a majority background of parental types.

In the cross presented, it was previously known that the markers *his-1*, *ino-1*, and *red-1* were linked (Hamlyn, 1982; Hamlyn *et al.*, 1985). Recovery of selectants on medium containing lysine selectively eliminates histidine- and inositol-requiring phenotypes (see Table III). Colonies with red morphology and no requirement for inositol arise following mitotic recombination within the linkage group concerned. Selection with the histidine medium is against inositol and lysine; thus, both his^+ and his^- types will be ino^+ lys^+. Mitotic recombination will give rise to his ino phenotypes. The data for this medium show clear linkage between *red-1* and *his-1*, and the low recovery of red and his phenotypes is due to linkage to ino, exemplifying the criteria referred to previously.

Analysis of segregants derived from heterozygotes produced in this cross is presented in Table IV and confirms the points already made. However, heterozygote analysis clearly provides the opportunity to map linkage groups.

Using both these approaches to genetic analysis, a tentative linkage map for *A. strictum* has been drawn up (Fig. 6) (Hamlyn, 1982; Hamlyn *et al.*, 1985). It is comprised of eight linkage groups, but a preliminary investigation of less frequently used markers suggests that this number may be an underestimate.

4. GENETICS OF CEPHALOSPORIN C PRODUCTION BY *ACREMONIUM STRICTUM* (*CEPHALOSPORIUM ACREMONIUM*)

The availability of parasexual genetic manipulation in *A. strictum* now provides an opportunity to explore the genetics of cephalosporin C biosynthesis and also possibilities of parasexual breeding.

4.1. Mutants That Affect Cephalosporin C Synthesis

Mutations that affect antibiotic production fall into two classes: (1) those that block synthesis and arise from changes in structural genes that code for enzymes involved in the biosynthetic pathway and (2) those that affect production in a regulatory fashion. These mutations might involve genes that directly regulate the antibiotic structural genes or other genes that have an indirect effect as a result of their pleiotropic action. Mutations of both types have been described in *A. strictum*.

The biosynthetic pathway for cephalosporin C is presented in Fig. 6, and only a brief account will be given here. For more detailed references, see

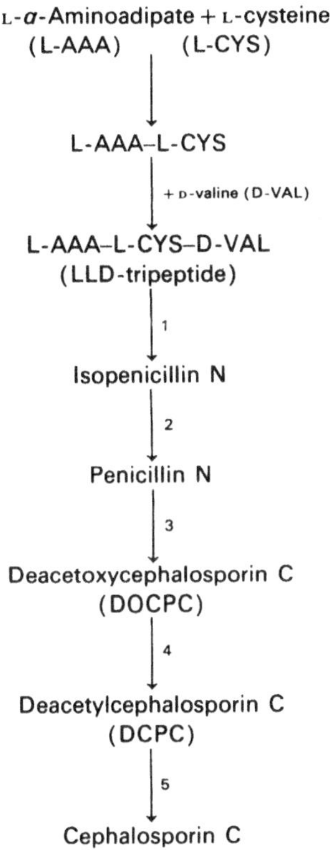

Reaction	Enzyme	Blocked mutants
1	Isopenicillin N synthetase (cyclase)	N2 (Fujisawa *et al.*, 1975)
2	"Racemase"	M8650-penN (Lemke and Nash, 1972)
3	DOCPC synthetase (expandase)	M-1443, M1836 (Yoshida *et al.*, 1978); RS85 (Queener *et al.*, 1974); 40-20 (Fujisawa *et al.*, 1975)
4	DCPC synthetase (dioxygenase)	20, 29, 26 (Fujisawa *et al.*, 1975; Fujisawa, 1977); MH63 (Queener *et al.*, 1974)
5	DCPC acetyl-transferase	40 (Fujisawa *et al.*, 1975)

Figure 6. Biosynthetic pathway for cephalosporin C showing location of mutants blocked in antibiotic synthesis.

Demain (1981) and Chapter 5. The antibiotic is synthesized from three precursor amino acids, L-aminoadipic acid, L-cysteine, and L-valine. In two steps, these compounds are converted to the tripeptide δ-(L-α-aminoadipyl)-L-cysteinyl-D-valine. The tripeptide undergoes a cyclization reaction to form isopenicillin N, a molecule with a β-lactam ring and a thiazolidine ring. This molecule then undergoes isomerization to form penicillin N. The final stages of cephalosporin C from penicillin N involve ring expansion with the formation of deacetylcephalosporin C, followed by deacetoxycephalosporin C, and leading finally to cephalosporin C. Each of the steps in the pathway is catalyzed by a specific enzyme; however, in a recent publication, it was suggested that the conversion of penicillin N to deacetylcephalosporin C and its conversion to deacetoxycephalosporin C are catalyzed by the same enzyme (Scheidegger *et al.*, 1984). Mutants, blocked at all the steps in this pathway, have been isolated and biochemically characterized (Fig. 6). Genetic studies involving these mutants have recently begun and are described in this chapter.

Regulatory genes that control cephalosporin C production are not as yet clearly defined; however, some of the raised-titer mutants that have been described undoubtedly include strains in which these genes have been mutated.

4.2. Genetic Studies on Cephalosporin-C-Blocked Mutants

Recent research at Nottingham has been carried out to provide a foundation for more extensive fundamental work. The emphasis of these studies to date has focused on the quantitative and regulatory aspects of cephalosporin C production and to a lesser extent on the characterization of the structural genes (Perez-Martinez, 1984). To establish these crosses, it was necessary to introduce suitable selective (auxotrophic) marker mutations, and in some strains, the antibiotic titer was depressed as a consequence (Elander *et al.*, 1977).

Crosses performed can be classified according to the genotypic relationship of strains as ancestral, divergent, or sister crosses (Ball, 1980) (Fig. 7).

Ancestral crosses were established between strains derived from M8650 and Co728 (Glaxo strain). Both haploid selectants and heterozygote progeny were recovered and analyzed. Progeny from several crosses performed gave mean antibiotic (penicillin N and cephalosporin C) potencies lower than the mean parental potencies; however, several of the progeny were better than the original Co728 strain in penicillin N production. In certain instances, the selective medium used to recover fusion products affected the antibiotic potency of the segregants recovered (Table V).

Analysis of the data obtained from these crosses for interactions of

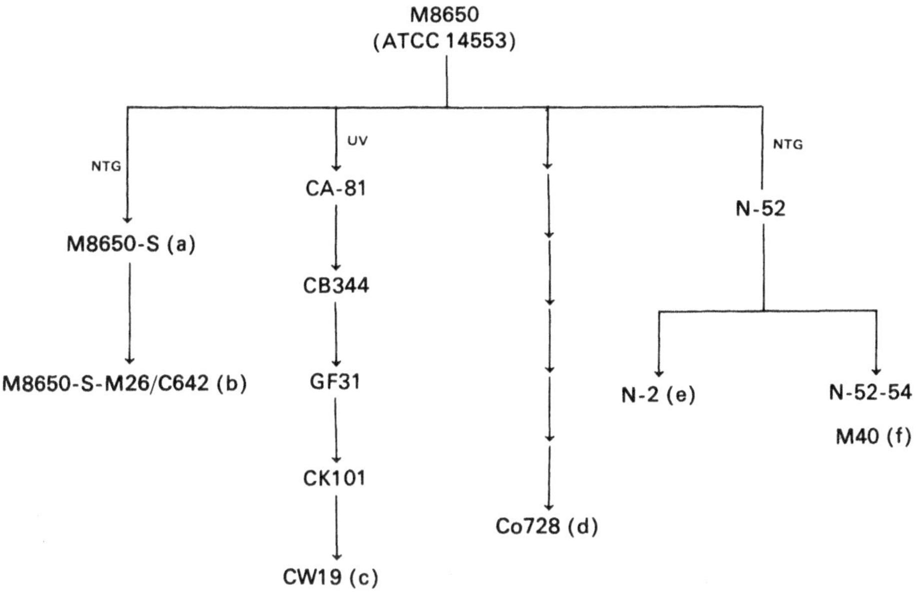

Figure 7. Genealogy of raised-titer strains of *Acremonium strictum* used in genetic studies. Definitions of crosses: Sister cross involves strains derived from a common progenitor; e.g., N2 × N-52-54 (NB N2 × M40) has been classified as a sister cross. Divergent cross involves strains from two lines of mutagenesis, e.g., CW19 × Co728. Ancestral cross involves a strain crossed with a progenitor, e.g., Co728 × M8650. References: (a) Nuesch *et al.* (1973); (b) Liersch *et al.* (1976); (c) Elander (1975); (d) Glaxo strain-improvement line; (e) Fujisawa and Kanzaki (1975); (f) Fujisawa *et al.* (1975).

Table V. Influence of Strain Markers on Mean Antibiotic Potencies of Segregants Derived from a Cross between M8650 *red-1, ino-1, azu-1, ben-1* × Co728 *arg-8*[a]

Selective medium			PenN	CPC
A	*red-1*	+	21.98	1.13
(+arginine)		−	6.69	0.00
	arg-8	+	21.98	1.10
		−	10.10	0.30
	azu-1	+	22.20	1.88
		−	19.53	0.79
C	*azu-1*	+	44.68	5.22
(+inositol)		−	32.39	2.23
	ben-1	+	52.42	3.77
		−	29.03	2.86

[a] See Table III for details of markers. (PenN) penicillin N; (CPC) cephalosporin C.

markers with antibiotic titer suggested evidence for factors or genes that interact with penicillin N and cephalosporin C production. Strains carrying the *ben* marker showed a raised penicillin titer, and strains wild-type for azauracil, i.e., sensitive to the inhibitor, showed enhanced cephalosporin C production. These possible linkage relationships are shown in the map in Fig. 8.

Divergent crosses involving C0728 (ex Glaxo) and M8650-S-M26/C462 (ex Ciba-Geigy) could also be achieved. As observed previously, the progeny displayed a spread of potencies of both penicillin N and cephalosporin C with means lower than the mean parental potency. Comparison of haploid selectant progeny and heterozygotic progeny revealed an interesting feature. Haploid progeny showed greater spread in total β-lactam potency, but generally had poor penicillin N/cephalosporin C ratios. Progeny from a heterozygote showed a much narrower range of potency that in all but a few

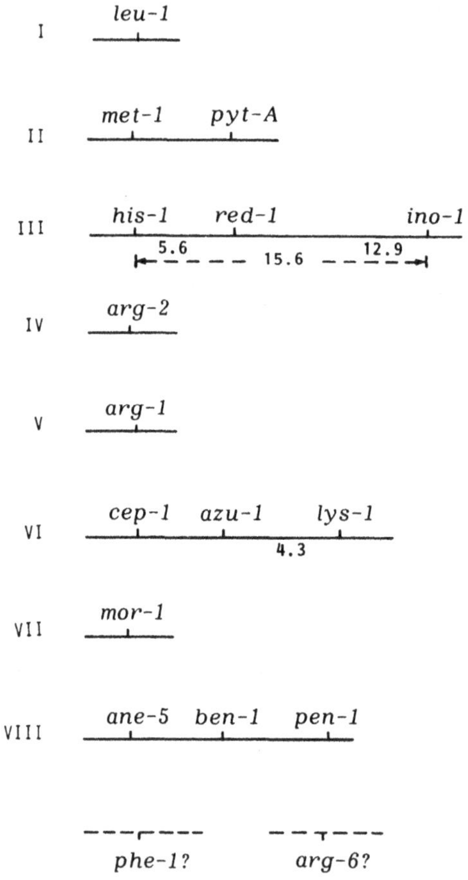

Figure 8. Linkage map of *Acremonium strictum*. Arrows under linkage group III indicate map distances between markers. From Hamlyn (1982) and Perez-Martinez (1984).

examples was lower than that of the parental strains. However, in these segregants, the spread of penicillin N/cephalosporin C ratios was very broad. These crosses suggested, further, that cephalosporin C overproduction was affected by other as yet unidentified genes.

A sister cross between strains of the Takeda line is an example of a cross designed to identify the individuality of structural genes. One strain had a block at ring cyclization, and the other was blocked at deacetylcephalosporin C. Recombination between these strains restored cephalosporin C synthesis, suggesting that these two steps in the pathway are subject to individual genetic control.

5. TOWARD A TRANSFORMATION AND GENE-CLONING SYSTEM IN *ACREMONIUM STRICTUM*

A complete understanding of the molecular genetics of fundamental processes in fungi such as *Acremonium* will depend on the opportunity to isolate the particular genes of interest. In many cases, this will require techniques for gene isolation and cloning to be made available in the fungus itself. Such a technology could be in extensive use within the next five years.

Transformation systems in yeasts and filamentous fungi involve the uptake of the vector into protoplasts in the presence of polyethylene glycol (Turner and Ballance, 1985). Clearly, this procedure has similarities to the well-established protoplast-fusion methods. In filamentous fungi, the vectors adopted for transformation are chimeric plasmids comprised of DNA from *Escherichia coli* and from the host fungus. Selection of the transformants is based on the repair of an auxotrophic deficiency by the wild-type gene carried on the plasmid. The restricted availability of such cloned genes has been the limiting factor in the expansion of different transformation systems in these organisms. Only one such gene from *A. strictum* has been isolated, that coding for β-isopropylmate dehydrogenase, a component of the leucine biosynthetic pathway, which was recovered by complementation in *E. coli*. Complementary *leu*$^-$ mutants of *A. strictum* have not been described.

In the absence of such auxotrophic selection, an alternative approach is to use resistance genes as selectable markers for transformants. Queener *et al.* (1984) have already developed such a transformation system in *A. strictum* using a resistance gene for the antibiotic hygromycin.

The general experience in all transformation systems in filamentous fungi is that the vector becomes integrated into the host genome with consequent low frequency of transformation. Integration may occur at several sites; however, integration at sites other than the homologous region tends to be unstable. To overcome these problems, two strategies have been adopted; the first has been to attempt to increase the transformation frequency by inserting fragments of ribosomal DNA into the vector (Skatrud *et al.*, 1984).

The aim of the second strategy is to produce an autonomously replicating vector; however, the value of such a vector in a filamentous fungus is not clear. A source of such a vector is a native fungal plasmid. *Acremonium strictum* is one of the species of filamentous fungi in which a plasmid has been described (Minuth *et al.*, 1982); however, until more information about the plasmid is available, it is not possible to comment on its suitability as a transformation and cloning vector. An alternative approach is to insert an autonomous replication sequence into the vector. Such sequences can be isolated as functional entities in *Saccharomyces cerevisiae* using a yeast integrative plasmid such as YIp5. Skatrud *et al.* (1984) obtained such a sequence from *A. strictum* in a 1.9-kilobase *Pst*I fragment. This sequence was part of the vector construction used to transform *A. strictum* (Queener *et al.*, 1984), but the plasmid did not replicate autonomously. The insertion of mitochondrial DNA (mtDNA) sequences into plasmids has also been suggested as a route to produce autonomously replicating fungal plasmids. Tudzynski and Esser (1982) constructed a hybrid plasmid of bacterial yeast and *A. strictum* mtDNA. Of the six plasmids obtained, one was found to behave as an autonomously replicating plasmid in *S. cerevisiae.*

As a first step in the study of the control of transcription and the expression of genes involved in the biosynthesis of cephalosporin C, one of these genes, which codes for isopenicillin N synthetase (IPS) (see Fig. 6), has been cloned and sequenced (Samson *et al.*, 1985). The gene was identified by determining the first 23 amino-terminal acids in the pure protein and using the sequence information to produce a set of synthetic oligonucleotides that were used to probe a cosmid library of *A. strictum* DNA. The IPS gene was isolated from a cosmid that hybridized with the probe and was sequenced and cloned in *E. coli.* A protein was isolated from the cells that comigrated with the authentic IPS protein. In the presence of crude extracts of the *E. coli*, the tripeptide δ-(L-α-aminoadipyl)-L-cysteinyl-D-valine was converted to a penicillinase-sensitive product that had antibacterial activity similar to that of isopenicillin N.

These experiments represent a major advance in fungal molecular genetics and are of obvious academic and economic interest. Cephalosporin C is a secondary metabolite, and so this research has clear significance with respect to our ultimate understanding of the regulation of secondary metabolism in fungi.

6. CONCLUDING REMARKS

In this review, I have attempted to show the relatively dramatic advance in the genetics of *A. strictum.* In the space of a decade, and with the application of the new genetic technologies, it has been possible to establish an experimental system for setting up crosses and in still more recent times the application of recombinant DNA techniques. We can therefore predict that in a short period of time, advances will be made in the molecular biology of this fungus. Together, these two fundamental aspects of *A. strictum* genetics will play an important part in the expansion of our understanding of, and our ability to manipulate, the synthesis of cephalosporin C.

REFERENCES

Anné, J., and Peberdy, J. F., 1976, Induced fusion of fungal protoplasts in polyethylene glycol, *J. Gen. Microbiol.* **92:**413–417.

Ball, C., 1980, Genetic modification of filamentous fungi, in: *Fungal Biotechnology* (J. E. Smith, D. R. Berry, and B. Kristiansen, eds.), Academic Press, New York and London, pp. 43–54.

Birkett, J. A., and Hamlyn, P. F., 1985, Protoplast fusion and genetic analysis in *Cephalosporium acremonium*, in: *Fungal Protoplasts: Applications in Biochemistry and Genetics* (J. F. Peberdy and L. Ferenczy, eds.), Marcel Dekker, New York, pp. 207–223.

Caten, C. E., 1981, Parasexual processes in fungi, in: *The Fungal Nucleus* (K. Gull and S. G. Oliver, eds.), Cambridge University Press, Cambridge, pp. 191–214.

Demain, A. L., 1981, Biosynthetic manipulations in the development of β-lactam antibiotics, in: *β-Lactam Antibiotics: Mode of Action, New Developments and Future Prospects* (M. R. J. Salton and G. D. Shockman, eds.), Academic Press, New York, pp. 567–583.

Elander, R. P., 1975, Genetic aspects of cephalosporin- and cephamycin-producing microorganisms, *Dev. Ind. Microbiol.* **16:**356–373.

Elander, R. P., Corum, C. J., De Valeira, H., and Wilgus, R. M., 1976, Ultraviolet mutagenesis and cephalosporin biosynthesis in strains of *Cephalosporium acremonium*, in: *Second International Symposium on the Genetics of Industrial Microorganisms* (K. D. Macdonald, ed.), Academic Press, London, pp. 253–271.

Elander, R. P., Chang, L. T., and Vaughan, R. W., 1977, Genetics of industrial microorganisms, in: *Annual Reports on Fermentation Processes*, Vol. 1 (D. Perlman, ed.), Academic Press, New York, pp. 1–40.

Fantini, A. A., 1962, Genetics and antibiotic production of *Emericellopsis* species, *Genetics* **47:**161–177.

Fawcett, P. A., Loder, P. B., Duncan, M. J., and Abraham, E. P., 1973, Formation and properties of protoplasts from antibiotic-producing strains of *Penicillium chrysogenum* and *Cephalosporium acremonium*, *J. Gen. Microbiol.* **79:**293–309.

Felix, H. R., Peter, H. H., and Treichler, H. J., 1981, Microbiological ring expansion of penicillin N, *J. Antibiot.* **34:**567–575.

Ferenczy, L., 1981, Microbial protoplast fusion, in: *Genetics as a Tool in Microbiology* (S. W. Glover and D. A. Hopwood, eds.), Cambridge University Press, Cambridge, pp. 1–34.

Fujisawa, Y., and Kanzaki, T., 1975, Role of acetyl-CoA:deacetylcephalosporin C acetyl transferase in cephalosporin C biosynthesis by *Cephalosporium acremonium*, *Agric. Biol. Chem.* **39:**2043–2048.

Fujisawa, Y., 1977, Studies on the biosynthesis of cephalosporin C, *Journal of Takeda Research Laboratory* **36:**271–276.

Fujisawa, Y., Kitano, K., and Kanzaki, T., 1975, Accumulation of deacetoxycephalosporin C by a deacetylcephalosporin C negative mutant of *Cephalosporium acremonium*, *Agric. Biol. Chem.* **39:**2049–2055.

Hamlyn, P. F., 1982, Protoplast fusion and genetic analysis in *Cephalosporium acremonium*, Ph.D. thesis, University of Nottingham.

Hamlyn, P. F., and Ball, C., 1979, Recombination studies with *Cephalosporium acremonium*, in: *Genetics of Industrial Microorganisms* (O. K. Sebek and A. I. Laskin, eds.), American Society for Microbiology, Washington, D.C., pp. 185–191.

Hamlyn, P. F., Bradshaw, R. E., Mellon, F., Santiago, C. M., Wilson, J. M., and Peberdy, J. F., 1981, Efficient protoplast isolation from fungi using commercial enzymes, *Enzyme Microbiol. Technol.* **3**:321–325.

Hamlyn, P. F., Birkett, J. A., Perez, G., and Peberdy, J. F., 1985, Parasexual recombination and genetic analysis in *Cephalosporium acremonium* using protoplast fusion, *J. Gen. Microbiol.* **131**:2813–2823.

Kanzaki, T., and Fujisawa, Y., 1976, Biosynthesis of cephalosporins, *Adv. Appl. Microbiol.* **20**:159–202.

Lemke, P. A., and Nash, C. H., 1971, Mutations that affect antibiotic synthesis by *Cephalosporium acremonium, Can. J. Microbiol.* **18**:255–259.

Liersch, M., Nuesch, J., and Treichler, H. J., 1976, Final steps in the biosynthesis of cephalosporin C, in: *Second International Symposium on the Genetics of Industrial Microorganisms* (K. Macdonald, ed.), Academic Press, London, pp. 179–198.

Matsumara, M., Imanaka, T., Yoshida, T., and Taguchi, H., 1980, Regulation of cephalosporin C production by endogenous methionine in *Cephalosporium acremonium, J. Ferment. Technol.* **58**:205–214.

Minuth, W., Tudzynski, P., and Esser, K., 1982, Extrachromosomal genetics of *Cephalosporium acremonium.* I. Characterization and mapping of mitochondrial DNA, *Curr. Genet.* **5**:227–231.

Nash, C. H., de la Higuera, N., Neuss, N., and Lemke, P. A., 1974, Application of biochemical genetics to the biosynthesis of β-lactam antibiotics, *Dev. Ind. Microbiol.* **15**:114–123.

Nuesch, J., Treichler, H. J., and Liersch, M., 1973, The biosynthesis of cephalosporin C, in: *Genetics of Industrial Microorganisms* (Z. Vanek, Z. Hostalek, and J. Cudlin, eds.), Academia, Prague, pp. 309–334.

Nuesch, J., Hinnen, A., Liersch, M., and Treichler, H. J., 1975, A biochemical approach to the biosynthesis of cephalosporin C, in: *Second International Symposium on the Genetics of Industrial Microorganisms* (K. Macdonald, ed.), Academic Press, London, pp. 451–472.

Peberdy, J. F., 1980, Protoplast fusion—a tool for genetic manipulation and breeding in industrial microorganisms, *Enzyme Microbiol. Technol.* **2**:23–29.

Perez-Martinez, G., 1984, Protoplast fusion and its consequences for cephalosporin C production in *Acremonium chrysogenum*, Ph.D. thesis, University of Nottingham.

Perez-Martinez, G., and Peberdy, J. F., 1987, Further studies on the genetics of *Acremonium strictum* (Gams: formerly *Cephalosporium acremonium*) using protoplast fusion methods, *J. Indust. Microbiol.*, in press.

Queener, S. W., and Ellis, L. F., 1975, Differentiation of mutants of *Cephalosporium acremonium* in complex medium: The formation of unicellular arthrospores and their germination, *Can. J. Microbiol.* **21**:1981–1996.

Queener, S. W., Capone, J. J., Radue, A. B., and Nagarajan, R., 1974, Synthesis of deacetoxycephalosporin C by a mutant of *Cephalosporium acremonium, Antimicrob. Agents Chemother.* **6**:334–337.

Queener, S. W., Ignolia, T. D., Skatrud, P. L., and Kaster, K. R., 1984, Recombinant DNA studies in *Cephalosporium acremonium*, in: *ASM Conference on the Genetics and Molecular Biology of Industrial Microorganisms*, Bloomington, Indiana, September/October 1984, Abstract No. 29.

Samson, S. M., Belagaje, R., Blakenship, D. T., Chapman, J. L., Perry, D., Skatrud, P. L., Van Frank, R. M., Abraham, E. P., Baldwin, J. E., Queener, S. W., and Ignolia, T. D.,

1985, Isolation, sequence determination and expression in *Escherichia coli* of the isopenicillin N synthetase gene from *Cephalosporium acremonium*, *Nature (London)* **318:**191–194.

Scheidigger, A., Kuenzi, M. T., and Nuesch, J., 1984, Partial purification and catalytic properties of a bifunctional enzyme in the biosynthetic pathway of β-lactams in *Cephalosporium acremonium*, *J. Antibiot.* **37:**522–531.

Skatrud, P. L., Carr, L., and Queener, S. W., 1984, Construction of plasmids containing *Cephalosporium acremonium* ribosomal DNA, in: *ASM Conference on the Genetics and Molecular Biology of Industrial Microorganisms*, Bloomington, Indiana, September/October 1984, Abstract No. 133.

Tudzynski, P., and Esser, K., 1982, Extrachromosomal genetics of *Cephalosporium acremonium*. II. Development of a mtDNA hybrid vector replicating in *Saccharomyces cerevisiae*, *Curr. Genet.* **6:**153–158.

Turner, G., and Ballance, D. J., 1985, Cloning and transformation in *Aspergillus*, in: *Gene Manipulations in Fungi* (J. W. Bennett and L. Lasure, eds.), Academic Press, New York, pp. 259–279.

Tuveson, R. W., and Coy, D. O., 1961, Heterokaryosis and somatic recombination in *Cephalosporium mycophilum*, *Mycologia* **53:**244–253.

Yoshida, M., Konomi, T., Kohsaka, M., Baldwin, J. E., Herchen, S., Singh, P., Hunt, N. A., and Demain, A. L., 1978, Cell-free ring expansion of penicillin N to deacetoxycephalosporin C by *Cephalosporium acremonium* CW19 and its mutants, *PNAS* **75:**6253–6257.

Chemistry and Biosynthesis of Penicillins and Cephalosporins

5

B. W. BYCROFT and R. E. SHUTE

1. INTRODUCTION

It has now been more than 50 years since Fleming's legendary observations on the inhibition of bacterial growth by *Penicillium notatum* laid the foundations of modern antibiotic chemotherapy.

Throughout the 1930s, the clinical potential of the active agents produced by this organism was recognized but not fully exploited, and it was to take the Allied Forces' need for antiinfective agents in the Second World War to stimulate an international research effort. This collaboration was directed first toward the production of penicillins on a large scale and second toward the chemistry of these substances. It is perhaps not generally appreciated that this joint industrial and academic endeavor was second in manpower and cost only to the Manhattan Project that led to the development of the atomic bomb.

The outcome of this immense effort was the provision of sufficient penicillin to supply the whole of the Allied Forces by the end of the war. In addition, the scientific program had established that the penicillins are a family of closely related natural products that possess a common nucleus based on the previously unknown fused β-lactam–thiazolidine system.

In the postwar period, the clinical success of the penicillins led to the wide-scale search for other microbial metabolites with potential antibacterial activity. This effort resulted in the isolation and characterization of a number of useful and important antibiotics. However, it was not until the late 1950s that the first of a new class of β-lactam antibiotics was isolated from an *Acremonium* species. Cephalosporin C was subsequently shown to be

B. W. BYCROFT and R. E. SHUTE • Department of Pharmacy, University of Nottingham, Nottingham NG7 2RD, England.

related to the penicillins both structurally and biosynthetically. Although it possesses broad-spectrum antibacterial activity, it is not in itself a particularly potent antibiotic, and indeed no naturally occurring cephalosporin has to date found clinical application. Cephalosporin C was, however, subsequently to serve as an important precursor for semi-synthetic cephalosporins.

Semisynthesis represented a milestone in the β-lactam antibiotic field. The initial isolation, also in the late 1950s, of the penicillin "nucleus," 6-aminopenicillanic acid, afforded a versatile intermediate for the construction of a virtually limitless range of semisynthetic penicillins. Since that time, modified penicillins and cephalosporins have occupied a preeminent position in modern-day therapeutics, and it is currently estimated that the annual worldwide sales of antibiotics approach $10 billion, of which β-lactam antibiotics account for approximately 60%. The fact that new penicillins and cephalosporins continue to be introduced into the clinic suggests that there is every likelihood that the research efforts in these areas, both industrial and academic, will proceed unabated into the 21st century. It is also important to note that to date, the basic ring systems of all clinically used penicillins and cephalosporins are still derived via fermentation, and they are likely to remain beyond the reach of economical total synthesis for the projectable future. In view of this, the last decade has seen a renewed interest in the biosynthesis of β-lactam antibiotics and the biochemical mechanism of construction of these highly strained systems.

Over the last half century, a vast wealth of knowledge has accumulated in the area of β-lactam antibiotics, and this chapter can only highlight aspects of the chemistry and biochemistry pertinent to the naturally occurring penicillins and cephalosporins from *Penicillium* and *Acremonium* spp. and, where appropriate, attempt to indicate their biotechnological significance.

2. PENICILLINS

2.1. Naturally Produced Penicillins

The original fungus isolated by Fleming was a strain of *Penicillium notatum*, and all the early work was carried out on this organism. The successful large-scale production of penicillin was due to the development of a submerged fermentaton process, the choice of an optimum nutrient medium (Moyer and Coghill, 1947), and the selection of suitable strains of *Penicillium* (Raper *et al.*, 1944). Later, penicillins were found to be produced by a wide variety of *Penicillia* and other species of fungi (Elander and Aoki, 1982). The increased productivity of the high-yielding strains led to the observation that the relative quantities of the various penicillins produced depended on the type of culture medium employed. The early penicillins produced were primarily F and K, whereas the principal penicillin obtained from the optimized medium was G (Table I). The latter was selected as the penicillin of choice because it possessed a high degree of biological activity and also because high titers could be obtained at the expense of other penicillins by supplementing the growth medium with phenylacetic acid (Moyer and Coghill, 1947). The recognition that a large variety of penicillins could be produced by the addition of potential side-chain precursors to the culture medium represents the first example of "directed biosynthesis." All side-chain precursors have proven to be aliphatic or aryl-aliphatic carboxylic acids, and to date, over 100 penicillins have been produced by this method of side-chain precursor supplementation (Behrens, 1949; Cole 1966a,b). The more significant of these from both the biochemical and the therapeutic standpoint are shown in Table I.

Penicillium chrysogenum, when grown in media deliberately deprived of side-chain precursors, was found to produce the "natural" aliphatic side-

Table I. Examples of Naturally Derived Penicillins

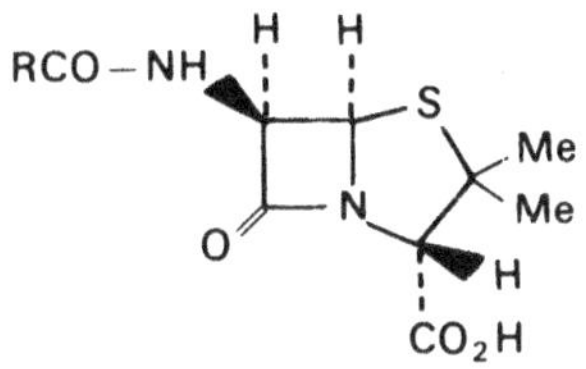

R group	Name
$CH_3CH_2CH=CHCH_2-$	Penicillin F (pentenylpenicillin)
$CH_3(CH_2)_4-$	Dihydropenicillin F (pentylpenicillin)
$CH_3(CH_2)_6-$	Penicillin K (heptylpenicillin)
$HO_2C\overset{D}{C}H(NH_2)(CH_2)_3-$	Penicillin N
$HO_2C\overset{L}{C}H(NH_2)(CH_2)_3-$	Isopenicillin N
$PhCH_2-$	Penicillin G
$C_6H_5O-CH_2-C(=O)-$	Penicillin V

chain penicillins K, F, and dihydro-F and, in addition, two previously unknown derivatives, namely, 6-aminopenicillanic acid (Batchelor *et al.*, 1959) and isopenicillin N (Cole and Batchelor, 1963). The significance of these latter two compounds is discussed in later sections.

2.2. Structure and General Chemistry

The early structural studies on penicillin were complicated by the fact that it was a mixture of closely related compounds, and meaningful investigations were to follow only when pure crystalline components were obtained during the war. However, the vast amount of accumulated chemical and physical data proved somewhat ambiguous, and the final confirmation of structure was to rest on the application of X-ray crystallography (Crowfoot *et al.*, 1949). This established beyond doubt the presence of a β-lactam ring within the penicillin molecule, a feature that had been proposed but considered too unlikely in a natural product. The detailed results of the theoretical and experimental studies carried out during the international collaboration were never published in the scientific literature, but are recorded in a unique volume published in the late 1940s (Clarke *et al.*, 1949).

The structural studies established that all the penicillins contain a common nucleus comprising a β-lactam ring, or azetidin-2-one ring, fused to a heterocyclic thiazolidine. This is now universally referred to as the "penam system" and is numbered as shown in Fig. 1. The array of naturally occurring penicillins therefore differ only in the nature of the side chain attached via an amide linkage to the penam nucleus.

The extraordinary chemical reactivity of the penicillins not only accounted for the problems relating to instability encountered in the initial investigations, but also gave an insight into their biological activity. The nucleus is sensitive to both acidic and basic conditions, but particularly to the latter, such that a solution of the sodium salt of penicillin G kept at

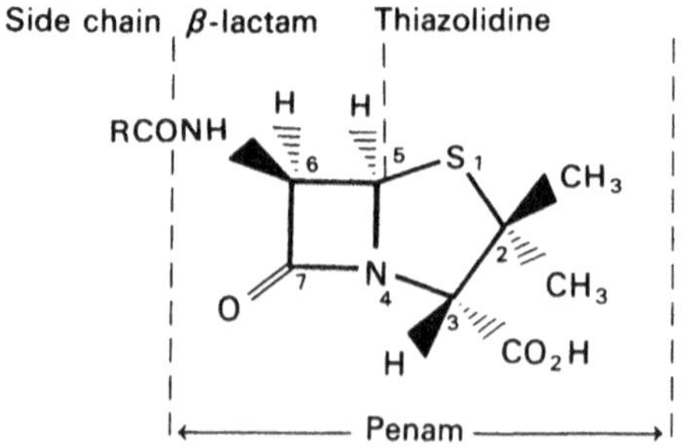

Figure 1. Nomenclature and numbering of the penicillin skeleton.

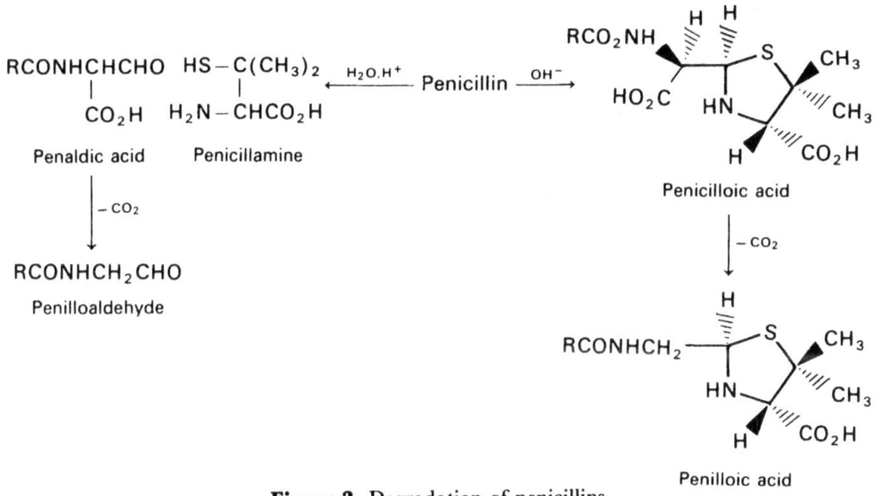

Figure 2. Degradation of penicillins.

pH 10 for 15 min at room temperature loses all biological activity. The reaction pathways in both cases are relatively complex, but essentially involve the opening of the β-lactam ring as shown in Fig. 2. This reactivity is ascribed to "ring strain" inherent in the bicyclic system arising from distortion of the amide bond both by the four-membered ring and by the nature of the ring fusion. Evidence for this strain is apparent from the bond angles and lengths obtained from X-ray crystallographic data as well as from the increased frequency in the infrared spectrum of the β-lactam carbonyl compared to normal amides and nonfused azetidin-2-ones (Sweet, 1972). It is noteworthy that all degradation products that lack the β-lactam ring are antibacterially inactive. However, D-penicillamine has found clinical application in the treatment of rheumatoid arthritis and is derived commercially from penicillin G.

Penicillins were found not to be as sensitive to redox conditions as they are to changes in pH. However, it was observed that reduction of penicillin G with Raney nickel removed the sulfur and afforded two products (Fig. 3). One was identified as phenylacetyl-L-alanyl-D-valine and was crucial in establishing the absolute stereochemistry of the molecule as well as hinting at its possible peptide origin. The structure of the other product was established as the related monocyclic azetidin-2-one shown. This compound proved to be relatively stable and lacked any notable biological activity. This finding once again emphasizes the fundamental importance of the intact bicyclic system for antibacterial activity.

The variable oxidation levels of sulfur render the penicillins particularly susceptible to oxidizing agents. Mild conditions such as peracids and peroxides afford the sulfoxides, whereas stronger oxidants, e.g., periodate,

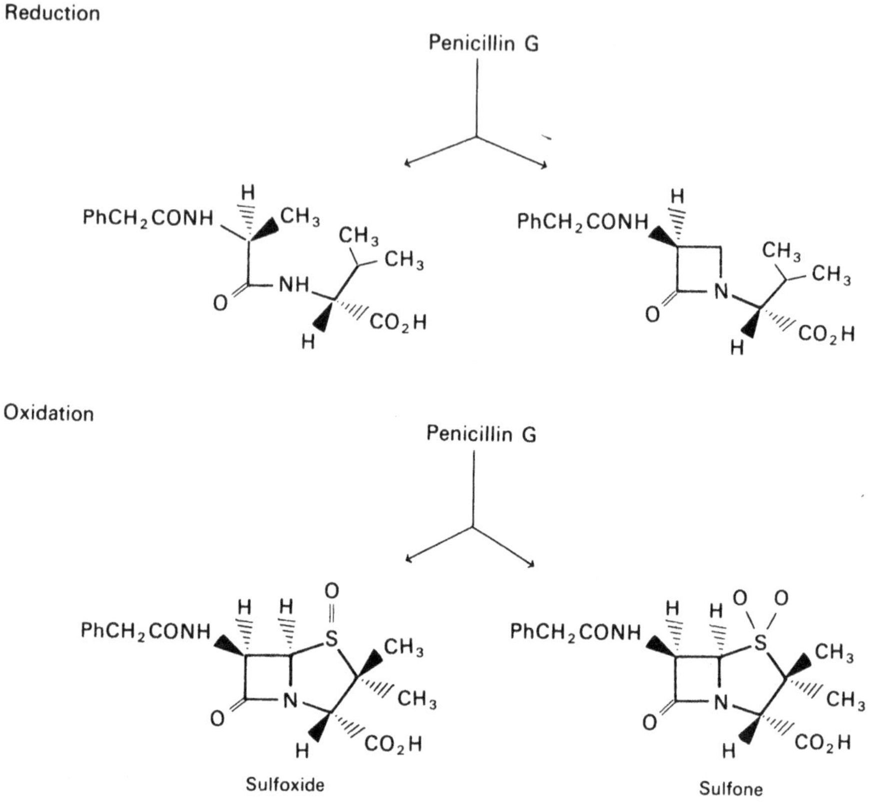

Figure 3. Mild reduction and oxidation of penicillins.

yield the sulfones (Fig. 3). More forcing conditions lead to total degradation. Although the penam system remains intact in the sulfoxide and sulfone derivatives, their antibacterial activity is considerably reduced.

2.3. Chemistry of 6-Aminopenicillanic Acid

The importance of 6-aminopenicillanic acid (6-APA) was realized with the initial observation that chemical acylation with phenylacetyl chloride yielded benzylpenicillin (Sheehan and Henery-Logan, 1962). This immediately led to the synthesis, from 6-APA and acid chlorides, of new penicillins otherwise unobtainable by the side-chain precursor supplementation method. However, obtaining 6-APA by direct fermentation is hampered by poor yields and the difficulties associated with its isolation; unlike the natural side-chain penicillins, 6-APA cannot be extracted into organic solvents at low pH. This source of 6-APA has now been replaced by processes based on the enzymatic or chemical removal of the acyl side-chain of either benzyl- or phenoxymethyl-penicillin.

Acylases capable of hydrolyzing off the side chain are produced by a variety of living organisms. In the main, those derived from actinomycetes and filamentous fungi hydrolyze aliphatic and aryloxymethylpenicillins, but cleave benzylpenicillin more slowly. Alternatively, those of bacterial origin, e.g., from the genus *Escherichia*, remove the phenylacetyl side chain from penicillin G, but hydrolyze aryloxymethylpenicillins less readily. The preparation and application of these enzymes have been reviewed (Huber *et al.*, 1972), and this procedure, using immobilized *E. coli* acylase, is currently the method of choice for the commercial production of 6-APA.

Initially, the likelihood of a method for chemical cleavage of the side chain appeared improbable, since it would involve the selective hydrolysis of a normal amide linkage in a molecule containing the reactive β-lactam system (see Section 2.2). The need for a chemical method was paramount in the case of the cephalosporins, since all attempts to find enzymes that could cleave cephalosporin C were unsuccessful (see Section 3.3). This was eventually achieved in both the penicillin and the cephalosporin series in high yield (Huber *et al.*, 1972) by converting the side-chain amide bond into the more highly reactive imino-chloride, which, on treatment with an excess of an alcohol followed by aqueous workup, afforded, in the case of penicillin, 6-APA and the side-chain ester (Fig. 4). Over the past 25 years, many thousands of semisynthetic penicillins have been synthesized and evaluated (Price, 1970; 1977). Unfortunately, only a very small number have reached the clinic; a selection of these is illustrated in Table II.

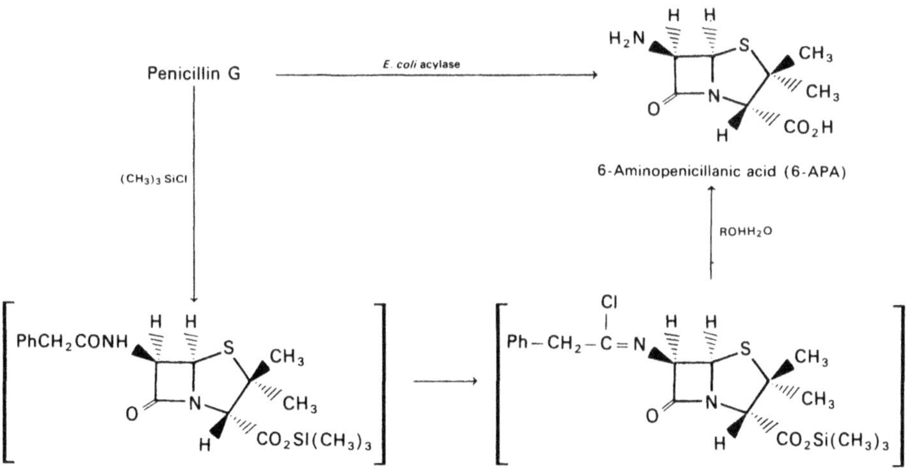

Figure 4. Preparation of 6-aminopenicillanic acid.

Table II. A Selection of Clinically Significant Semisynthetic Penicillins Derived from 6-Aminopenicillanic Acid

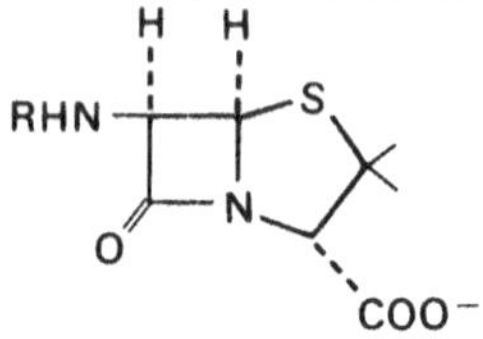

R group	Name
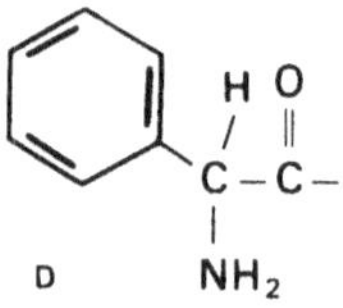	Ampicillin
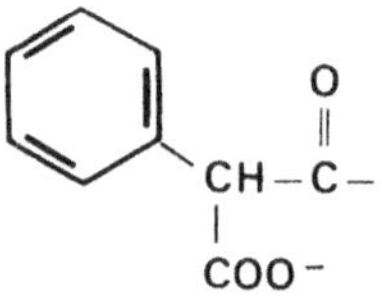	Carbenicillin
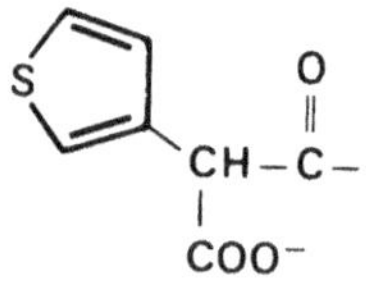	Ticarcillin
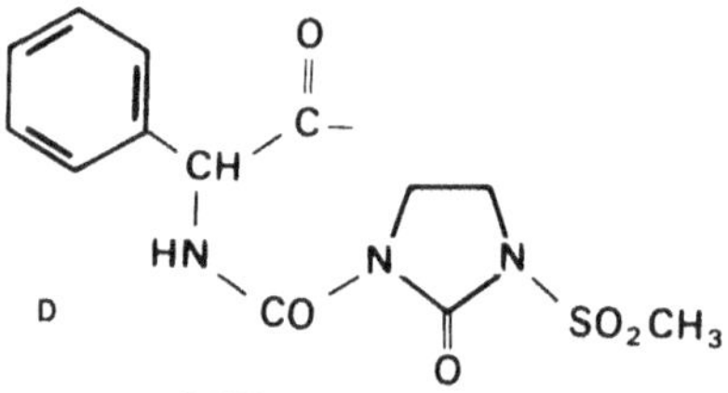	Mezlocillin
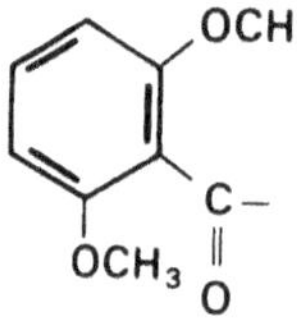	Methicillin
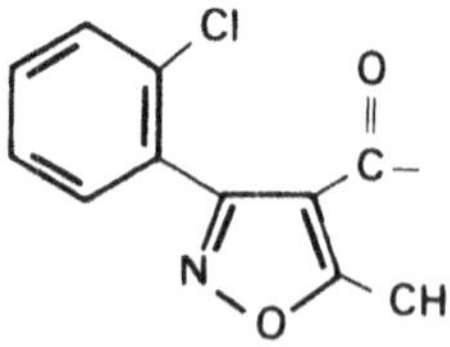	Cloxacillin

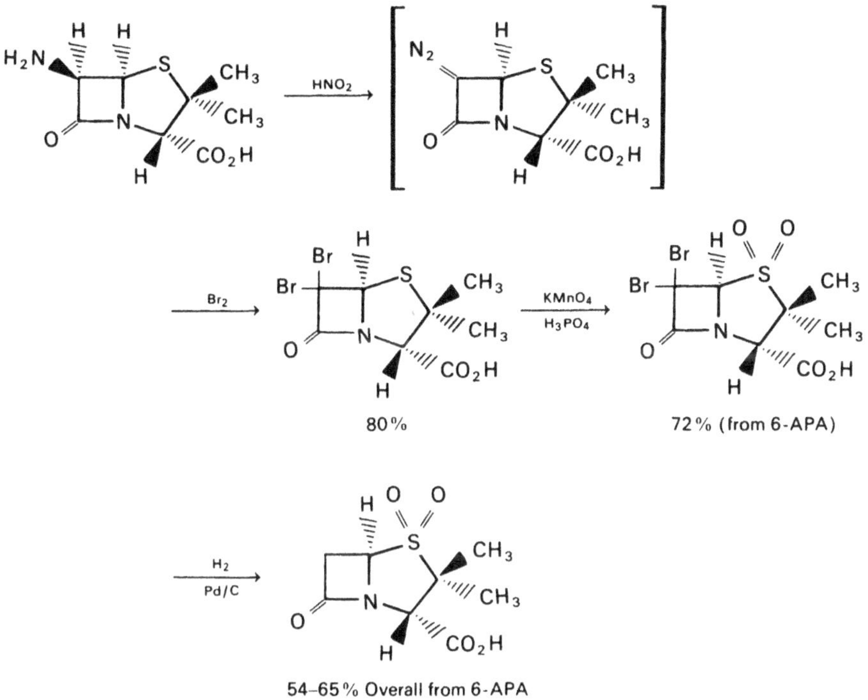

Figure 5. Synthesis of Sulbactam.

In view of the efficiency of the commercial production, 6-APA has become not only the precursor of semisynthetic penicillins but also a cheap and versatile starting material for the medicinal chemist (Jung *et al.*, 1980). The level of sophisticaton of chemical modification is illustrated by the synthesis of penicillanic acid sulfone (Sulbactam) (Volkmann *et al.*, 1982) (Fig. 5). This compound and related compounds have attracted considerable interest in recent years. They have virtually no antibacterial activity themselves, but act as selective inhibitors of the β-lactam-inactivating enzymes. β-Lactamases are largely responsible for the resistance of an increasing number of pathogenic organisms to antibiotic therapy (see Section 6).

2.4. Total Synthesis of Penicillins

Once the chemical structure of penicillin had been established, the chemists optimistically considered that developing a commercial synthesis would pose few problems. This proved to be far from the truth, and intensive research through to the mid-1950s failed to achieve this objective. The β-lactam ring proved too sensitive to the chemical techniques available at

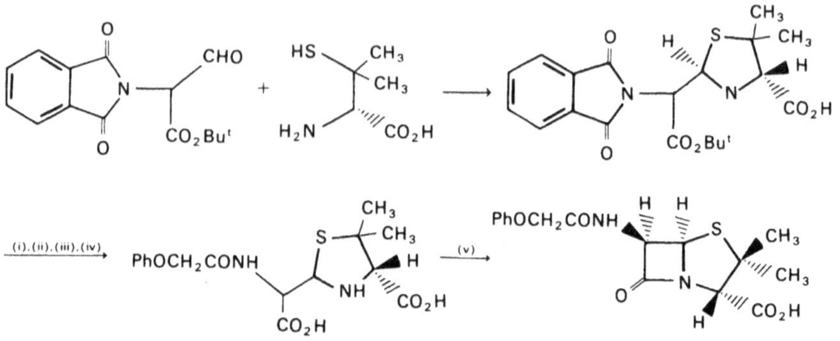

Figure 6. Synthesis of a penicillin. *Reagents*: (i) N_2H_4-HCL; (ii) $PhOCH_2COCl$-Et_3N; (iii) HCl, pyridine; (iv) KOH (1 equivalent); (v) DCC-aqueous dioxan.

the time, and it required the development of new methodology, particularly from the emerging field of peptide chemistry, to allow protection and deprotection of the various reactive groups and the construction of the β-lactam ring. The classic achievement of the first total synthesis is outlined in Fig. 6 (Sheehan and Henery-Logan, 1959). Although of great academic interest, this route and other routes subsequently developed (Holden, 1982) have found no commercial utility, and it is likely that fermentation will remain the predominant source of penicillins for the projectable future.

3. CEPHALOSPORINS

3.1. Naturally Occurring Cephalosporins

The discovery and development of the cephalosporins began in the immediate postwar period with the observation that an organism isolated from seawater near a sewage outlet at Cagliari in Sardinia was active against both gram-negative and gram-positive bacteria (Brotzu, 1948). This fungus was initially identified as a strain of *Cephalosporium acremonium* but later reclassified as *Acremonium chrysogenum* (Lemke and Brannon, 1972) and more recently as *A. strictum* (see Chapter 1). Advances in this area were slow because the activity was associated with a number of different types of compounds. A connection with the penicillins was established when the so-called "cephalosporin N" was shown to be a new penicillin (penicillin N) with a δ-D-α-aminoadipyl side chain (see Table I) (Abraham and Newton, 1954). Although the isomeric compound, isopenicillin N, with the α-l-configuration in the side chain was shortly after isolated from *Penicillium* fermentations

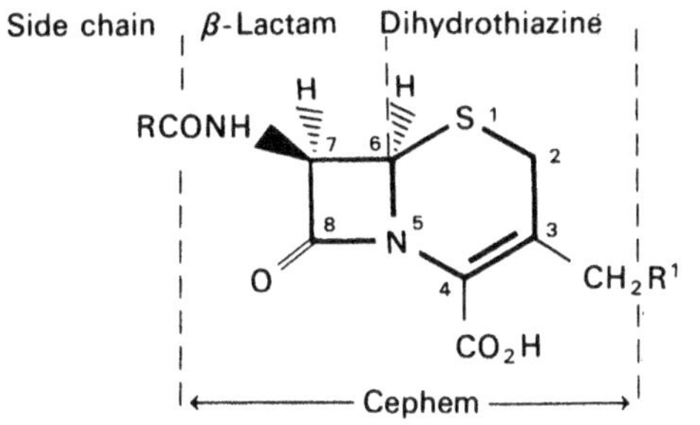

Figure 7. Nomenclature and numbering of the cephalosporin skeleton.

when side-chain precursors were limited (see Section 2.1), penicillin N has not to date been observed as a metabolite of any *Penicillium* sp. The significance of penicillin N and isopenicillin N to the biosynthesis of both groups is discussed in Section 7.

Cephalosporin C was first recognized as a contaminant of partially purified penicillin N. The isolation of this new antibiotic in quantity proved to be a formidable undertaking. Nevertheless, small amounts of the pure antibiotic were accumulated for structural studies (Abraham and Newton, 1961), and the structure proposed on the basis of this work was later confirmed by X-ray crystallographic analysis (Hodgkin and Maslen, 1961). Cephalosporin C differs from penicillin N by substitution of a dihydrothiazine ring for the thiazolidine ring and by the presence of an acetoxymethyl group attached to the C-3 position of the cephalosporin (cephem) ring system (Fig. 7).

Unlike the naturally occurring penicillins, all the cephalosporins thus far isolated possess the δ-D-α-aminoadipyl residue or a side chain that is derived by biochemical transformation of this group (Table III). In addition, the production of other cephalosporins cannot be induced by the addition of carboxylic acid derivatives to the culture media (Ott *et al.*, 1962), nor has 7-aminocephalosporanic acid ever been observed in any fermentation.

A further important difference from the penicillins relates to modifications of the bicyclic system. Deacetoxy- and deacetyl cephalosporin C have been observed in fermentation broths, and a number of derivatives involving substitution on the C-3 methyl have been reported. Table III lists the cephalosporins isolated from *Acremonium* spp. (W. B. Turner and Aldridge, 1983).

3.2. Structure and General Chemistry of Cephalosporins

The structural studies on cephalosporin C were considerably facilitated by the foreknowledge of the properties of the penicillins. Like penicillin N,

Table III. Naturally Occurring Cephalosporins

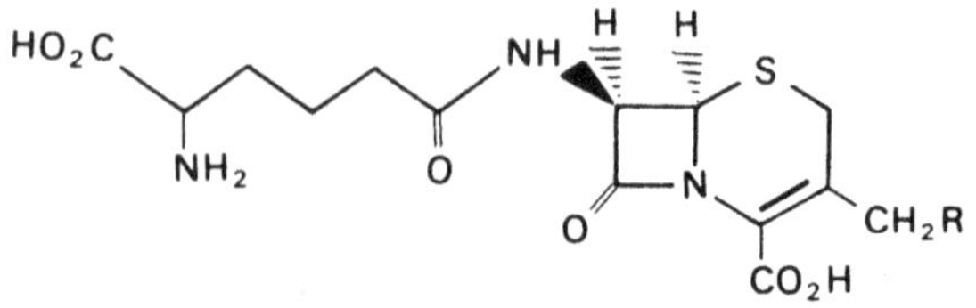

R group	Name
$-H$	Deacetoxycephalosporin C
$-OH$	Deacetylcephalosporin C
$-SMe$	—
$-SC(Me)_2-CH(NH_2)CO_2H$	—

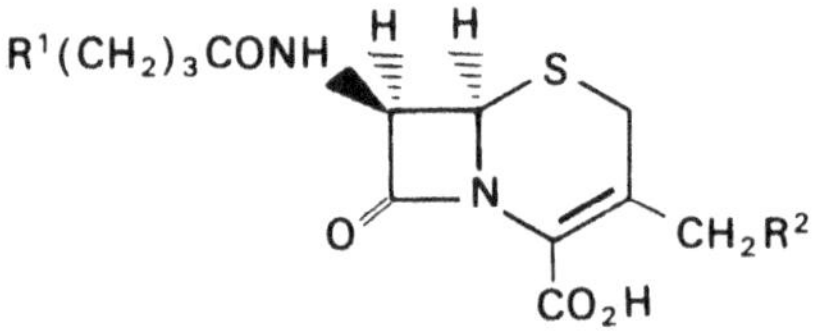

R^1 group	R^2 group
$-CO_2H$	$-OH$
$-COCO_2H$	$-OH$
$-CO_2H$	$-OAc$
$-COCO_2H$	$-OAc$
$-CO_2H$	$-H$
$-COCO_2H$	$-H$
$-CH(NH_2AC)-CO_2H$	$-H$

cephalosporin C behaved as a monoaminodicarboxylic acid and contained a δ-D-α-aminoadipyl residue that linked to the rest of the molecule through its δ-carboxyl group. The infrared absorption spectrum of its sodium salt showed a strong band at 1782 cm^{-1} characteristic of the β-lactam carbonyl and similar to the stretching frequency of the corresponding carbonyl in penicillins. Despite these similarities, cephalosporin C showed some marked differences from penicillin N. On acid hydrolysis, it did not yield D-penicillamine, but rather gave a number of sulfur-containing fragments that lacked nitrogen. Also, the valine obtained from cephalosporin C after treatment with Raney nickel was racemic, whereas the valine-containing fragment from penicillin N had the D-configuration (see Section 2.2). This result, and further spectroscopic data obtained principally from ultraviolet absorption and nuclear magnetic resonance (NMR) spectra, were thus not consistent with the presence of a fused β-lactam–thiazolidine ring system, and at this stage the structure containing the dihydrothiazine ring system was proposed (Fig. 7) (Abraham and Newton, 1961). This structure was shown to be consistent with the known chemical and physical properties and subsequently confirmed by X-ray crystallography (Hodgkin and Maslen, 1961). The presence of the allylacetoxy functionality provides an additional reactive site within the molecule, complicating the degradative pathways for acid and neutral conditions; some of these are illustrated in Fig. 8 (Abraham and Loder, 1972). As expected, the instability of the β-lactam ring to basic

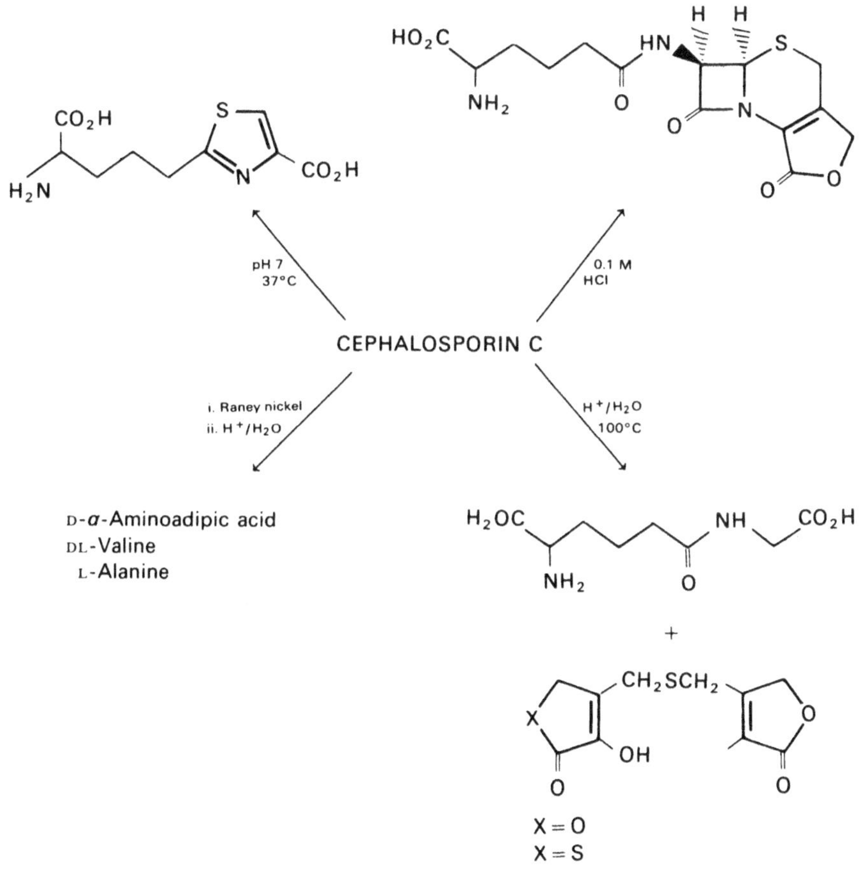

Figure 8. Chemical degradation of cephalosporin.

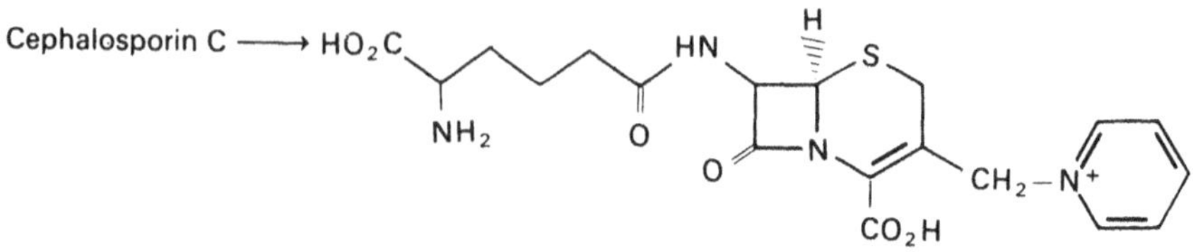

Figure 9. C-3 methyl substitution of cephalosporin C.

conditions coupled with the presence of a further reactive site results in complete degradation of the molecule at high pH. It is possible, however, to displace the acetoxy group on the C-3 methyl by nucleophiles with surprising ease and without disrupting the rest of the molecule. During the purification of cephalosporin C in pyridine acetate buffer, the formation of a pyridinium betaine was observed (Fig. 9) (Hale *et al.*, 1961). This was subsequently shown to be a general reaction with a variety of weak heterocyclic bases and was later to find extensive application in the production of semisynthetic cephalosporins.

3.3. Chemistry of 7-Aminocephalosporanic Acid and Related Compounds

In the light of the success of the semisynthetic penicillins, it was axiomatic that the development of similar methods for the production of 7-aminocephalosporanic acid (7-ACA) would be painstakingly pursued. Surprisingly, the first reported conversion of cephalosporin C to 7-ACA was achieved by mild acid hydrolysis (Loder *et al.*, 1961), but in a yield of less than 1%. Nevertheless, material obtained in this way was used to prepare semisynthetic cephalosporins with simple modified side chains. The antibacterial potential of these early compounds demanded the development of more practical procedures for the synthesis of 7-ACA. Unfortunately, it appears that nature has not yet developed, or is carefully guarding, acylases capable of cleaving off the aminoadipyl side chain, and extensive evaluation of bacterial, fungal, and other natural sources (Demain *et al.*, 1963; Claridge *et al.*, 1963) has failed to reveal any enzyme that can effect this transformation. This failure can clearly be ascribed to the structure of the side chain and not to the cephem nucleus, since other semisynthetic cephalosporins have been successfully cleaved enzymatically. For example, acylases of *Nocardia* and *Proteus* cleave 7-phenoxyacetyl cephalosporin C to 7-ACA in appreciable yields (Sjoeberg *et al.*, 1967).

The first practical and selective deacylation method involved the treatment of cephalosporin C with nitrosyl chloride in formic acid at room temperature followed by quenching with aqueous methanol (Morin *et al.*, 1962). The proposed mechanism for this transformation involves the formation, by

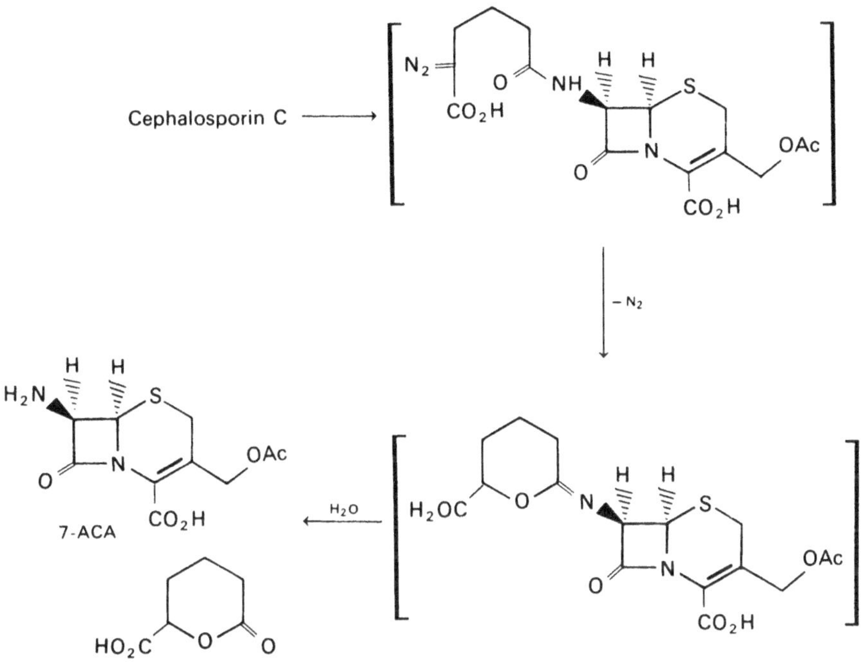

Figure 10. Preparation of 7-aminocephalosporanic acid (7-ACA).

an intramolecular cyclization, of an intermediate iminolactone that is more susceptible than the β-lactam ring aqueous hydrolysis (Fig. 10). The importance of 7-ACA to the pharmaceutical industry has meant that over the years, a variety of further sophisticated chemical procedures for side-chain removal have been developed, although in essence all proceed through similar reactive intermediates (Hatfield *et al.*, 1981).

With the ready availability of 7-ACA, the synthesis of a large number of cephalosporin derivatives became possible. Initially, such syntheses involved only the acylation of the 7-amino function, but the potential for further structural modification at the C-3 position to optimize biological activity was soon realized (Kukolja and Chauvette, 1982). Displacement of the acetoxy group with heterocyclic tertiary bases and then later with sulfur nucleophiles widened the scope for the medicinal chemist, and over the past 20 years, a vast number of semisynthetic derivatives of an ever-increasing complexity have been and continue to be synthesized (Sassiver and Lewis, 1977; Webber and Ott, 1977). It is perhaps noteworthy that few cephalosporins are active after oral administration due to their relative acid instability, and research has in the main been directed toward developing broad-spectrum injectable agents. A number of these are exemplified in Table IV.

Table IV. A Selection of Clinically Significant Semisynthetic Cephalosporins Derived from 7-Aminocephalosporanic Acid

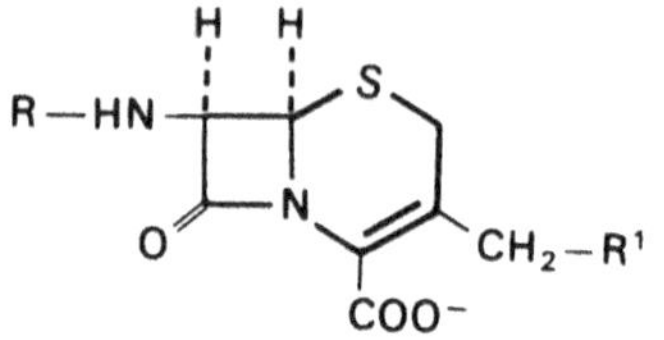

Side-chain group (R)	C-3 substituent (R^1)	Name
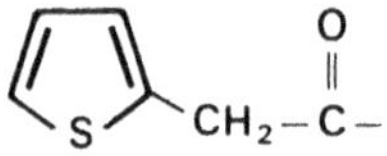	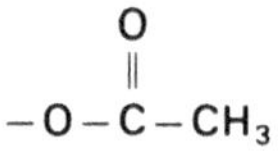	Cephalothin
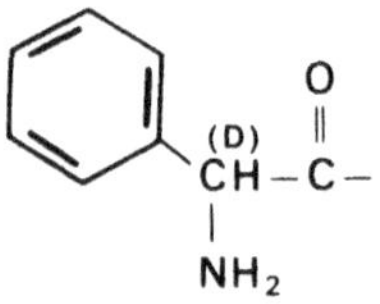	–H	Cephalexin
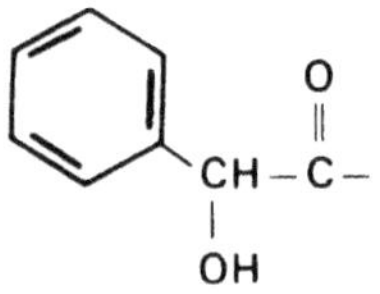	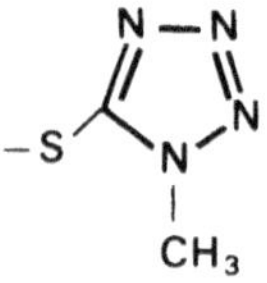	Cefamandole
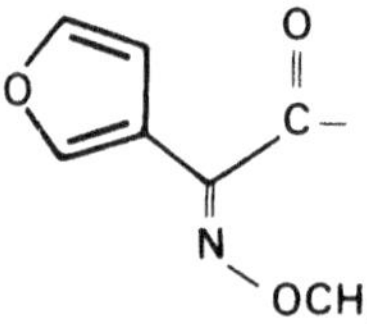	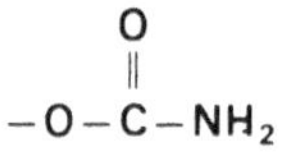	Cefuroxime
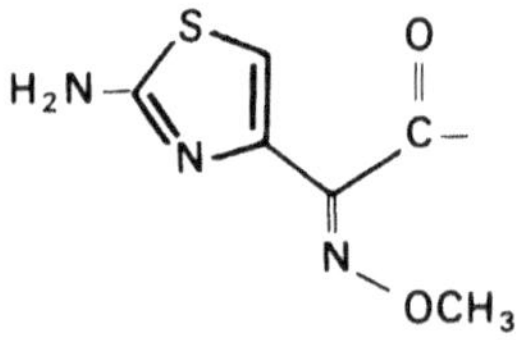	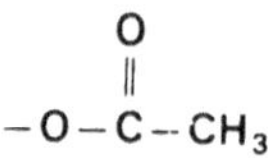	Cefotaxime
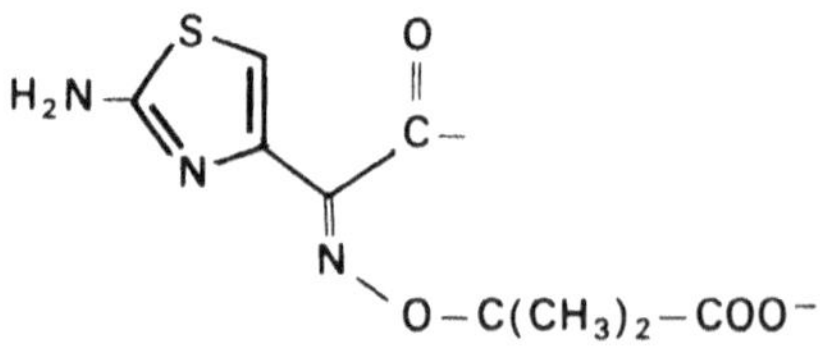	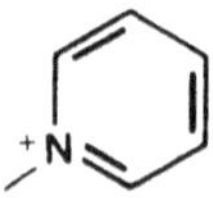	Ceftazidime

3.4. Total Synthesis of Cephalosporins

The difficulties experienced with the penicillins strongly mitigated against total synthesis, contributing significantly to the medicinal chemistry of the cephalosporins. Nevertheless, the complexity of the structure represented a formidable challenge, and it is a tribute to the developments that had occurred in synthetic organic chemistry that an elegant total synthesis was achieved within five years of the structure being reported. Moreover, the strategy involved the initial construction of the β-lactam ring, the very

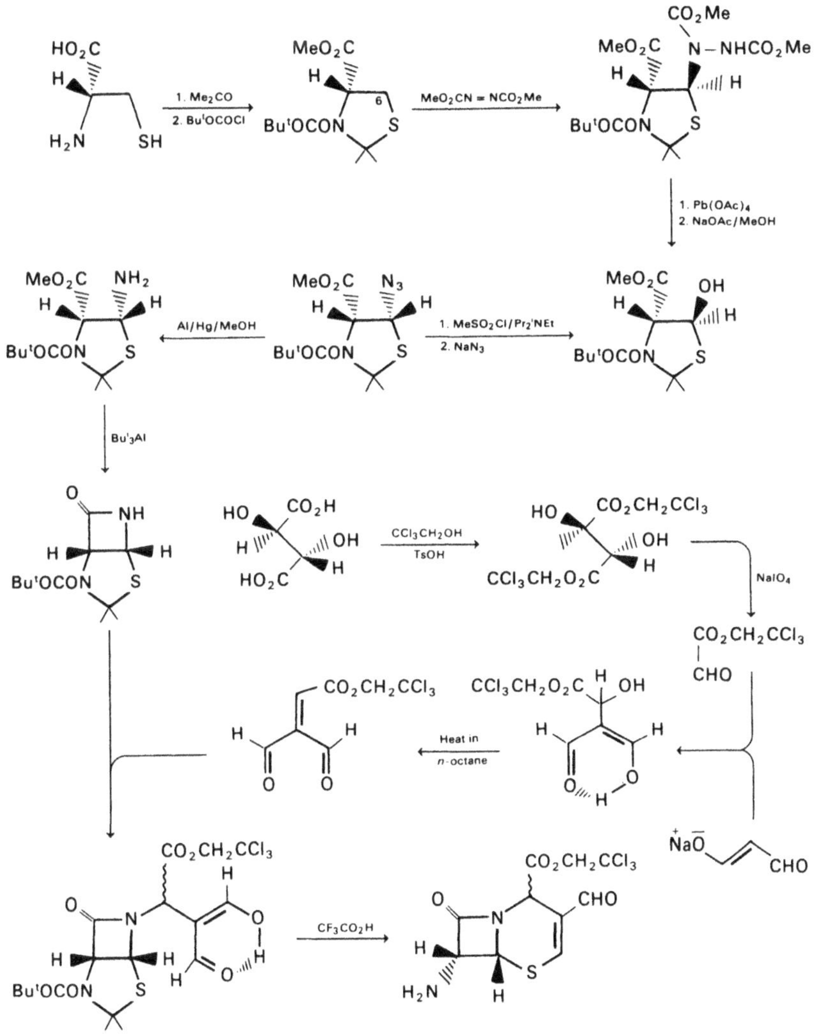

Figure 11. Total synthesis of cephalosporin C.

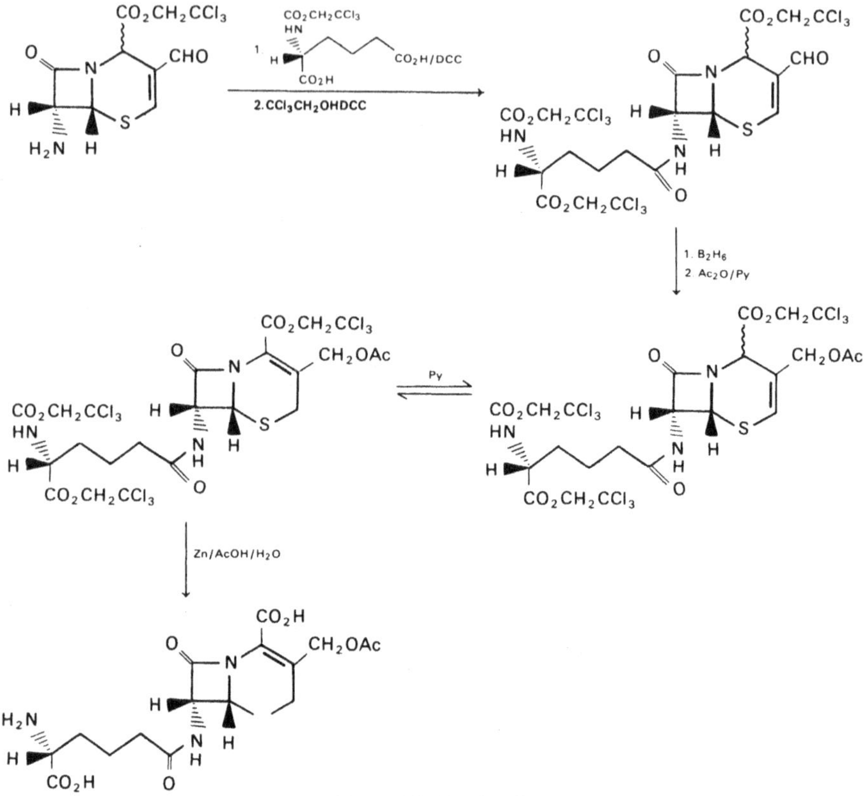

Figure 11. (*Continued*)

feature that had been left until the end in the penicillin synthesis. A brief outline of the synthesis, which was the subject of the Nobel Prize address in 1965, is shown in Fig. 11 (Woodward, 1966). Although no totally synthetic cephalosporin has, or is likely to have, any clinical significance, it is important to note that the methodology developed in this synthesis was later to be applied to novel β-lactam systems unrelated to cephalosporins and penicillins. A number of these compounds are potent antibiotics with considerable clinical potential; at present, total synthesis represents the most economical approach to their production (Brown and Roberts, 1985).

4. CHEMICAL INTERRELATIONSHIP BETWEEN PENICILLINS AND CEPHALOSPORINS

The similarities between the structures of penicillins and cephalosporins coupled with the co-occurrence of penicillin N and cephalosporin C in

Acremonium spp. were used very early on to suggest that cephalosporins were derived biochemically via an oxidative ring expansion (see Section 7). However, there was no chemical precedence for opening the thiazolidine ring while keeping the more sensitive β-lactam ring intact. Nonetheless, it had been noted that penicillins were readily oxidized to the corresponding sulfoxide derivatives (Section 2.2). These derivatives were considerably more stable to acidic and basic conditions than the parent penicillin, but their chemistry had received only scant attention.

During a reexamination of these compounds in the early 1960s, it was observed, with some excitement, that penicillin sulfoxide esters undergo a thermal opening of the thiazolidine ring and rearrangement to deacetoxycephalosporin and 2-methyl-substituted penicillin derivatives (Fig. 12) (Morin *et al.*, 1963). This unique type of reaction, the details of which are beyond the scope of this review, has now been extensively investigated (Cooper and Spry, 1972; Cooper and Koppel, 1982) and shown

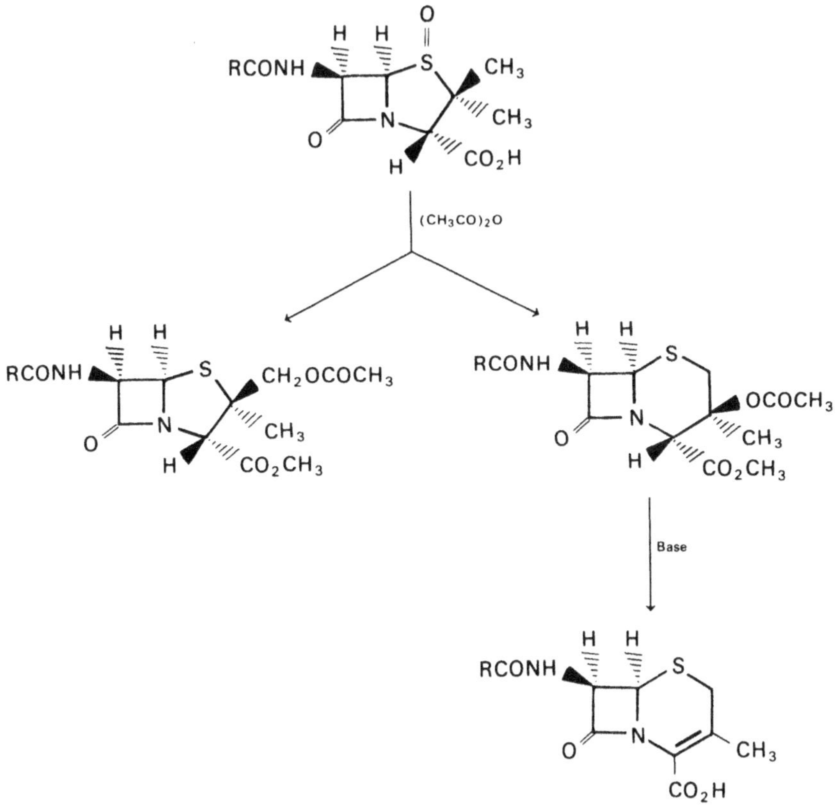

Figure 12. Chemical conversion of penicillin to cephalosporin.

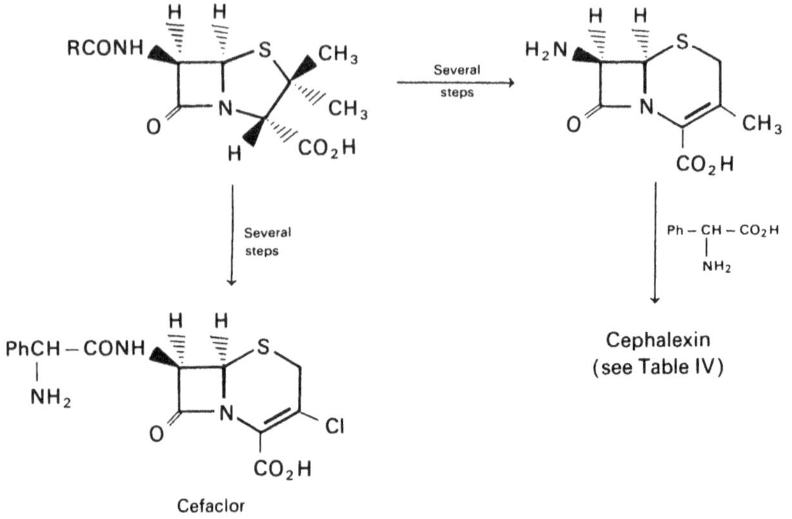

Figure 13. Commercial synthesis of Cephalexin.

to proceed through an alkene-sulfenic acid intermediate (Cooper *et al.*, 1973). The application of this chemistry for the preparation of semisynthetic deacetoxycephalosporins from the readily available penicillins was soon to be recognized. Indeed, it has resulted in the development of an industrial process for the manufacture of Cephalexin, an orally active broad-spectrum antibiotic (Fig. 13) (Chauvette *et al.*, 1971). It is perhaps noteworthy that this process depended on the selective protection of the carboxyl group as a trichloroethyl ester, a strategy that was pioneered in the total synthesis of cephalosporin C (see Section 3.4). This is necessary because acid-catalyzed rearrangement of penicillin sulfoxides to cephalosporins results in extensive decarboxylation if the carboxyl is unprotected.

5. RECENT ADVANCES IN SEMISYNTHESIS OF PENICILLINS AND CEPHALOSPORINS

It was not until the 1970s that more esoteric modifications of the basic skeletons could be considered. This possibility was due first to the enormous advancements that were being made in synthetic methodology and second to improvements in fermentation technology leading to cheaper β-lactam precursors.

Such modification was demonstrated in the first instance by refining and extending the penicillin sulfoxide rearrangement to give a range of 3-substituted cephems (Kukolja and Chauvette, 1982). An electronegative heteroatom attached directly to the 3-position was shown to impart enhanced biological properties, e.g., Cefaclor (Fig. 13).

Stimulated by the isolation and characterization of a family of closely related 7-methoxycephalosporins, the cephamycins (Nagarajan *et al.*, 1971), from various *Streptomyces* spp., a number of methods were immediately developed for the introduction of a methoxy group into the 6- and 7-positions of penicillins and cephalosporins, respectively (Gordon and Sykes, 1982; Morin and Gorman, 1982). These compounds are noteworthy because of their increased stability to a wide range of gram-negative β-lactamases. A number of the more clinically significant examples are shown, together with the cephamycins, in Table IV.

The interest in the semisynthesis of novel modified β-lactam antibiotics was stimulated not only by the isolation of the cephamycins but also by a range of new types of β-lactam systems, the latter being predominantly from actinomycetes and bacterial species and therefore beyond the scope of this review. However, the structure of these compounds considerably influenced the direction of research with respect to chemical modifications of penicillins and cephalosporins. So that these developments may be appreciated, the structures of the major groups are presented in Table V.

The molecular structure of clavulanic acid (Table V), represented a completely novel oxa-fused β-lactam system, and the corresponding oxa-cephems, i.e., with oxygen replacing sulfur in the dihydrothiazine ring, a system that has yet to be found in nature, became an obvious target for both total and semisynthesis (Cooper, 1980). The pinnacle of the research in this

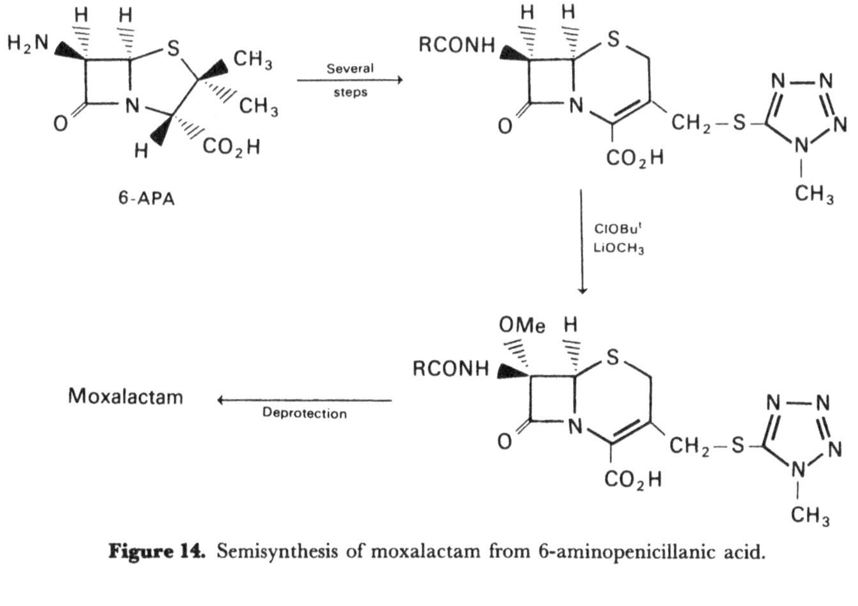

Figure 14. Semisynthesis of moxalactam from 6-aminopenicillanic acid.

Table V. Structures of the Principal Antibiotics Isolated from Actinomycetes and Bacterial Species

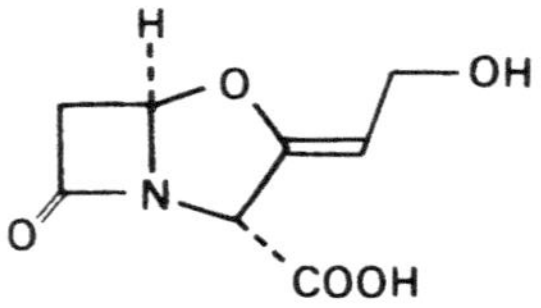

Clavulanic acid

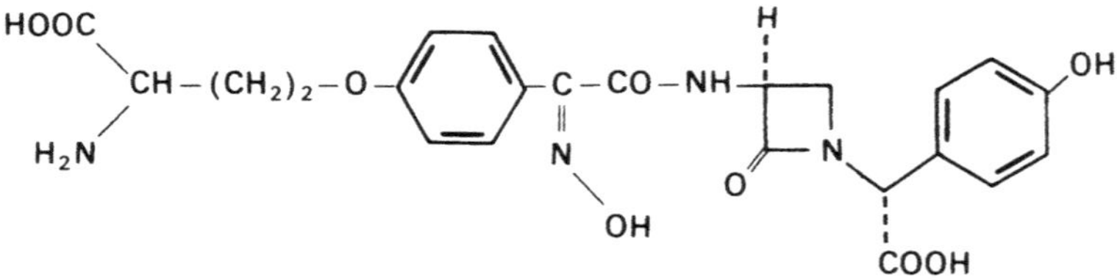

Nocardicin

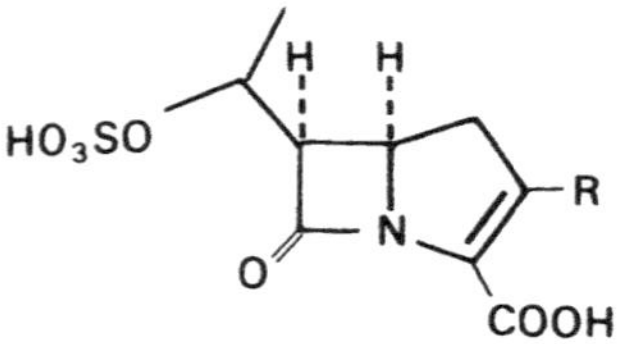

Olivanic acids

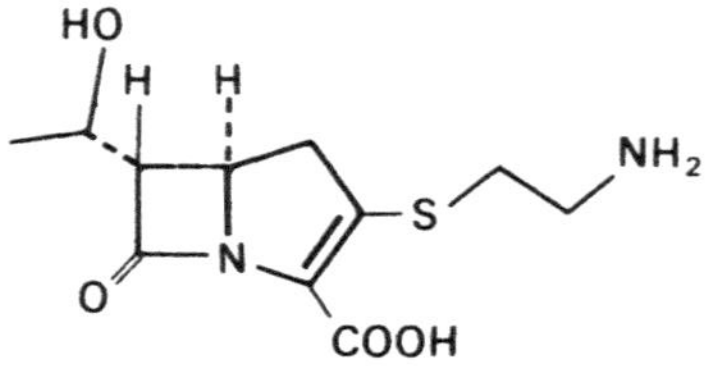

Thienamycin

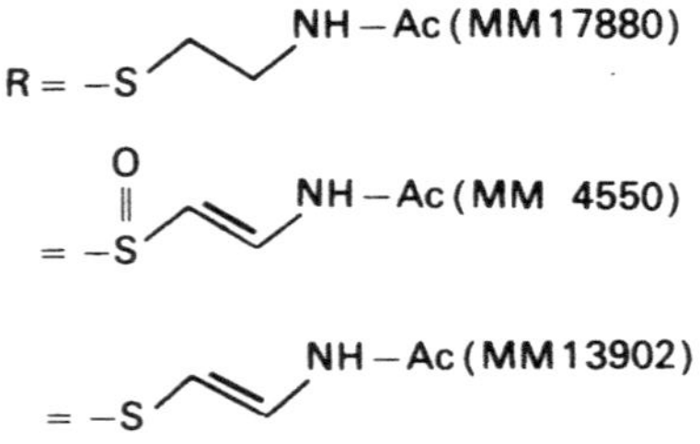

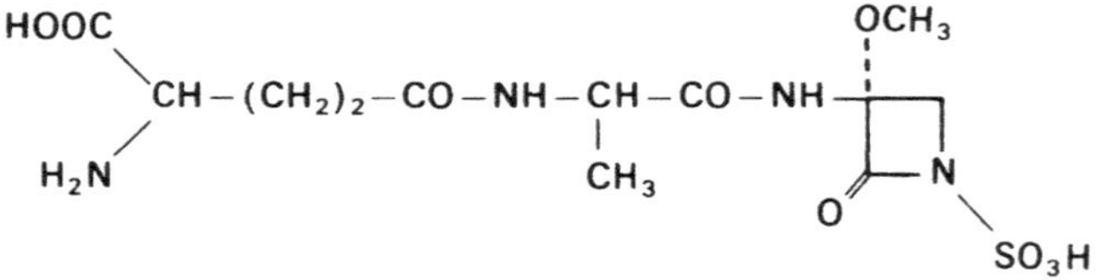

Sulfazecin

area is epitomized by moxalactam (Table VI), a potent gram-negative antibacterial agent. The synthesis of this compound, for which penicillin G is the starting material, involves not only the opening of the thiazolidine ring followed by the construction of the oxa–cephem system, but also semisynthetic modifications at C-3 and the side chain, together with the introduction of a 7-methoxy substituent as outlined in Fig. 14 (Hamashima *et al.*, 1981). Thus, moxalactam, as an important clinical antibiotic, embodies the most complex series of chemical modifications of a naturally occurring β-lactam precursor yet achieved.

As a consequence of the success with the oxacephems, attention was

Table VI. Semisynthetic 7-Methoxycephalosporins Related to Cephamycin C

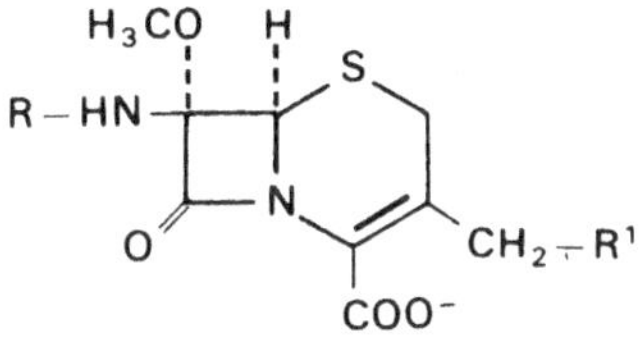

Side-chain group (R)	C-3 substituent (R^1)	Name
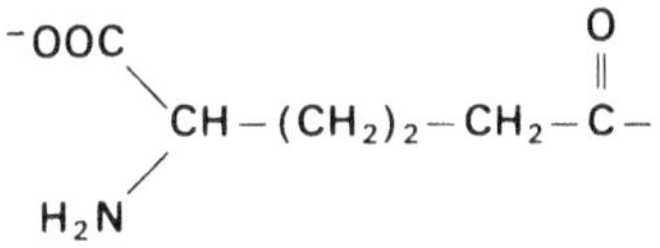	$-O-\overset{O}{\overset{\|\|}{C}}-NH_2$	Cephamycin C
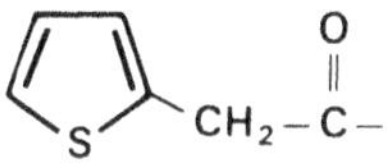	$-O-\overset{O}{\overset{\|\|}{C}}-NH_2$	Cefoxitin
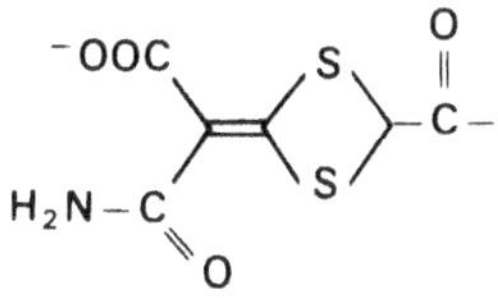	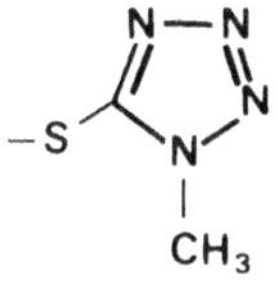	Cefotetan
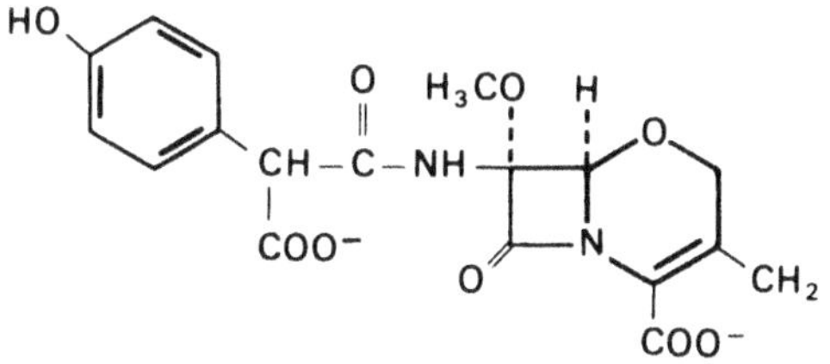	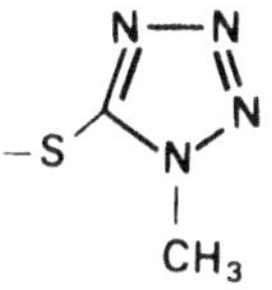	Moxalactam

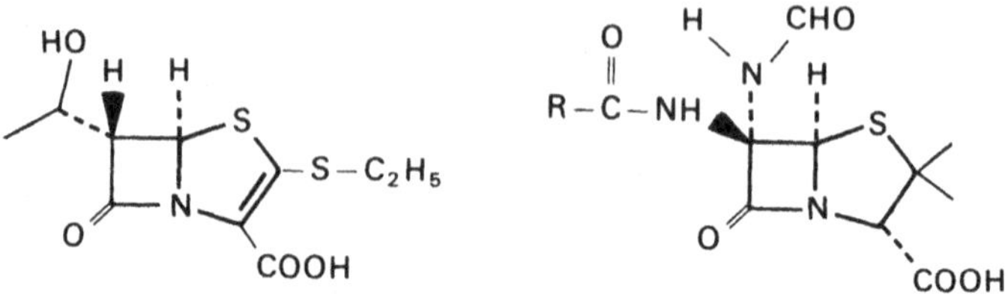

Sch 29482 6-Formamidopenicillin

Figure 15. Synthetic hybrid penicillins.

redirected toward hybrids of the penicillin–cephalosporin systems, namely, the penems (Woodward, 1977; Ernest, 1982). These systems, too, have not yet been observed as natural products and, because of the substantially increased ring strain, represent a considerable synthetic challenge. Yet, again, penicillins proved to be useful, inexpensive, chiral starting materials for synthesis in this area, and currently the most promising clinical candidate is Sch 29482 (Fig. 15) (Ganguly *et al.*, 1982).

Since almost all the fundamental advances have stemmed from investigations into the secondary metabolites of microorganisms, it is reasonable to suppose that more detailed examination of nature's resources will provide further quantum leaps in terms of new *β*-lactam systems that will then require fine-tuning by medicinal chemists. In this respect, it is interesting to note that very recently, attention has again centered on modification of the penam and cephem ring systems, this attention being prompted by the isolation of 7-formamidocephalosporins from a *Flavobacterum* sp. (Singh *et al.*, 1984) and the independent semisynthesis of antibacterially active 6-formamidopenicillins (Fig. 15) (Ponsford, 1985).

6. BIOCHEMICAL MODE OF ACTION OF β-LACTAM ANTIBIOTICS AND MECHANISMS OF RESISTANCE

The mode of action of *β*-lactam antibiotics and the mechanisms by which organisms acquire resistance are areas that have been and continue to be intensively investigated because of their obvious relevance to the synthesis and clinical importance of semisynthetic penicillins and cephalosporins.

The evolution of the understanding of the fundamental principles in this area has occurred in three discrete steps. Early work on the morphological and lytic effects of penicillins on sensitive organisms led to the conclusion that they selectively inhibit bacterial cell-wall biosynthesis (Gardner, 1940).

The second stage involved the unraveling of the complex chemical structure of bacterial cell walls and the detailed mechanisms by which they

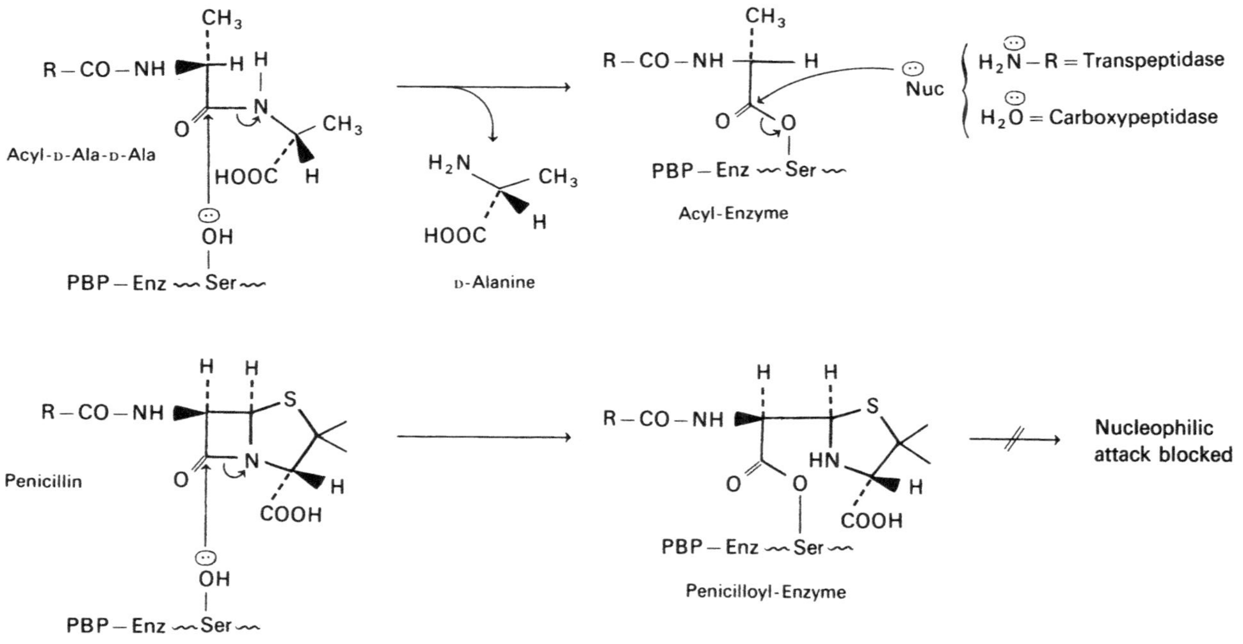

Figure 16. Proposed mechanism for interaction of acyl-D-Ala-D-Ala (top) and penicillin (bottom) with PBPs showing participation of an active-site serine residue, involvement of an acyl- or penicilloyl-enzyme intermediate, and possible subsequent reactions.

are biosynthesized (Ghuysen, 1977; Mirelman, 1979). This phase also led to the demonstration that one of the terminal steps in the biosynthesis, catalyzed by the enzyme peptidoglycan transpeptidase, is the penicillin-sensitive reaction (Tipper and Strominger, 1965). Furthermore, it was shown that penicillin behaves as a substrate analogue for the dipeptide terminus (acyl-D-Ala-D-Ala) crucially involved in the final stages of cell-wall cross-linking (Fig. 16).

The discovery that several proteins within all bacterial cell walls so far examined bind penicillin specifically suggested multiple targets for penicillin action (Suginaka *et al.*, 1972; Blumberg and Strominger, 1972). These so-called "penicillin-binding proteins" (PBPs) represent the sites at which all β-lactam antibiotics exert their bactericidal effect by interacting with different subsets of PBPs. The current stage of research is concentrating on detailed studies of PBPs, the complex role they play in bacterial cell-wall construction, and their interaction with the various types of β-lactam antibiotics (Waxman and Strominger, 1982; Bycroft and Shute, 1985).

It has been known for the last 40 years that bacteria that were initially susceptible to β-lactam antibiotics can suddenly or progressively develop resistance (Abraham and Chain, 1940). A number of factors have been advanced (Ghuysen, 1980) to account for bacterial resistance, the most significant of which is the production of β-lactamases, and these enzymes now represent the most serious threat to the continuing therapeutic utility of β-lactam antibiotics. β-Lactamases catalyze the hydrolysis of the azetidin-2-one ring of a variety of antibiotics, but more particularly of the penicillins and cephalosporins (Fig. 17). The enzymes are most conveniently detected by using the chromogenic cephalosporin nitrocefin (O'Callaghan *et al.*, 1972). The intact substrate molecule is yellow, but becomes pink when the β-lactam bond is broken due to conjugation of the

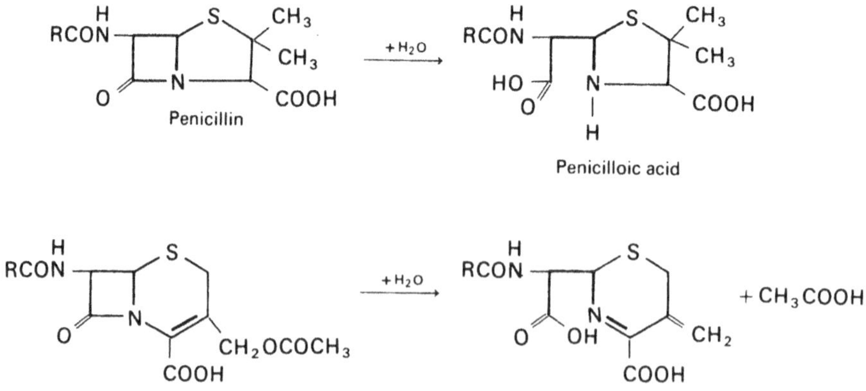

Figure 17. Hydrolysis of β-lactams by β-lactamases.

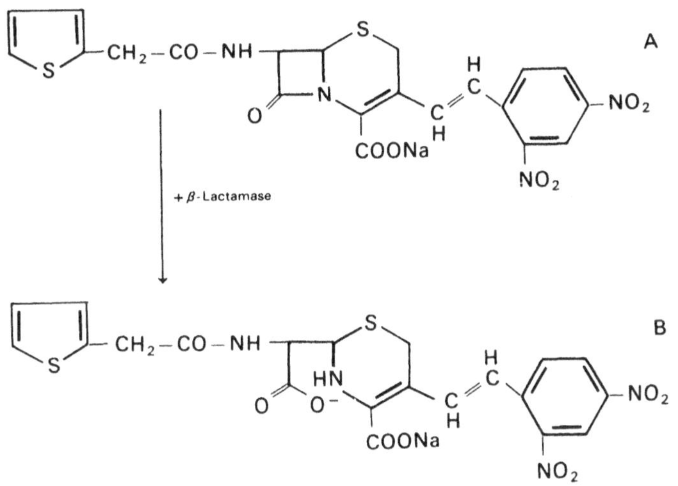

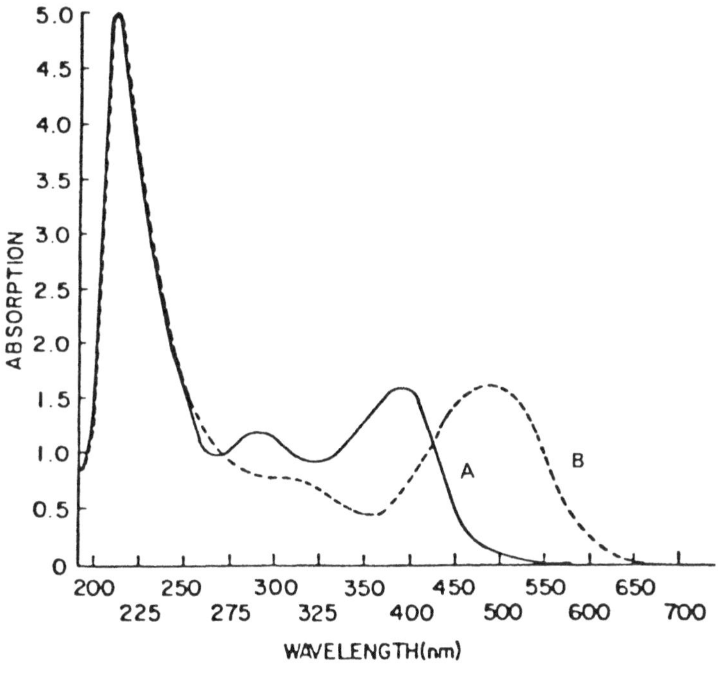

Figure 18. Absorption spectrum of the chromogenic cephalosporin nitrocefin before (A) and after (B) treatment with β-lactamase.

nitrogen with the chromophore (Fig. 18). This system has found widespread application not only for detection but also for the study of the enzyme kinetics.

β-Lactamases are widely distributed throughout microorganisms (matthew and Harris, 1976), and the different types have been classified on the basis of their substrate profiles (Sykes and Matthew, 1976). Although extensive physical and chemical studies have been conducted on a number of these enzymes (Hamilton-Miller and Smith, 1979), the considerable information that would accrue from a fully refined X-ray crystallographic analysis on even one enzyme is still eagerly awaited. However, the studies that have been carried out to date indicate the intermediacy of an acyl–enzyme complex involving the hydroxyl group of an active-site serine residue (Anderson and Pratt, 1981). Thus, it would appear that β-lactamases are a subgroup of the serine proteases and have probably evolved from the enzymes of the cell-wall biosynthetic machinery.

7. BIOSYNTHESIS OF PENICILLINS AND CEPHALOSPORIN C

7.1. Introduction

Biosynthetic studies on penicillin were started in the 1940s. The main objectives of this work were improvement of the yield of penicillins from fermentation together with the production of new penicillins by addition of appropriate precursors to the culture medium. Both these objectives were realized to some extent, yields were improved, and a large number of penicillins containing modified side chains were produced (see Section 2.1).

Structural examination and degradative studies (see Sections 2.2 and 3.2) of both the penicillins and the cephalosporins firmly suggested that their ring systems are derived from cyst(e)ine and valine (see Fig. 19). As a result of early radiochemical labeling studies using whole cells of the producing fungi, it was soon established that both the penam and the cephem nuclei are built up from these amino acids. The isolation of the tripeptide δ-(α-aminoadipyl)-cysteinyl valine from *P. chrysogenum* in 1960 (Arnstein and Morris, 1960) led to the proposal that this substance is the precursor of the penicillins—the so-called "tripeptide theory." The "Arnstein" tripeptide was subsequently identified as a metabolite of *Acremonium*. However, because of the permeability constraints imposed by experiments using intact cells, i.e., to the uptake of amino acids and dipeptides, investigations up until the early 1970s were limited to establishing the involvement of valine and cyst(e)ine as precursors of penicillin and cephalosporin C. These investigations were followed by studies directed toward determining which atoms of the amino acids are incorporated into

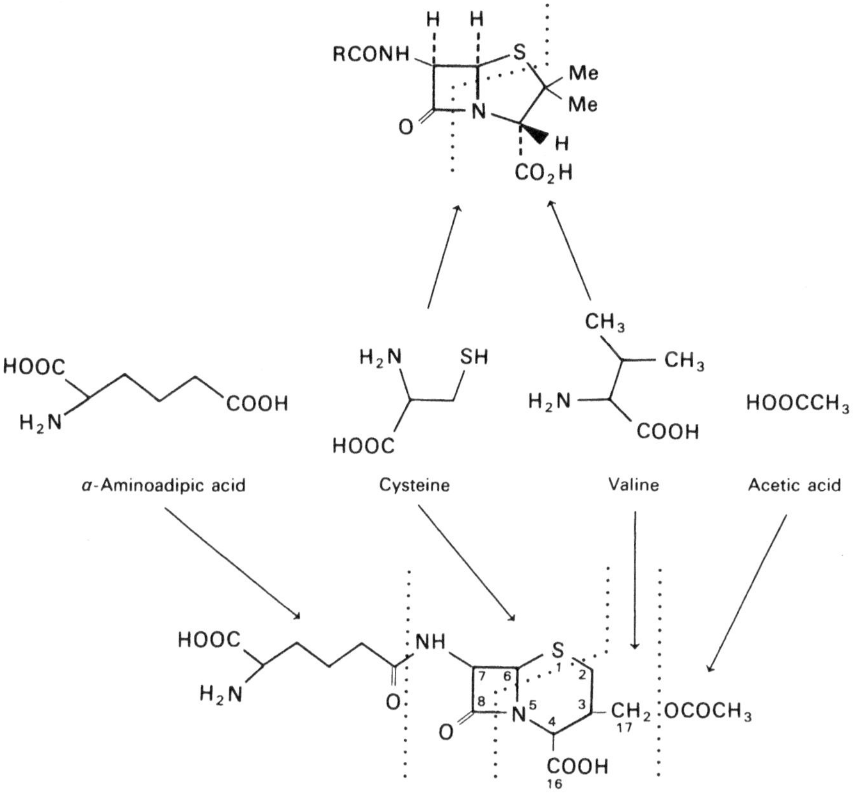

Figure 19. Biochemical building blocks of penicillins and cephalosporins.

the antibiotics and the stereochemical fate of the various chiral and prochiral centers involved.

The development in the late 1970s of a cell-free system capable of converting isopenicillin N into cephalosporin C provided the quantum leap needed for further cell-free studies. These studies were to establish the involvement of the Arnstein tripeptide in the pathways and provide a first insight into what appears to be a unique biochemical mechanism for the formation of the bicyclic systems.

The results obtained up until the mid 1970s have been the subject of a number of reviews (Behrens, 1949; Abraham *et al.*, 1965; Abraham, 1974; Demain, 1974; Aberhart, 1977). These are largely concerned with the incorporation of cyst(e)ine and valine into both the penam and the cephem skeleton in whole organisms. Developments associated with the cell-free investigations are covered in later reviews (O'Sullivan and Abraham, 1981; Queener and Neuss, 1982).

7.2. Incorporation of Valine and Cyst(e)ine into Penicillins and Cephalosporin C in Whole Cells

The feeding of [4 ^{14}C]-DL-valine to *P. chrysogenum* first demonstrated the incorporation of this amino acid into benzylpenicillin (Arnstein and Grant, 1954). Later examination of the relative rates of incorporation of D- and L-[1-^{14}C] valine indicated that L-valine was more rapidly taken up by the mycellium and used for penicillin biosynthesis (Stevens and De Long, 1958; Bycroft *et al.*, 1976). Similar studies showed that [1-^{14}C]-DL-valine served as a precursor of penicillin N and cephalosporin C in *Acremonium* (Trown *et al.*, 1963b) and that L-valine more efficiently labeled cephalosporin C (Demain, 1963). Because the relative rates of uptake of L- and D-valine by whole cells differ, rates of incorporation of the isotopically labeled isomers cannot be compared directly. This point was of some considerable significance and has been extensively investigated, since the corresponding center in the penicillins has the D-configuration.

Neither the tritium label from D- nor that from L-[2-^{3}H] valine is incorporated into penicillin by *P. chrysogenum* cells (Bycroft *et al.*, 1975b; Adriaens *et al.*, 1975), suggesting that the D-isomer is not a direct precursor of penicillin. The loss of label from the α-position of the L-isomer was to be expected with the inversion of configuration in the final product. This must also occur without loss of the nitrogen, since a mixture of L-[^{15}N] valine and L-[U ^{14}C]-valine was converted to benzylpenicillin without appreciable change in the isotopic ratio. These results suggested that L-valine is not initially converted intracellularly to free D-valine via α-ketoisovalerate before incorporation (Booth *et al.*, 1976).

The fate of the diastereotopic methyl groups of L-valine in its conversion to penicillin was a question of some considerable interest, since it would provide important information concerning the stereochemistry associated with the formation of the carbon–sulfur bond in the thiazolidine moiety within the penam nucleus. To this end, valines, prochirally labeled at the β-center with ^{2}H or ^{13}C, were synthesized. The synthetic [^{13}C] valines were then converted by *P. chrysogenum* into penicillin V and by *Acremonium* into penicillin N and cephalosporin C. The ^{13}C-NMR spectra of the penicillins revealed enhanced signals corresponding to only one of the methyl groups attached to the thiazolidine ring in each case, demonstrating that the biosynthesis had proceeded completely stereospecifically. Since the assignment of the methyl signals in the ^{13}C spectrum of penicillins had been unequivocally established, it was possible to conclude that the incorporation of valine into penicillin proceeds with overall retention of configuration at C-3. Furthermore, in the biosynthesis of cephalosporin C, it was shown that C-2 of the dihydrothiazine ring was specifically labeled by 3R-[4 ^{13}C] valine, while the C-3 methylene carbon was labeled by the 3S-isomer

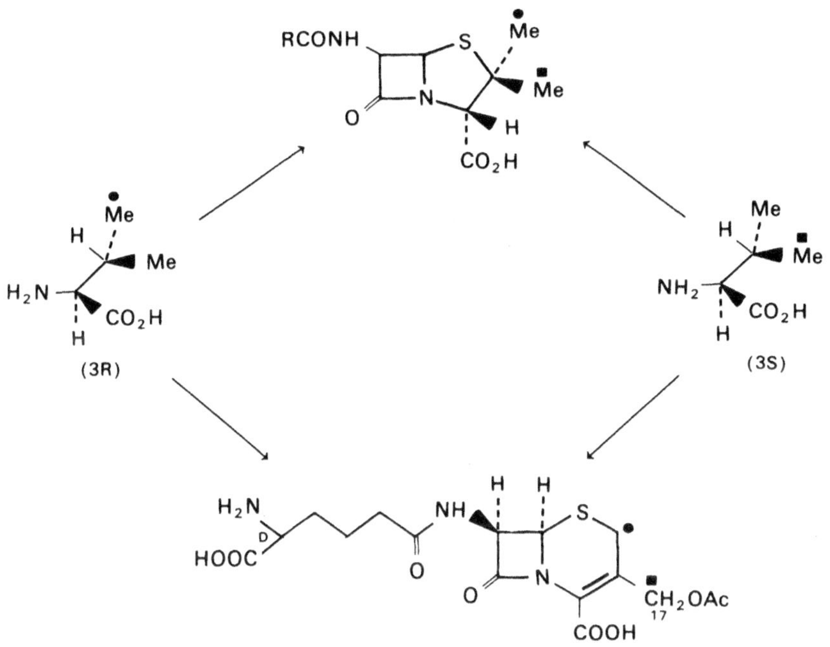

Figure 20. Incorporation of prochirally labeled valine into penicillin and cephalosporin.

(Fig. 20) (Neuss *et al.*, 1973). Similar experiments using the deuterium-labeled chiral valines and mass spectrometric analysis of penicillin N supported the conclusions described above, but in addition established that all the original methyl hydrogens of valine remain intact in penicillin biosynthesis (Kluender *et al.*, 1974).

The first significant radiochemical experiments to demonstrate that the β-lactam ring is derived from cyst(e)ine were conducted in the early 1950s (Arnstein and Grant, 1954). To establish whether cyst(e)ine was incorporated intact, triply labeled [3-^{14}C]-, [^{15}N]-, [^{35}S]-D- and L-cystine were fed to *P. chrysogenum.* Since the stereochemistry at the C-6 position in penicillins possesses the L-configuration, it was not surprising that D-cystine was a relatively poor precursor. Although the resultant penicillin had very nearly the same ^{15}N/^{14}C ratio as the precursor, the ^{35}S content was considerably enriched above that of the ^{14}C. It was concluded that this enrichment was due to the incorporation into penicillin of [^{35}S] cyst(e)ine biosynthesized *de novo* from the catabolic products of the triply labeled cystine. Later, it was shown that cyst(e)ine acted as a similar precursor for penicillin N and cephalosporin C in *Acremonium* (Trown *et al.*, 1963a,b).

Since a hypothetical pathway for the formation of the β-lactam ring involving an intermediate α-β-dehydrocyst(e)ine had been presented, it was important to ascertain whether the α-hydrogen of cyst(e)ine is retained or lost. Surprisingly, this determination proved difficult, and initial experiments proved somewhat ambiguous. Later work with high-yielding strains of *P. chrysogenum* established beyond doubt that the α-hydrogen is indeed retained (Bycroft *et al.*, 1975a) and thus excluded a dehydro-intermediate.

The inevitable loss of one of the C-3 hydrogens from L-cyst(e)ine in the formation of the β-lactam ring in penicillin biosynthesis posed the question as to the stereochemical fate of the diastereotopic hydrogens at this position. Both the C-3 stereospecifically tritium-labeled cyst(e)ines were fed concurrently with ^{14}C-labeled cyst(e)ine in separate experiments to *P. chrysogenum* and the benzylpenicillin isolated. It was observed that for the 3R-[^{3}H] isomer, most of the tritium was retained, whereas for the corresponding 3S-isomer, most of the tritium was lost. Thus, it was concluded that the ring closure to the β-lactam ring occurs with retention of configuration (Fig. 20) (Morecombe and Young, 1975; Aberhart *et al.*, 1975).

It is perhaps significant to note that 2-alkyl-, 3-alkyl-, and 3,3-dialkyl-cyst(e)ines cannot substitute for cyst(e)ine in the biosynthetic pathway, and to date only cyst(e)ine has been shown to serve as a direct penicillin precursor in whole cells. Similarly, attempts to induce directed biosynthesis using analogues of valine have been equally unsuccessful.

7.3. Tripeptide Theory

The isolation from the mycelium of *P. chrysogenum* of small quantities of the linear tripeptide, δ-(α-aminoadipyl)-cysteinyl valine, the stereochemistry of which was not defined, initiated the tripeptide theory for penicillin biosynthesis (Arnstein and Morris, 1960). Although it was not until later that a penicillin possessing an α-aminoadipyl side chain was observed in the culture filtrates of a *Penicillium* sp. (Flynn *et al.*, 1962), it was proposed that all other penicillins might be obtained from such a penicillin molecule by an enzymatic side-chain transfer. The α-aminoadipyl side chain of penicillin N from *Acremonium* was known to possess the D-configuration (Newton and Abraham, 1953). However, the free α-aminoadipic acid observed in *P. chrysogenum* was of the L-configuration (Arnstein and Morris, 1960), and it was subsequently shown that this was also the configuration of the side chain within the penicillin from *Penicillium*, i.e., isopenicillin N (Cole and Batchelor, 1963). It is noteworthy that even though cephalosporin C and penicillin N both possess a D-α-aminoadipyl side chain, radiolabeling experiments have clearly established that L-α-aminoadipic acid is the direct precursor of this unit (Warren *et al.*, 1967). It was not until considerably later that the stereochemistry of this important putative intermediate was established beyond doubt. The absolute configurations of the component amino acids were determined by circular dichroism, and it was thus established that the tripeptide from *Acremonium* is δ-L-(α-aminoadipyl)-L-cysteinyl-D-valine (Loder and Abraham, 1971). The same conclusions were later drawn for the *Penicillium* tripeptide (Adriaens *et al.*, 1975; Chan *et al.*, 1976; Fawcett *et al.*, 1976). *Acremonium* spp. have also yielded simple derivatives of the Arnstein tripetide (Fujisawa, 1977; Queener and Neuss, 1982) and two tetrapeptides containing a carboxy-terminal glycine (Loder and Abraham, 1971). More recently, α-aminoadipyl-alanyl-valine, α-aminoadipyl-serinyl-valine, and α-aminoadipyl-serinyl-isodehydrovaline have been isolated from culture filtrates of *P. chrysogenum* (Neuss *et al.*, 1980). So far, only the α-aminoadipyl-serinyl-valine has been shown to correspond with the LLD-configuration of the tripeptide.

The theory itself was simple and attractive, but attempts to demonstrate conversion of the tripeptide into β-lactam antibiotics by whole cells of either *Penicillium* or *Acremonium* were unsuccessful, no doubt due to the permeability constraints already mentioned. Nevertheless, considerable circumstantial evidence was accumulating throughout the 1970s that the tripeptide is an obligatory precursor of penicillins and cephalosporins. However, this hypothesis and the mechanism of the biochemical construction of the tripeptide were not to be substantiated until the availability of reliable cell-free preparations (see Section 7.5).

7.4. Biosynthesis of Penicillins and Cephalosporin C and Its Relationship to Primary Metabolism

The principal β-lactams, isopenicillin N, penicillin N, and cephalosporin C, are all secondary metabolites derived from the amino acids L-α-aminoadipic acid, L-cyst(e)ine, and L-valine, all of which are constituents of cellular metabolism, α-Aminoadipic acid is not a universal metabolite, but is a common intermediate for the synthesis of lysine in higher fungi (Sinha and Bhattacharjee, 1970). Data from the studies on *Penicillium* and *Acremonium* indicated that the α-aminoadipic acid of the β-lactam antibiotics is indeed derived as an intermediate of lysine biosynthesis. The demonstration in the late 1950s that lysine could inhibit benzylpenicillin biosynthesis in *P. chrysogenum* (Demain, 1957) was later followed by the observation that this inhibition could be reversed by the addition of α-aminoadipic acid (Somerson *et al.*, 1961). Similar results were obtained for isopenicillin N (Cole and Batchelor, 1963). Lysine, therefore appeared to inhibit general penicillin biosynthesis by feedback regulation of the lysine pathway, thereby limiting the endogenous supply of α-aminoadipic acid. These results were consistent with the tripeptide theory and implied that an α-aminoadipylpenicillin might be a crucial intermediate in the synthesis of all other penicillins. Genetic evidence that α-aminoadipic acid is an obligatory intermediate for the biosynthesis of penicillins stemmed from mutations that specifically blocked the lysine pathway. These fall into two phenotypic classes: Mutants blocked prior to the synthesis of α-aminoadipic acid are unable to synthesize penicillins, although they grow well when supplemented with lysine. The addition of an exogenous supply of α-aminoadipic acid to the culture medium, however, restores their capacity to synthesize penicillins. Alternatively, mutants that are blocked after α-aminoadipic acid are capable of synthesizing β-lactams if the medium contains added lysine (Goulden and Chattaway, 1968). A lysine-requiring mutant of *Acremonium* has also been obtained (Lemke and Nash, 1972; Friedrich and Demain, 1977; Luengo *et al.*, 1979).

Both penams and cephems possess a sulfur atom derived from L-cyst(e)ine. The carbon skeleton of cyst(e)ine is derived from serine, either by the interaction of reduced sulfur catalyzed by cysteine synthetase or alternatively via the condensation of serine and homocyst(e)ine mediated by β-cystathionase to form cystathione, the precursor of cyst(e)ine via trans-sulfuration. In fungi, sulfur metabolism involves either the sulfate reduction or the reverse transsulfuration pathway. In *Penicillium*, the sulfur for penicillin is derived efficiently via the sulfate reduction pathway, but can also originate via the trans-sulfuration pathway (Segel and Johnson, 1963). In *Acremonium*, the sulfur for cephalosporin C is readily derived from methionine via peverse trans-sulfuration. Although methionine and sulfate labeled with ^{35}S can be incorporated into penicillin in *P. chrysogenum* fermentation, neither stimulates antibiotic production. In contrast, methionine has a marked stimulatory effect on the synthesis of penicillin N and cephalosporin C in *Acremonium* (Nüesch *et al.*, 1973).

It has already been established that L-valine is the direct precursor for the penicillins and cephalosporins (see Section 7.2). The first step in the biosynthesis of valine in fungi is the conversion of pyruvate to acetolactate by acetohydroxyacid synthetase. This enzyme is sensitive to feedback inhibition by L-valine, and the accumulation of endogenous valine can apparently influence the level of production of penicillin by *Penicillium*. This is borne out by the fact that in high-penicillin-producing mutants, this enzyme does not seem to be subject to the same degree of feedback inhibition as in the low producers (Goulden and Chattaway, 1969).

7.5. Biosynthetic Studies Using Cell-Free Systems

By the early 1970s, it was becoming apparent that no further substantial advances in the understanding of penicillin biosynthesis could be achieved using whole cells. Experiments using particulate fractions of sonicated cells of *Acremonium* were the first to show the potential of cell-free systems. These studies indicated that the tripeptide could be synthesized from the constituent amino acids, probably via the initial formation of L-α-aminoadipyl-L-cysteine followed by the addition of L-valine. In this step, the configuration of the valine is in some way inverted (Abraham, 1974). No synthesis of β-lactam antibiotics occurred with this system, and attempts to convert the tripeptide into β-lactam antibiotics with either intact or broken cell preparations were unsuccessful at this time. More recently, further studies have substantiated these findings and demonstrated that the enzyme of enzymes involved are dependent on divalent magnesium ions and ATP (Adlington *et al.*, 1983b). Except for the possible involvement of a racemase for covalently bound valine, the biosynthesis of the tripeptide is probably analogous to that of glutathione (γ-L-glutamyl-L-cysteinyl-glycine). Glutathione synthetase has been purified to homogeneity (Meister and Tate, 1976), but the relationship of this system to the tripeptide synthetase must await further experimentation.

The conversion of synthetic singly radiolabeled tripeptide into penicillin by cell-free extracts of *Acremonium* tended to confirm its long-suspected role in the biosynthetic pathway (Fawcett *et al.*, 1976; Bost and Demain; 1977). This was established beyond doubt in an experiment whereby the tripeptide labeled with ^{14}C in the aminoadipic acid and a ^{3}H label in the valine methyl groups was added to the cell-free system. The specific radioactivities of the isolated isopenicillin N and the tripeptide precursor were the same, as were the relative specific activities of the aminoadipyl and penicillamine fragments derived from the isopenicillin N (O'Sullivan *et al.*, 1979).

The initial cell-free systems for the conversion of the tripeptide to isopenicillin N were obtained by the lysis of enzymatically prepared protoplasts, a method that was not applicable to the production of reasonable quantities of extract. Hence, the recognition that active preparations could be obtained by mechanically breaking *Acremonium* vegetative cells marked an important advance (Abraham *et al.*, 1981). The isopenicillin N synthetase has since been purified to homogeneity and has a molecular weight of 40,000 (Pang *et al.*, 1984). The enzyme uses molecular oxygen as a cosubstrate for the oxidative cyclization of the tripeptide, and the activity is stimulated by ferrous ions, ascorbic acid, and dithiothreitol. It has also been demonstrated that one molecule of oxygen is required for the conversion of one molecule of tripeptide to isopenicillin N (Fig. 21) (White *et al.*, 1982). In the enzymatic process, the four hydrogens indicated (Fig. 21) are lost from the tripeptide during the oxidative steps. The stereochemical consequence with respect to the β-positions in both the cysteine and valine residues corresponds, as expected, to that observed from the labeling studies using whole organisms (see Section 7.3) (Baldwin *et al.*, 1981a; Adlington *et al.*, 1983a).

Investigations designed to explore the specificity of the synthetase enzyme have revealed a degree of flexibility that would perhaps not have

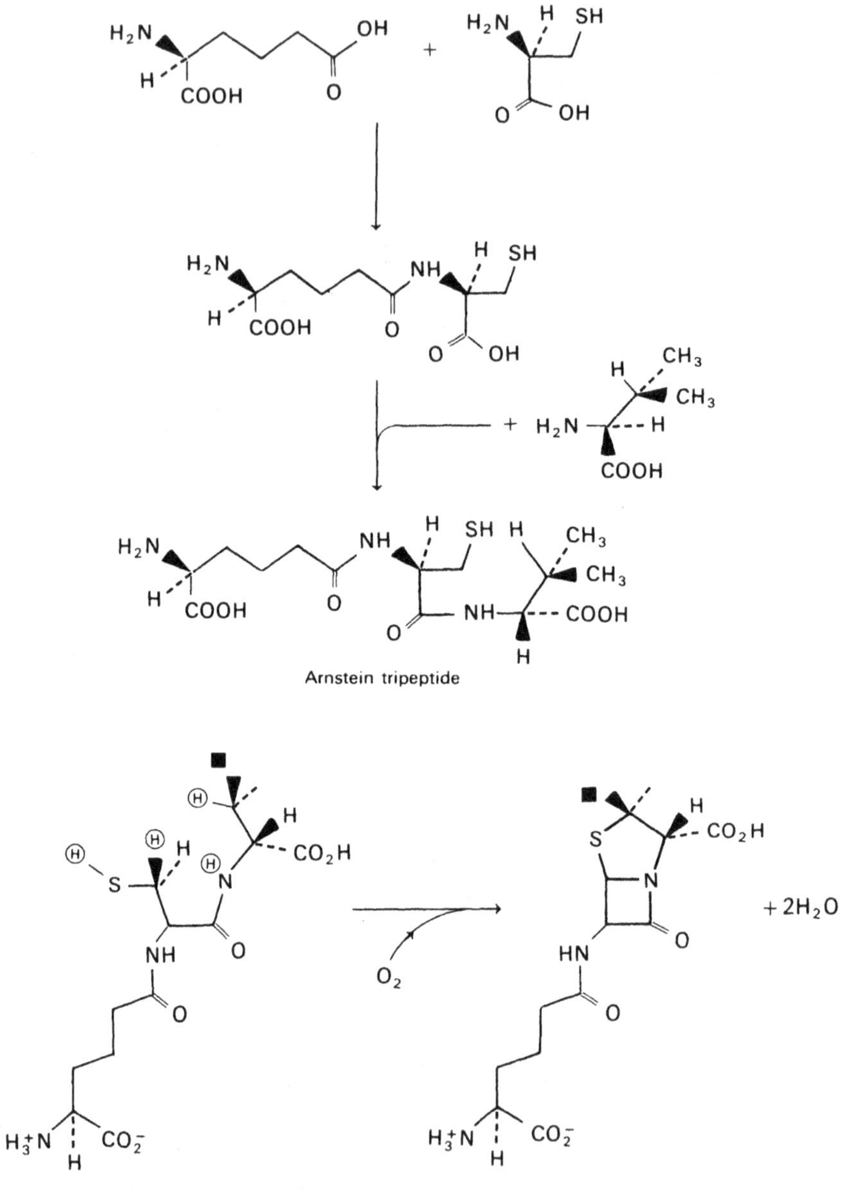

Figure 21. The tripeptide precursor of penicillins.

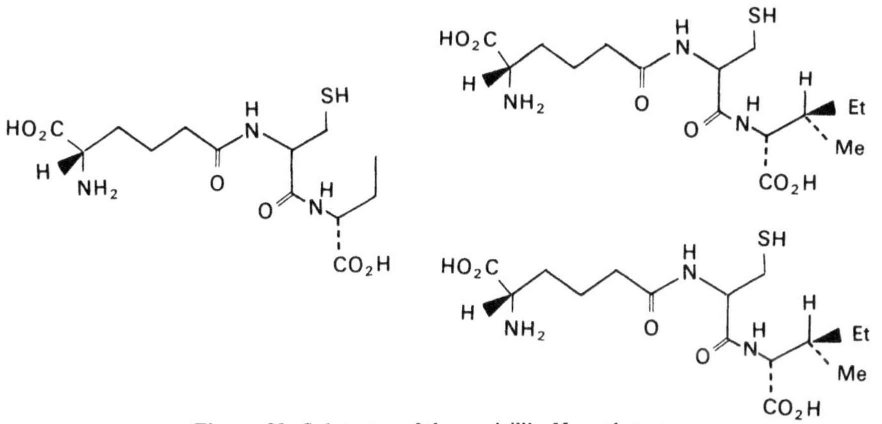

Figure 22. Substrates of the penicillin N synthetases.

been predicted from the whole cell studies. Tripeptide analogues containing modifications of the α-aminoadipic acid residue (Baldwin *et al.*, 1984a) and also those with the valine replaced by either D-α-aminobutyric acid or D-isoleucine (Baldwin *et al.*, 1983) have so far been reported to be substrates for the enzyme (Fig. 22). To date, however, the L-cysteine residue appears to be invariant.

Considerable amounts of kinetic data using deuterated and non-deuterated tripeptide and analogues (Baldwin, 1985) suggest that the reaction sequence is stepwise and involves an enzyme-bound monocyclic β-lactam (Baldwin *et al.*, 1984b). For the tripeptide, cyclization of the enzyme-bound intermediate to isopenicillin N occurs with retention of configuration at the β-position of the valine residue (Fig. 23). However, in the tripeptide

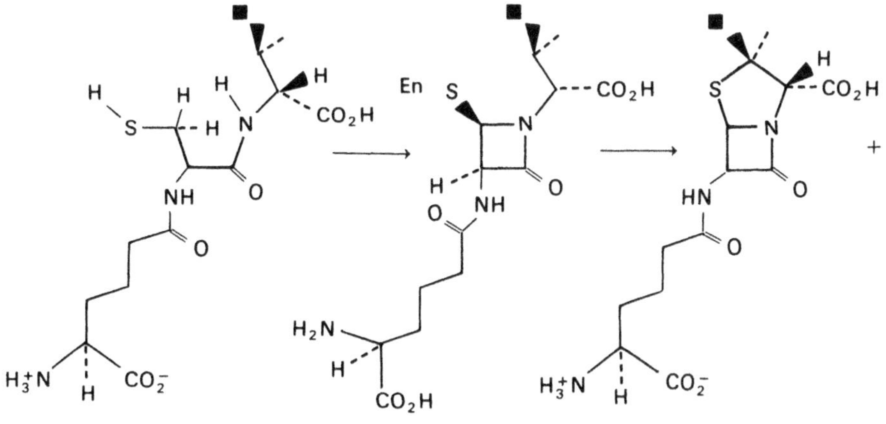

Figure 23. Mechanism of tripeptide cyclization.

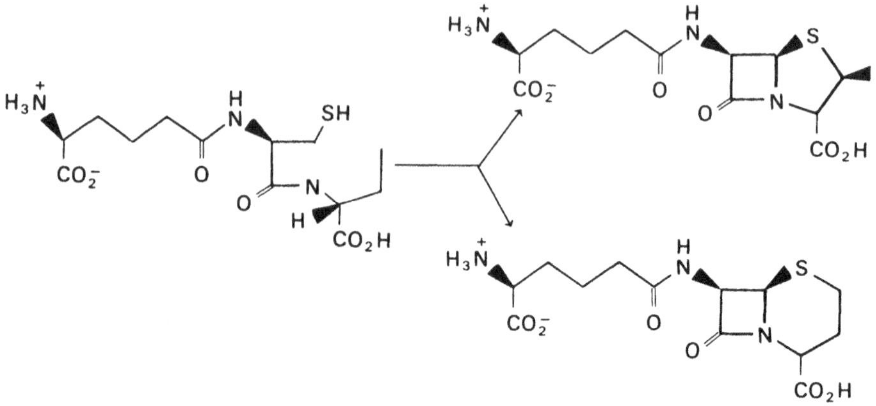

Figure 24. Cyclization of the α-aminobutyric acid analogue of the tripeptide with penicillin N synthetase.

analogue that contains α-aminobutyric acid, cyclization occurs with both retention and inversion of configuration; furthermore, the cyclization also yields six-membered cepham products (Fig. 24) (Baldwin *et al.*, 1983). To date, the foregoing evidence would suggest that the isopenicillin N synthetase is a nonheme, iron-dependent dehydrogenase that probably utilizes an iron–hydrogen peroxide complex and that the ring closure itself involves a free-radical mechanism.

Recently, cell-free extracts of *P. chrysogenum* have also been shown to convert radiolabeled tripeptide to isopenicillin N. It has been suggested that in the course of the conversion, the monocyclic β-lactam (see Fig. 23) is formed as a free intermediate (Meesschaert *et al.*, 1980). These data conflict with the aforementioned studies in that this azetidinone is extremely unstable and not converted by the synthetase into isopenicillin N (Baldwin, 1985).

It had been suspected for some time that isopenicillin N is an obligatory precursor of the "natural" penicillins produced by *P. chrysogenum.* However, it was not until the mid 1970s that a partially purified enzyme (isopenicillin N acyltransferase), capable of exchanging the L-α-aminoadipyl moiety with an acyl group derived from an acyl-CoA, was reported (Fawcett *et al.*, 1975). The enzyme isopenicillin N amidolyase, which is yet to be purified and may be the same as isopenicillin N acyltransferase, is claimed to catalyze the hydrolysis of isopenicillin N to 6-APA and L-α-aminoadipic acid (Fig. 25) (Meesschaert *et al.*, 1980). Surprisingly, in view of their commercial potential, these enzymes have received rather scant attention.

More or less simultaneous with the demonstration of the cell-free conversion of the tripeptide to isopenicillin N, extracts of *Acremonium* were shown to catalyze the transformation of "penicillin N" to deacetoxycephalosporin C (Fig. 26) (Kohsaka and Demain, 1976;

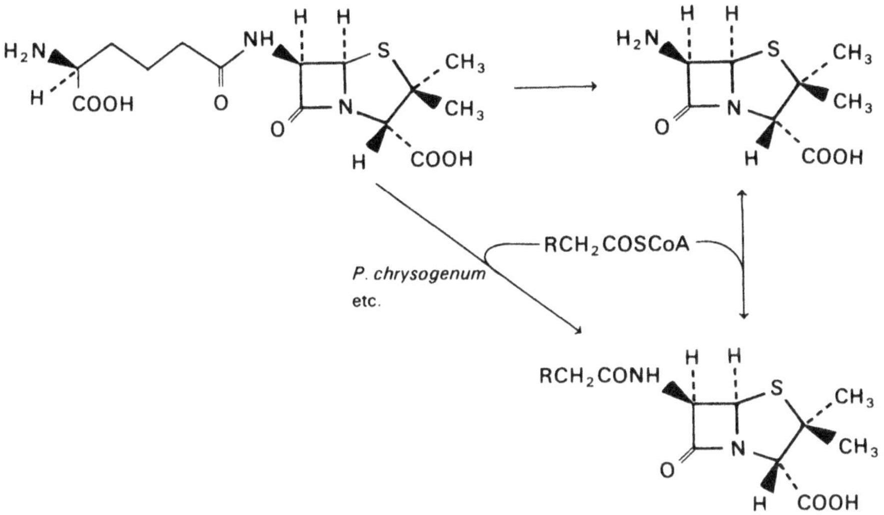

Figure 25. Side-chain exchange of isopenicillin N.

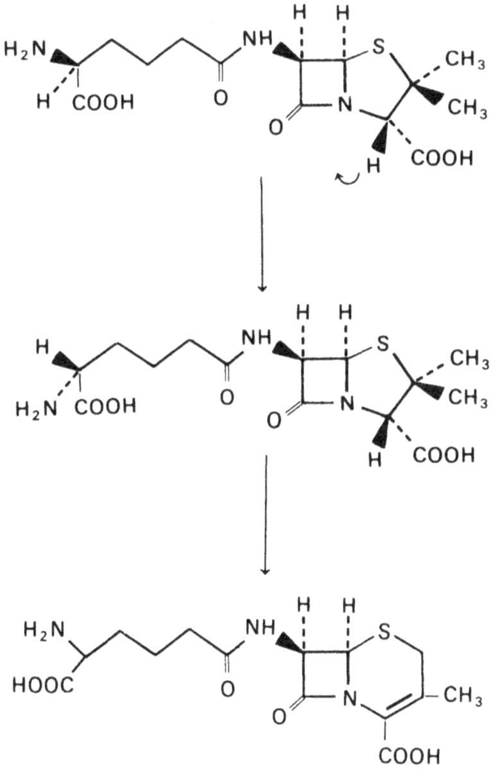

Figure 26. Derivation of deacetoxycephalosporin from isopenicillin N.

Yoshida *et al.*, 1978; Baldwin *et al.*, 1980). The isomerization of the L-α-aminoadipyl side chain of isopenicillin N to the D-configuration observed in penicillin N has recently been demonstrated in a cell-free extract, but there is evidence to suggest that the racemase responsible is extremely labile (Baldwin *et al.*, 1981b; Jayatilake *et al.*, 1981). The ring-expansion enzyme has been partially purified and been shown to be an iron- and α-ketoglutarate-dependent enzyme similar to the isopenicillin N synthetase (Baldwin *et al.*, 1981b). The precise mechanism, as well as the substrate specificity, of this enzyme remain to be elucidated.

The isolation of the 3-β-hydroxycepham (Fig. 27) as a metabolite of *Acremonium* (Miller *et al.*, 1981) supported the claim that this compound was an intermediate in the ring expansion; however, cell-free extracts of *Acremonium* known to contain the ring-expansion enzyme failed to convert this metabolite into deacetoxy- or deacetyl cephalosporin C. This would clearly indicate that the cepham is a shunt metabolite, rather than an intermediate in the ring expansion (Queener and Neuss, 1982).

The conversion of deacetoxy- to deacetylcephalosporin C by cell-free systems has been investigated by a number of groups (Brewer *et al.*, 1980), and the reaction has been shown to require the presence of oxygen, ferrous ions, ascorbic acid, and α-ketoglutarate. The oxygen attached to the C-3 methylene group is derived, as expected, from molecular oxygen. Recently, it has been claimed that this enzyme and the ring-expansion enzyme are one and the same (Schneidegger *et al.*, 1984).

Deacetylcephalosporin C is a well-established biosynthetic precursor of cephalosporin C in *Acremonium*, since it accumulates in cephalosporin-C-negative mutants (Fujisawa *et al.*, 1975; Liersch *et al.*, 1976). Buffered cell-free extracts of the parent strains of these mutants were capable of effecting the acetylation in the presence of acetyl-CoA and magnesium ions (Fujisawa and Kanzaki, 1975). This O-acetyltransferase has been partially purified and shown to have a broad specificity for compounds containing a 3-hydroxymethylcephem nucleus, irrespective of the nature of the C-7 side chain, as well as for DL-serine and DL-homoserine (Liersch *et al.*, 1976).

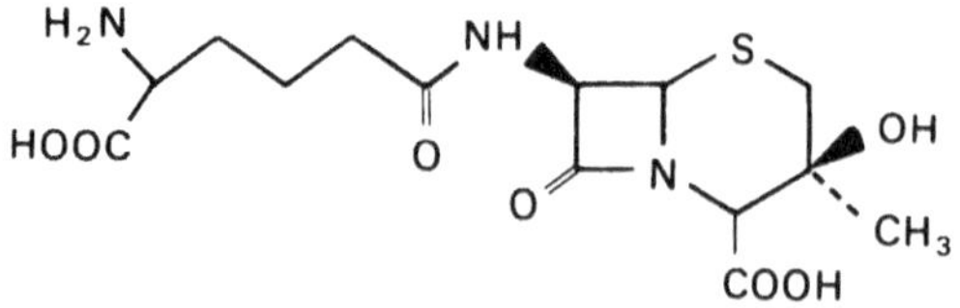

Figure 27. 3-β-hydroxycepham derivative from *Acremonium* spp.

7.6. Possible Future Developments and Direction for Biosynthetic Studies on Penicillins and Cephalosporin C

The past five years have witnessed dramatic and substantial advances in the basic understanding of the biochemistry and enzymology associated with the penicillin and cephalosporin biosynthetic pathways. However, many fundamental questions—e.g., the precise mechanism of action of both the isopenicillin N synthetase and the ring-expansion enzyme—remain to be answered. High-resolution NMR and X-ray studies should provide detailed three-dimensional structural information that, when coupled with substrate-specificity studies, should indicate which amino acid residues are essential for the catalytic processes and metal binding. Such studies will require significant amounts of enzyme material, which will no doubt be made more readily available by the application of cloning techniques. In the long term, alteration of the amino acid sequence within the binding site of appropriate enzymes via "site-specific mutagenesis" offers the possibility of modifying substrate specificity with the ultimate prospect of the cell-free production of novel β-lactam antibiotics. This biochemical approach to the construction of "unnatural" penicillins and cephalosporins, with side chains other than α-aminoadipic acid, may in the future complement and possibly supersede the semisynthetic methods for the production of clinically significant antibiotics described in this chapter.

REFERENCES

Aberhart, D. A., 1977, Biosynthesis of β-lactam antibiotics, *Tetrahedron* **33:**1545–1559.

Aberhart, D. J., Chu, J. Y. R., and Lin, L. J., 1975, Studies on the biosynthesis of β-lactam antibiotics: Synthesis and incorporation into penicillin G of (2RS, 2′RS, 3R, 3′R)-3,3′-^{3}H2-cystine and (2RS, 2′RS, 3S, 3′S)-3,3′-^{3}H2-cystine, *J. Chem. Soc. Perkin Trans. 1* **1975:**2517–2523.

Abraham, E. P., 1974, *Biosynthesis and Enzymic Hydrolysis of Penicillins and Cephalosporins*, University of Tokyo Press, Tokyo.

Abraham, E. P., and Chain, E. B., 1940, An enzyme from bacteria able to destroy penicillin, *Nature (London)* **146:**837.

Abraham, E. P., and Loder, P. B., 1972, Cephalosporin C, in: *Cephalosporins and Penicillins: Chemistry and Biology* (E. H. Flynn, ed.), Academic Press, New York, pp. 2–26.

Abraham, E. P., and Newton, G. G. F., 1954, Synthesis of D-δ-amino-δ-carboxyvalerylglycine (a degradation product of cephalosporin N) and of DL-δ-amino-δ-carboxyvaleramide, *Biochem. J.* **58:**226–268.

Abraham, E. P., and Newton G. G. F., 1961, The structure of cephalosporin C, *Biochem. J.* **79:**377–393.

Abraham, E. P., Newton, G. G. F., and Warren, S. C., 1965, Problems relating to the biosynthesis of peptide antibiotics, in: *Biogenesis of Antibiotic Substances* (Z. Vanek and Z. Hostalek, eds.), Academic Press, New York, pp. 169–194.

Abraham, E. P., Huddlestone, J. A., Jayatilake, G. S., O'Sullivan, J., and White, R. L., 1981,

Conversion of δ-(L-α-aminoadipyl)-L-cysteinyl-D-valine to isopenicillin N in cell-free extracts of *Cephalosporium acremonium*, in: *Recent Advances in the Chemistry of β-Lactam Antibiotics: 2nd International Symposium* (G. I. Gregory, ed.), Royal Society of Chemistry, London, pp. 125–134.

Adlington, R. M., Aplin, R. T., Baldwin, J. E., Chakravarti, B., Field, L. D., John, E. M. M., Abraham, E. P., and White, R. L., 1983a, Conversion of $^{17}O/^{18}O$-labelled δ-(L-α-aminoadipyl)-L-cysteinyl-D-valine to $^{17}O/^{18}O$-labelled isopenicillin N in a cell-free extract of *Cephalosporium acremonium*, *Tetrahedron* **39:**1061–1068.

Adlington, R. M., Baldwin, J. E., Lopez-Nieto, M., Murphy, J. A., and Patel, N., 1983b, A study of the biosynthesis of the tripeptide δ-(L-α-aminoadipyl)-L-cysteinyl-D-valine in a β-lactam-negative mutant of *Cephalosporium acremonium*, *Biochem. J.* **213:**573–576.

Adriaens, P., Vanderhaege, B., Meesschaert, B., and Eyssen, H., 1975, Incorporation of double-labelled L-cystine and DL-valine in penicillin, *Antimicrob. Agents Chemother.* **8:**15–17.

Anderson, E. G., and Pratt, R. F., 1981, Pre-steady state β-lactamase kinetics: Observation of a covalent intermediate during turnover of a fluorescent cephalosporin by the β-lactamase of *Staphylococcus aureus* PC1, *J. Biol. Chem.* **256:**11,401–11,404.

Arnstein, H. R. V., and Grant, P. T., 1954, The biosynthesis of penicillin: The incorporation of some amino acids into penicillin, *Biochem. J.* **57:**353–359.

Arnstein, H. R. V., and Morris, D., 1960, The structure of a peptide containing α-aminoadipic acid, cystine and valine, present in the mycelium of *Penicillium chrysogenum*, *Biochem. J.* **76:**357–361.

Baldwin, J. E., 1985, Recent studies on the biosynthesis of penicillins, in: *Recent Advances in the Chemistry of β-Lactam Antibiotics: 3rd International Symposium* (A. G. Brown and S. M. Roberts, eds.), Royal Society of Chemistry, London, pp. 62–85.

Baldwin, J. E., Singh, P. D., Yoshida, M., Sawada, Y., and Demain, A. L., 1980, Incorporation of ^{3}H and ^{14}C from 6-α-^{3}H-penicillin N and 10-^{14}C-6-α-^{3}H-penicillin N into deacetoxycephalosporin C, *Biochem. J.* **186:**889–895.

Baldwin, J. E., Jung, M., Usher, J. J., Abraham, E. P., Huddleston, J. A., and White, R. L., 1981a, Penicillin biosynthesis: Conversion of deuterated δ-(L-α-aminoadipyl)-L-cysteinyl-D-valine into isopenicillin N by a cell-free extract of *Cephalosporium acremonium*, *J. Chem. Soc. Chem. Commun.* **1981:**246–247.

Baldwin, J. E., Keeping, J. W., Singh, P. D., and Vallejo, C. A., 1981b, Cell-free conversion of isopenicillin N into deacytoxycephalosporin C by *Cephalosporium acremonium* M-0198, *Biochem. J.* **194:**649–651.

Baldwin, J. E., Abraham, E. P., Adlington, R. M., Chakravati, B., Derome, A. E., Murphy, J. A., Field, L. D., Green, N. B., Ting, H.-H., and Usher, J. J., 1983, Penicillin biosynthesis: Dual pathways from a modified substrate, *J. Chem. Soc. Chem. Commun.* **1983:**1317–1319.

Baldwin, J. E., Abraham, E. P., Adlington, R. M., Bahadur, G. A., Chakravati, B., Domayne-Hayman, B. P., Field, L. D., Flitsch, S. L., Jayatilake, G. S., Spakovskis, A., Ting, H.-H., Turner, N. J., White, R. L., and Usher, J. J., 1984a, Penicillin biosynthesis: Active site mapping with aminoadipylcysteinylvaline variants, *J. Chem. Soc. Chem. Commun.* **1984:**1225–1227.

Baldwin, J. E., Abraham, E. P., Lovel, C. G., and Ting, H.-H., 1984b, Inhibition of penicillin biosynthesis by δ-(L-α-aminoadipyl)-L-cysteinylglycine: Evidence for initial β-lactam ring formation, *J. Chem. Soc. Chem. Commun.* **1984:**902–903.

Batchelor, F. R., Doyle, F. P., Nayler, J. H. C., and Rolinson, G. N., 1959, Synthesis of penicillin-6-aminopenicillanic acid in penicillin fermentations, *Nature (London)* **183:**257–258.

Behrens, O. K., 1949, Biosynthesis of penicillins, in: *The Chemistry of Penicillin* (H. T. Clarke, J. R. Johnson, and R. Robinson, eds.), Princeton University Press, Princeton, New Jersey, pp. 657–679.

Blumberg, P. M., and Strominger, J. L., 1972, Five penicillin-binding components occur in *Bacillus subtilis* membranes, *J. Biol. Chem.* **247:**8107–8113.

Booth, H., Bycroft, B. W., Wels, C. M., Corbett, K., and Maloney, A. P., 1976, Application of ^{15}N pulsed Fourier transform nuclear magnetic resonance spectroscopy to biosynthesis: Incorporation of L-^{15}N-valine into penicillin G, *J. Chem. Soc. Chem. Commun.* **1976:**110–111.

Bost, P. E., and Demain, A. L., 1977, Studies on cell-free biosynthesis of β-lactam antibiotics, *Biochem. J.* **161:**681–687.

Brewer, S. J., Farthing, J. E., and Turner, M. K., 1977, Oxygenation of the 3-methyl group of 7-β-(5-D-aminoadipamido)-3-methylceph-3-em-4-carboxylic acid (deacetoxycephalosporin C) by extracts of *Acremonium chrysogenum, Biochem. Soc. Trans.* **5:**1024–1026.

Brotzu, G., 1948, *Lavori dell'Istituto d'Igiene di Cagliari,* Cagliari.

Brown, A. G., and Roberts, S. M. (eds.), 1985, *Recent Advances in the Chemistry of β-Lactam Antibiotics: 3rd International Symposium,* Royal Society of Chemistry, London.

Bycroft, B. W., and Shute, R. E., 1985, The molecular basis for the mode of action of β-lactam antibiotics and mechanisms of resistance, *Pharm. Res.* **1985:**3–14.

Bycroft, B. W., Wels, C. M., Corbett, K., and Lowe, D. A., 1975a, Incorporation of α-^{2}H- and α-^{3}H-L-cystine into penicillin G and the location of the label using isotope exchange and ^{2}H nuclear magnetic resonance, *J. Chem. Soc. Chem. Commun.* **1975:**123.

Bycroft, B. W., Wels, C. M., Corbett, K., and Maloney, A. P., 1975b, Biosynthesis of penicillin G from D- and L-^{14}C- and α-^{3}H-valine, *J. Chem. Soc. Chem. Commun.* **1975:**923–924.

Bycroft, B. W., Wels, C. M., Corbett, K., and Maloney, A. P., 1976, Studies on the biosynthesis of penicillin G in a high-producing strain of *Penicillium chrysogenum,* in: *Recent Advances in the Chemistry of β-Lactam Antibiotics: 1st International Symposium* (J. Elks, ed.), Royal Society of Chemistry, London, pp. 12–19.

Chan, J. A., Huang, F. C., and Sih, C. J., 1976, The absolute configuration of the amino acids in δ-(α-aminoadipyl)cysteinylvaline from *Penicillium chrysogenum, Biochemistry* **15:**177–180.

Chauvette, R. R., Pennington, P. A., Ryan, C. W., Cooper, R. D. G., Jose, F. L., Wright, I. G., Van Heyningen, E. M., and Huffman, G. W., 1971, Chemistry of cephalosporin antibiotics: Conversion of penicillins to cephalexin, *J. Org. Chem.* **36:**1259–1267.

Claridge, C. A., Luttinger, J. R., and Lein, S., 1963, Specificity of penicillin amidases, *Proc. Soc. Exp. Biol. Med.* **113:**1008–1012.

Clarke, H. T., Johnson, J. R., and Robinson, R. (eds.), 1949, *The Chemistry of Penicillin,* Princeton University Press, Princeton, New Jersey.

Cole, M., 1966a, Microbial synthesis of penicillins. I, *Process Biochem.* **1:**334–338.

Cole, M., 1966b, Microbial synthesis of penicillins. II, *Process Biochem.* **1:**373–377.

Cole, M., and Batchelor, F. R., 1963, Aminoadipoylpenicillin in penicillin fermentations, *Nature (London)* **198:**383–384.

Cooper, R. D. G., 1980, New β-lactam antibiotics, in: *Topics in Antibiotic Chemistry,* Vol. III (P. G. Sammes, ed.), Wiley and Sons, New York, pp. 39–199.

Cooper, R. D. G., and Koppel, G. A., 1982, The chemistry of penicillin sulfoxide, in: *Chemistry and Biology of β-Lactam Antibiotics,* Vol. I (R. B. Morin, and M. Gorman, eds.), Academic Press, New York, pp. 1–92.

Cooper, R. D. G., and Spry, D. O., 1972, Rearrangements of cephalosporins and penicillins, in: *Cephalosporins and Penicillins: Chemistry and Biology* (E. H. Flynn, ed.), Academic Press, New York, pp. 183–254.

Cooper, R. D. G., Hatfield, L. D., and Spry, D. O., 1973, Chemical interconversion of the β-lactam antibiotics, *Acc. Chem. Res.* **6:**32–40.

Crowfoot, D., Bunn, C. W., Rogers-Low, B. W., and Turner-Jones, A., 1949, The X-ray crystallographic investigation of the structure of penicillin, in: *The Chemistry of Penicillin* (H. T. Clarke, J. R. Johnson, and R. Robinson, eds.), Princeton University Press, Princeton, New Jersey, pp. 310–366.

Demain, A. L., 1957, Inhibition of penicillin formation by lysine, *Arch. Biochem. Biophys.* **67**:244–246.

Demain, A. L., 1963, Synthesis of cephalosporin C by resting cells of *Cephalosporium acremonium*, *Clin. Med.* **70**:2045–2051.

Demain, A. L., 1974, Biochemistry of penicillin and cephalosporin fermentations, *Lloydia* **37**:147–167.

Demain, A. L., Walton, R. B., Newkirk, J. F., and Miller, I. M., 1963, Microbial degradation of cephalosporin C, *Nature (London)* **199**:909–910.

Elander, R. D., and Aoki, H., 1982, β-Lactam-producing micro-organisms: Their biology and fermentation behavior, in: *Chemistry and Biology of β-Lactam Antibiotics*, Vol. III (R. B. Morin and M. Gorman, eds.), Academic Press, New York, pp. 84–183.

Ernest, I., 1982, The penams, in: *Chemistry and Biology of β-Lactam Antibiotics*, Vol. II (R. B. Morin and M. Gorman, eds.), Academic Press, New York, pp. 315–360.

Fawcett, P. A., Usher, J. J., and Abraham, E. P., 1975, Behavior of tritium-labelled isopenicillin N and 6-aminopenicillanic acid as potential precursors in an extract of *Penicillium chrysogenum*, *Biochem. J.* **151**:741–746.

Fawcett, P. A., Usher, J. J., Huddleston, J. A., Bleany, R. C., Nisbet, J. J., and Abraham, E. P., 1976, Synthesis of δ-(α-aminoadipyl)cysteinylvaline and its role in penicillin biosynthesis, *Biochem. J.* **157**:651–660.

Felix, H. R., Nuesch, J., and Wehrli, W., 1980, Investigation of the two final steps in the biosynthesis of cephalosporin C using permeabilized cells of *Cephalosporium acremonium*, *FEMS Microbiol. Lett.* **8**:55–58.

Flynn, E. H., McCormick, M. H., Stamper, M. C., De Valeria, H., and Godzeskii, C. W., 1962, A new natural penicillin from *Penicillium chrysogenum*, *J. Am. Chem. Soc.* **84**:4594–4595.

Friedrich, C. G., and Demain, A. L., 1977, Homocitrate synthetase as the crucial site of the lysine effect on penicillin biosynthesis, *J. Antibiot.* **30**:760–761.

Fujisawa, Y., 1977, Studies on the biosynthesis of cephalosporin C, *J. Takeda Res. Lab.* **36**(3/4):295–356.

Fujisawa, Y., and Kanzaki, T., 1975, Role of acetyl CoA-deacetylcephalosporin C acetyltransferase in cephalosporin C biosynthesis by *Cephalosporium acremonium*, *Agric. Biol. Chem.* **39**:2043–2048.

Fujisawa, Y., Shirafuji, H., Kida, M., Nara, K., Yoneda, M., and Kanzaki, T., 1975, Accumulation of deacetylcephalosporin C by cephalosporin C-negative mutants of *Cephalosporium acremonium*, *Agric. Biol. Chem.* **39**:1295–1301.

Ganguly, A. K., Girijavallabhan, J. M., McCombie, S., Pinto, P., Rizvi, R., Jeffrey, P. D., and Lin, S., 1982, Synthesis of Sch 29482 [5R, 6S, 8R] 6-(1-hydroxyethyl)-2-ethylthiopenem-3-carboxylic acid sodium salt): A novel penem antibiotic, *J. Antimicrob. Chemother.* **9**(Suppl. C):1–6.

Gardner, A. D., 1940, Morphological effects of penicillin on bacteria, *Nature (London)* **146**:837–838.

Ghuysen, J.-M., 1977, Biosynthesis and assembly of bacterial cell walls, *Cell Surfaces Rev.* **4**:463–595.

Ghuysen, J.-M., 1980, Antibiotics and peptidoglycan metabolism, in: *Topics in Antibiotic Chemistry*, Vol. V (P. G. Sammes, ed.), Wiley and Sons, New York, pp. 9–117.

Gordon, E. M., and Sykes, R. B., 1982, Cephamycin antibiotics, in: *Chemistry and Biology of β-Lactam Antibiotics*, Vol. I (R. B. Morin and M. Gorman, eds.), Academic Press, New York, pp. 199–370.

Goulden, S. A., and Chattaway, F. W., 1968, Lysine control of α-aminoadipate and penicillin synthesis in *Penicillium chrysogenum*, *Biochem. J.* **110**:55P–56P.

Goulden, S. A., and Chattaway, F. W., 1969, End-product control of acetohydroxyacid synthetase by valine in *Penicillium chrysogenum* Q176 and a high penicillin-yielding mutant, *J. Gen. Microbiol.* **59:**111–118.

Hale, C. W., Newton, G. G. F., and Abraham, E. P., 1961, Derivatives of cephalosporin C formed with certain heterocyclic tertiary bases: The cephalosporin Ca family, *Biochem. J.* **79:**403–408.

Hamashima, Y., Matsumura, H., Matsumura, S., Nagata, W., Narisada, M., and Yoshida, T., 1981, Synthesis and structure–activity relationships of 1-oxacephem derivatives, in: *Recent Advances in the Chemistry of β-Lactam Antibiotics: 2nd International Symposium* (G. I. Gregory, ed.), Royal Society of Chemistry, London, pp. 57–79.

Hamilton-Miller, J. M. T., and Smith, J. T. (eds.), 1979, *The β-Lactamases*, Academic Press, New York.

Hatfield, L. D., Lunn, W. H. W., Jackson, B. G., Peters, L. R., Blaszczak, L. C., Fisher, J. W., Gardner, J. P., and Dunigan, J. M., 1981, Application of phosphorus–halogen compounds in cleavage of the 7-amide group of cephalosporins, in: *Recent Advances in the Chemistry of β-Lactam Antibiotics: 2nd International Symposium* (G. I. Gregory, ed.), Royal Society of Chemistry, London, pp. 109–124.

Hodgkin, D. C., and Maslen, E. N., 1961, The X-ray structure of cephalosporin C, *Biochem. J.* **79:**393–402.

Holden, K. G., 1982, Total synthesis of penicillins, cephalosporins, and their nuclear analogs, in: *Chemistry and Biology of β-Lactam Antibiotics*, Vol. II (R. B. Morin and M. Gorman, eds.), Academic Press, New York, pp. 99–164.

Huber, F. M., Chauvette, R. R., and Jackson, B. G., 1972, Preparative methods for 7-aminocephalosporanic acid and 6-aminopenicillanic acid, in: *Cephalosporins and Penicillins: Chemistry and Biology* (E. H. Flynn, ed.), Academic Press, New York, pp. 27–72.

Jayatilake, G. S., Huddleston, J. A., and Abraham, E. P., 1981, Conversion of isopenicillin N into penicillin N in cell-free extracts of *Cephalosporium acremonium*, *Biochem. J.* **194:**645–646.

Jung, F. A., Pilgrim, W. R., Poyser, J. P., and Sirett, P. J., 1980, The chemistry and antimicrobial activity of new synthetic β-lactam antibiotics, in: *Topics in Antibiotic Chemistry*, Vol. IV (P. G. Sammes, ed.), Wiley and Sons, New York, pp. 11–265.

Kluender, H., Huang, F. C., Fritzberg, A., Schones, H., Sih, C. J., Fawcett, P., and Abraham, E. P. 1974, Studies on the incorporation of (2S,3R)-4,4,4-^{2}H3-valine and (2S,3S)-4,4,4-^{2}H3-valine into β-lactam antibiotics, *J. Am. Chem. Soc.* **96:**4054–4055.

Kohsaka, M., and Demain, A. L., 1976, Conversion of penicillin N to cephalosporin(s) by cell-free extracts of *Cephalosporium acremonium*, *Biochem. Biophys. Res. Commun.* **70:**465–473.

Kukolja, S., and Chauvette, R. R., 1982, Cephalosporin antibiotics prepared by modifications at the C-3 position, in: *Chemistry and Biology of β-Lactam Antibiotics*, Vol. I (R. B. Morin and M. Gorman, eds.), Academic Press, New York, pp. 93–198.

Lemke, P. A., and Brannon, D. R., 1972, Microbial syntheses of cephalosporin and penicillin compounds, in: *Cephalosporins and Penicillins: Chemistry and Biology* (E. H. Flynn, ed.), Academic Press, New York, pp. 370–437.

Lemke, P. A., and Nash, 1972, Mutations that affect antibiotic synthesis by *Cephalosporium acremonium*, *Can. J. Microbiol.* **18:**255–259.

Liersch, J., Nüesch, J., and Treichler, H. J., 1976, Final steps in the biosynthesis of cephalosporin C, in: *Second International Symposium on the Genetics of Industrial Microorganisms* (K. D. MacDonald, ed.), Academic Press, New York, pp. 179–195.

Loder, P. B., and Abraham, E. P., 1971, Isolation and nature of intracellular peptides from a cephalosporin C-producing *Cephalosporium* sp., *Biochem. J.* **123:**471–476.

Loder, P. B., Newton, G. G. F., and Abraham, E. P., 1961, The cephalosporin C nucleus (7-aminocephalosporanic acid) and some of its derivatives, *Biochem. J.* **79:**408–416.

Luengo, J. M., Revilla, G., Villanueva, J. R., and Martin, J. E., 1979, Lysine regulation of penicillin biosynthesis in low-producing and industrial strains of *Penicillium chrysogenum*, *J. Gen. Microbiol.* **115:**207–211.

Matthew, M., and Harris, A. M., 1976, Identification of β-lactamases by analytical isoelectric focusing: Correlation with bacterial taxonomy, *J. Gen. Microbiol.* **94:**55–67.

Meesschaert, B., Adriaens, P., and Eyssen, H., 1980, Studies on the biosynthesis of isopenicillin N with a cell-free preparation of *Penicillium chrysogenum*, *J. Antibiot.* **33:**722–730.

Meister, A., and Tate, S. S., 1976, Glutathione and related γ-glutamyl compounds, in: *Annual Review of Biochemistry*, Vol. 45 (E. D. Snell, ed.), Annual Reviews, Palo Alto, California, pp. 559–604.

Miller, R. D., Huckstep, L. L., McDermott, J. P., Queener, S. W., Kukolja, S., Spry, D. O., Elzey, T. K., Lawrence, S. W., and Neuss, N., 1981, High performance liquid chromatography (HPLC) of natural products: The use of HPLC in biosynthetic studies of cephalosporin C in the cell-free system, *J. Antibiot.* **34:**984–993.

Mirelman, D., 1979, Biosynthesis and assembly of cell wall peptidoglycan, in: *Bacterial Outer Membranes: Biosynthesis and Functions* (M. Inouye, ed.), Wiley and Sons, New York, pp. 115–166.

Morecombe, D. J., and Young, D. W., 1975, Synthesis of chirally labelled cysteines and the steric origin of C-5 in penicillin biosynthesis, *J. Chem. Soc. Chem. Commun.* **1975:**198–199.

Morin, R. B., and Gorman, M., 1982, Introduction of a 6(7)-methoxyl group into penicillins and cephalosporins, in: *Chemistry and Biology of β-Lactam Antibiotics*, Vol. III (R. B. Morin and M. Gorman, eds.), Academic Press, New York, pp. 395–401.

Morin, R. B., Jackson, B. G., Flynn, E. H., and Roeske, R. W., 1962, Chemistry of cephalosporin antibiotics: 7-Aminocephalosporanic acid from cephalosporin C, *J. Am. Chem. Soc.* **84:**3400–3401.

Morin, R. N., Jackson, B. G., Mueller, R. A., Lavagnino, E. R., Scanlon, W. B., and Andrews, S. L., 1963, Chemistry of cephalosporins: Chemical correlation of penicillin and cephalosporin antibiotics, *J. Am. Chem. Soc.* **85:**1896–1897.

Moyer, A. J., and Coghill, R. D., 1947, Penicillin: The effect of phenylacetic acid on penicillin production, *J. Bacteriol.* **53:**329–341.

Nagarajan, R., Boeck, L. D., Gorman, M., Hamill, R. C., Higgens, C. E., Hoehn, M. M., Stark, W. M., and Whitney, J. G., 1971, β-Lactam antibiotics from *Streptomyces*, *J. Am. Chem. Soc.* **93:**2308–2310.

Neuss, N., Nash, C. H., Baldwin, J. E., Lemke, P. A., and Grutzner, J. B., 1973, Incorporation of (2RS,3S)-4 ^{13}C-valine into cephalosporin C, *J. Am. Chem. Soc.* **95:**3797–3798.

Neuss, N., Miller, R. D., Affolder, C. A., Nakatsukasa, W., Mabe, J., Huckstep, L. L., De La Higuera, N., Hunt, A. H., Occolowitz, J. L., and Gilliam, J. H., 1980, High performance liquid chromatography of natural products: Isolation of new tripeptides from the fermentation broth of *P. chrysogenum*, *Helv. Chim. Acta* **63:**1119–1129.

Newton, G. G. F., and Abraham, E. P., 1953, Isolation of penicillaminic acid and D-α-aminoadipic acid from cephalosporin N, *Nature (London)* **172:**395.

Nüesch, J., Treichler, H. J., and Liersch, M., 1973, The biosynthesis of cephalosporin C, in: *Genetics of Industrial Micro-organisms*, Vol. 2 (Z. Vanek, Z. Hostalek, and J. Cudlin, eds.), Elsevier, Amsterdam, pp. 309–334.

O'Callaghan, C. H., Morris, A., Kirby, S., and Shingler, A. H., 1972, Novel method for the detection of β-lactamases by using a chromogenic cephalosporin substance, *Antimicrob. Agents Chemother.* **1:**283–288.

O'Sullivan, J., and Abraham, E. P., 1981, Biosynthesis of β-lactam antibiotics, in: *Antibiotics*, Vol. IV (J. W. Corcoran, ed.), Springer-Verlag, Berlin, pp. 101–122.

O'Sullivan, J., Bleany, R. C., Huddleston, J. A., and Abraham, E. P., 1979, Incorporation of ^{3}H from δ-(L-α-4,5-^{3}H2-adipyl)-L-cysteinyl-D-4,4-^{3}H2-valine into isopenicillin N, *Biochem. J.* **184:**421–426.

Ott, J. L., Godzeski, C. W., Pavey, D. E., Farran, J. D., and Horton, D. R., 1962, Biosynthesis of cephalosporin C: Factors affecting the fermentation, *Appl. Microbiol.* **10:**515–523.

Pang, C.-P., Chakravarti, B., Adlington, R. M., Ting, H.-H., White, R. L., Jayatilake, G. S., Baldwin, J. E., and Abraham, E. P., 1984, Purification of isopenicillin N synthetase, *Biochem. J.* **222:**789–495.

Ponsford, R. J., 1985, Recent advances in the chemistry and biology of penicillins, in: *Recent Advances in the Chemistry of β-Lactam Antibiotics: International Symposium* (A. G. Brown and S. M. Roberts, eds.), Royal Society of Chemistry, London, pp. 32–51.

Price, K. E., 1970, Structure–activity relationships of semisynthetic penicillins, *Adv. Appl. Microbiol.* **11:**17–75.

Price, K. E., 1977, Structure–activity relationships of semisynthetic penicillins (supplement), in: *Structure–Activity Relationships among the Semi-Synthetic Antibiotics* (D. Perlman, ed.), Academic Press, New York, pp. 61–86.

Queener, S. W., and Neuss, N., 1982, The biosynthesis of β-lactam antibiotics, in: *Chemistry and Biology of β-Lactam Antibiotics,* Vol. III (R. B. Morin and M. Gorman, eds.), Academic Press, New York, pp. 1–81.

Raper, K. B., Alexander, D. F., and Coghill, R. D., 1944, *Penicillin notatum* and allied species, *J. Bacteriol.* **48:**639–659.

Sassiver, M. L., and Lewis, A., 1977, Structure–activity relationships among semi-synthetic cephalosporins: The first generation compounds, in: *Structure–Activity Relationships among the Semi-Synthetic Antibiotics* (D. Perlman, ed.), Academic Press, New York, pp. 87–160.

Schneidegger, A., Kuenzi, M. T., and Nüesch, J., 1984, Partial purification and catalytic properties of a bifunctional enzyme in the biosynthetic pathway of β-lactams in *Cephalosporium acremonium, J. Antibiot.* **37:**522–531.

Segel, I. H., and Johnson, M. J., 1963, Intermediates in inorganic sulfate utilization by *Penicillium chrysogenum, Arch. Biochem. Biophys.* **103:**216–226.

Sheehan, J. C., and Henery-Logan, K. R., 1959, The total synthesis of penicillin V, *J. Am. Chem. Soc.* **81:**3089–3094.

Sheehan, J. C., and Henery-Logan, K. R., 1962, The total and partial general synthesis of the penicillins, *J. Am. Chem. Soc.* **84:**2983–2989.

Singh, P. D., Young, M. G., Johnson, J. H., Cimarusti, C. M., and Sykes, R. B., 1984, Bacterial production of 7-formamidocephalosporins: Isolation and structure determination, *J. Antibiot.* **37:**773–780.

Sinha, A. K., and Bhattacharjee, J. K., 1970, Control of a lysine-biosynthetic step by two unlinked genes of *Saccharomyces, Biochem. Biophys. Res. Commun.* **39:**1205–1210.

Sjoeberg, B., Nathorst-Westfelt, L., and Ortengren, B., 1967, Enzymatic hyrolysis of some penicillins and cephalosporins by *Escherichia coli* acylase, *Acta Chem. Scand.* **21:**547–551.

Somerson, N. L., Demain, A. L., and Nunheimer, R. D., 1961, Reversal of lysine inhibition of penicillin production by α-aminoadipic acid, *Arch. Biochem. Biophys.* **93:**238–241.

Stevens, C. M., and De Long, C. W., 1958, Valine metabolism and penicillin biosynthesis, *J. Biol. Chem.* **230:**991–999.

Suginaka, H., Blumberg, P. M., and Strominger, J. L., 1972, Peptidoglycan of bacterial cell walls: Multiple penicillin-binding components in *Bacillus subtilis, Bacillus cereus, Staphylococcus aureus* and *Escherichia coli, J. Biol. Chem.* **247:**5279–5288.

Sweet, R. M., 1972, Chemical and biological activity: Inferences from X-ray crystal structures, in: *Cephalosporins and Penicillins: Chemistry and Biology* (E. H. Flynn, ed.), Academic Press, New York, pp. 280–309.

Sykes, R. B., and Matthew, M., 1976, The β-lactamases of gram-negative bacteria and their role in resistance to β-lactam antibiotics, *J. Antimicrob. Chemother.* **2:**115–157.

Tipper, D. J., and Strominger, J. L., 1965, Mechanism of action of penicillins: A proposal based on their structural similarity to acyl D-alanyl-D-alanine, *Proc. Natl. Acad. Sci. U.S.A.* **54:**1133–1141.

Trown, P. W., Sharp, M., and Abraham, E. P., 1963a, α-Oxoglutarate and a precursor of D-α-aminoadipic acid residue in cephalosporin C, *Biochem. J.* **86:**280–284.

Trown, P. W., Smith, B., and Abraham, E. P., 1963b, Biosynthesis of cephalosporin C from amino acids, *Biochem. J.* **86:**284–291.

Turner, M. K., Farthing, J. E., and Brewer, S. J., 1978, Oxygenation of 3-^{3}H-methyldeacetoxylosporin C (7-β-D-aminoadipamido-3-methylceph-3-em-4-carboxylic acid) to 3-^{3}H-hydroxymethyldeacetylcephalosporin C by 2-oxoglutarate-linked dioxygenases from *Acremonium chrysogenum* and *Streptomyces clavuligerus*, *Biochem. J.* **173:**839–850.

Turner, W. B., and Aldridge, D. C., 1983, *Fungal Metabolites*, Vol. II, Academic Press, New York, pp. 424–429.

Volkmann, R. A., Carrol, R. D., Drolet, R. B., Elliott, M. L., and Moore, B. S., 1982, Efficient preparation of 6,6-dihalopenicillanic acids: Synthesis of penicillanic acid *S,S*-dioxide (Sulbactam), *J. Org. Chem.* **47:**3344–3345.

Warren, S. C., Newton, G. G. F., and Abraham, E. P., 1967, Use of α-aminoadipic acid for the biosynthesis of penicillin N and cephalosporin C by a *Cephalosporium* sp., *Biochem. J.* **103:**891–901.

Waxman, D. J., and Strominger, J. L., 1982, β-Lactam antibiotics: Biochemical modes of action, in: *Chemistry and Biology of β-Lactam Antibiotics*, Vol. III (R. B. Morin and M. Gorman, eds.), Academic Press, New York, pp. 209–285.

Webber, J. A., and Ott, J. L., 1977, Structure–activity relationships in the cephalosporins: Recent developments, in: *Structure–Activity Relationships among the Semi-Synthetic Antibiotics* (D. Perlman, ed.), Academic Press, New York, pp. 161–237.

White, R. L., John, E. M., Baldwin, J. E., and Abraham, E. P., 1982, Stoichiometry of oxygen consumption in the biosynthesis of isopenicillin N from a tripeptide, *Biochem. J.* **203:**791–793.

Woodward, R. B., 1966, Recent advances in the chemistry of natural products, *Science* **153:**487–493.

Woodward, R. B., 1977, Recent advances in the chemistry of β-lactam antibiotics, in: *Recent Advances in the Chemistry of β-Lactam Antibiotics: 1st International Symposium* (J. Elks, ed.), Royal Society of Chemistry, London, pp. 167–180.

Yoshida, M., Konomi, T., Kohsaka, M., Baldwin, J. E., Herchen, S., Singh, P., Hunt, N. A., and Demain, A. L., 1978, Cell-free ring expansion of penicillin N to deacetoxycephalosporin C by *Cephalosporium acremonium* CW-19 and its mutants, *Proc. Natl. Acad. Sci. U.S.A.* **75:**6253.

Secondary Metabolites of *Penicillium* and *Acremonium*

6

P. G. MANTLE

1. SECONDARY METABOLISM IN FUNGI

Fungi elaborate an almost bewildering array of metabolites that, being additional to those normally regarded as having a specific role in the living process, are generally termed *secondary metabolites*. It is now clear that all the compounds are biosynthesized from a few key metabolic intermediates that are themselves either intermediates or end products of primary metabolism, which can be thought of as all the integrated metabolic processes the products of which are ultimately involved in structure, replication, differentiation, communication, and homeostasis at the cellular or organismic level.

The most important intermediate of primary metabolism that may be diverted into secondary products is acetate. This may be formed into a β-ketomethylene chain, or polyketide, of 3–10 acetate residues in length, by a process that is in part identical to steps of fatty acid biosynthesis, usually using acetyl-CoA as a starter with subsequent additions of malonyl-CoA. The longer the chain, the more folding permutations are theoretically possible to cyclize the chain. Once the chain is folded, there are sometimes rearrangements involving ring cleavage that confuse the original sequence of acetate units.

An alternative fate of acetate into secondary metabolites is via β-hydroxy-β-methylglutarate and mevalonate, which give rise to the C5 compounds isopentenyl pyrophosphate and dimethylallylpyrophosphate. Both isomers condense to form the first step of terpenoid biosynthesis, and dimethylallylpyrophosphate may sometimes alone contribute an isoprene unit to add structural variety to metabolites.

The other principal source of structural components for secondary

P. G. MANTLE • Department of Biochemistry, Imperial College of Science & Technology, London SW7 2AY, England.

metabolites is amino acids. Since they are frequently incorporated whole, or nearly so, nitrogen is thereby inserted and may contribute to heterocyclic variety. Chlorine and sulfur are other elements that are occasionally involved, in addition to oxygen that may be added through a number of steps gradually elevating the molecule to a higher oxidation state.

Various combinations of ketide, isoprenoid, and amino acid moieties therefore constitute the basis of structural diversity, and if secondary metabolism is initially approached in this rather simple way, its practicalities and philosophy assume more comprehensible proportions.

2. *PENICILLIUM* METABOLITES

Rationalization of the structural diversity of fungal secondary metabolites has been admirably achieved by Turner (1971) and Turner and Aldridge (1983). Their mammoth task has been to arrange the metabolites according to either their known biosynthetic origin or their deduced origin based either on analogies with compounds the pedigree of which has been experimentally verified or on theoretical prediction integrating the small number of possible precursors with fundamental principles in reactions of organic molecules. In the event, a small proportion of fungal metabolites have continued to defy even the latter approach and await assignment to a category based on biosynthesis until definitive radiolabeling experiments are performed. This is the context in which a view of the metabolite productivity of the *Penicillia* must be set. Ideally, it must also be integrated within a taxonomy of the genus that embraces the widest possible range of metabolic processes—not only those that are expressed in morphological characters but also those that are revealed only by explorations at the molecular level. This objective is not made any easier by the rich morphological diversity in the genus *Penicillium* and the attendant continuing evolution of a rational taxonomy. It is convenient for the purposes of this chapter that the important recent revision by Pitt (1979a) is available. Consequently, the taxonomic framework for the present compilation of *Penicillium* metabolites is Pitt's speciation. The organisms that produce compounds discovered or studied since about the mid-1970s have often been named following direct consultation with Pitt. The majority, however, are reported to be produced by *Penicillium* species the validity of which is less well standardized, but for the purpose of this chapter, the names have been transposed to accord with Pitt's listing of synonyms. It is recognized that this approach may compound some errors, but the original species assignment is usually evident in the literature cited. The references have been selected to justify the producing organism and to provide entry to the literature on structure and biosynthesis.

In the attempt to construct a rational, though simple, framework for the 380 or so *Penicillium* metabolites that have been reported in the literature up to the end of 1984, there has been some deviation from Turner's approach. Whereas the acetate-derived fatty acid metabolites and polyketides and the acetate mevalonate-derived isoprenoids are irrefutably valid groupings, compounds derived from shikimate are essentially from an integral part of aromatic amino acid biosynthesis and are placed alongside those from a later step (anthranilic acid) leading to tryptophan. In some instances, rationalized listing of some groups of related metabolites has obscured stereochemical features.

The alkaloids, compounds that contain nitrogen, have almost all been listed according to the biosynthetic origin of the nitrogen, usually from an amino acid. The exception is the ochratoxins (e.g., *72*), which are phenylalanine-derived. However, since the other moiety is essentially the polyketide citrinin (*70*), ochratoxins have been placed with citrinin in the pentaketides. Sometimes alkaloids contain, for example, an isoprenoid moiety, but their amino acid derivation is considered to be of prime importance because it involves utilization of an end product of primary metabolism, rather than an intermediate. This invokes the attempt to resolve the philosophical dilemma surrounding the contention that all metabolic products might be expected to have a function for the producing organism but that, in practice, very few seem to play any direct role. Bu'Lock (1980) offered the perceptive explanation that the value of a secondary metabolite to the organism that produces it is not so much, or often at all, in the product itself, but in the process of metabolism involved. By this means, any or all of acetate, mevalonate, and amino acids are removed from the metabolic pool in which, as their concentration increases in certain circumstances, they exert feedback regulation of their own synthesis. Removal of excess metabolites therefore keeps open important primary metabolic pathways during times of environmental or nutritional stress. Some alkaloids such as penitrems have a large terpenoid moiety and therefore could reasonably be regarded as adding to the isoprenoid type of compounds, which are otherwise a rather minor group of the *Penicillium* metabolites. The frequency of distribution of metabolites in the various biosynthetic categories is summarized in Tables Ia and Ib. The species listed account for 65% of those described by Pitt. Of course, it is possible that omission of some is due as much to lack of investigation as to any intrinsic paucity of secondary metabolism potential. Nevertheless, among the species that produce secondary metabolites, versatility is widely distributed, though the subgenus *Penicillium* is particularly rich and a few species have somewhat flamboyant biosynthesis, forming several compounds that happen to display potent biological activity in other organisms. These activities constitute an aspect of the biotechnological relevance of *Penicillia*. The penicillins must be the most

Table Ia. Distribution Frequency of Secondary Metabolites among *Penicillia* That Do Not Have a Perfect State[a]

Subgenus *Aspergilloides*

Metabolite type	*P. glabrum*	*P. purpurescens*	*P. spinulosum*	*P. lividum*	*P. thomii*	*P. sclerotiorum*	*P. implicatum*	*P. restrictum*	*P. roseopurpureum*	*P. resedanum*	*P. citreonigrum*	*P. turbatum*	*P. cyaneum*	*P. decumbens*
Tri- and tetraketides	—	—	4	—	1	—	—	—	—	—	3	—	—	—
Pentaketides	—	2	1	1	2	—	1	—	—	—	4	—	—	—
Hexaketides	—	—	3	—	—	3	—	—	—	—	—	—	—	—
Heptaketides	3	—	—	—	—	—	—	—	—	—	4	—	—	—
Octaketides	4	—	—	—	—	3	—	1	2	—	—	2	—	—
Nonaketides	—	—	—	—	—	—	—	—	—	1	3	—	—	—
Decaketides	—	—	—	—	—	—	—	—	—	—	—	—	—	—
Fatty-acid-derived	—	—	3	—	—	—	—	—	—	—	—	—	1	6
Monoterpenes	—	—	—	—	—	—	—	—	—	—	2	—	—	—
Sesquiterpenes	—	—	—	—	—	—	—	—	—	—	—	—	—	2
Triterpenes	—	—	—	—	—	—	—	—	—	—	—	—	—	—
Shikimic- and anthranilic-derived	—	—	—	—	—	—	—	—	—	—	—	—	—	—
Aliphatic-amino-acid-derived	1	1	—	—	2	1	—	—	-	—	1	—	—	—
Aromatic-amino-acid-derived	—	—	—	—	—	—	—	—	—	—	—	—	—	—
Diketopiperazines	—	—	6	—	—	—	—	—	—	—	—	3	—	—
Derived from two amino acids (nondiketopiperazine)	—	—	—	—	—	—	—	—	—	—	—	—	—	—
Derived from several amino acids	—	—	—	—	—	—	—	—	—	—	—	—	—	—
Miscellaneous	—	—	—	—	—	1	—	2	—	—	—	—	—	—

Subgenus *Furcatum*

Metabolite type	*P. janthinellum*	*P. velutinum*	*P. griseoroseum*	*P. canescens*	*P. janczewskii*	*P. melinii*	*P. fellutanum*	*P. jensenii*	*P. oxalicum*	*P. simplicissimum*	*P. madriti*	*P. raistrickii*	*P. novae-zeelandiae*	*P. citrinum*	*P. corylophyllum*	*P. miczynskii*	*P. herquei*	*P. paxilli*
Tri- and tetraketides	1	—	1	—	—	1	—	1	—	1	2	—	1	—	—	—	—	1
Pentaketides	2	1	—	2	1	—	1	1	—	—	—	—	1	9	—	—	—	—
Hexaketides	—	—	—	—	—	—	—	—	—	—	—	—	—	—	—	—	—	—
Heptaketides	2	—	—	—	2	1	1	—	—	1	—	—	—	1	—	—	6	—
Octaketides	—	—	1	1	—	—	—	—	1	—	—	—	—	3	—	—	—	—
Nonaketides	—	—	—	—	—	—	1	2	—	2	—	—	—	4	—	2	—	—
Decaketides	—	—	—	—	—	—	—	—	—	—	—	—	—	—	—	—	—	—
Fatty-acid-derived	1	—	—	—	—	—	8	1	—	1	—	—	—	—	—	—	—	—
Monoterpenes	—	—	—	—	—	—	—	—	—	—	—	—	—	—	—	—	—	—
Sesquiterpenes	—	—	—	—	—	—	—	—	—	—	—	—	—	—	—	—	—	—
Triterpenes	—	—	—	—	—	—	—	—	—	—	—	—	—	1	—	—	—	—
Shikimic- and anthranilic-derived	—	—	—	—	—	—	—	—	1	—	—	—	—	—	—	—	—	—
Aliphatic-amino-acid-derived	—	—	—	—	—	1	—	—	—	—	—	—	—	—	—	—	—	—
Aromatic-amino-acid-derived	7	—	—	—	3	—	—	—	—	—	—	—	—	6	2	—	1	1
Diketopiperazines	2	—	—	—	3	—	1	—	1	3	—	1	—	—	1	—	—	1
Derived from two amino acids (nondiketopiperazine)	—	—	—	—	—	—	—	—	—	—	—	—	—	—	—	—	—	—
Derived from several amino acids	—	—	—	—	—	—	—	—	—	—	—	—	—	—	—	—	—	—
Miscellaneous	1	—	—	1	—	—	—	—	—	1	—	—	—	—	1	—	—	—

[a] From Pitt (1979a). Species are bracketed to indicate affiliations in distinct series.

Table Ib. Distribution Frequency of Secondary Metabolites among *Penicillia* That Do Not Have a Perfect State[a]

Subgenus *Penicillium*

Metabolite type	*P. expansum*	*P. chrysogenum*	*P. atromentosum*	*P. viridicatum*	*P. crustosum*	*P. roquefortii*	*P. hirsutum*	*P. aurantiogriseum*	*P. camembertii*	*P. griseofulvum*	*P. puberulum*	*P. olivicolor*	*P. brevicompactum*	*P. granulatum*	*P. verrucosum*	*P. italicum*	*P. fennelliae*	*P. arenicola*
Tri- and tetraketides	2	—	—	1	—	4	1	7	—	22	4	1	5	1	—	1	1	1
Pentaketides	1	2	—	4	—	1	1	1	—	—	1	—	4	—	1	—	—	—
Hexaketides	—	1	—	—	—	—	—	—	—	—	—	—	—	—	—	—	—	—
Heptaketides	—	—	—	6	—	—	—	2	—	12	—	—	—	—	—	—	—	—
Octaketides	—	—	1	—	—	—	—	4	—	—	—	—	—	—	—	—	—	—
Nonaketides	1	1	—	—	—	—	—	—	—	—	—	—	4	—	—	—	—	—
Decaketides	—	—	—	—	—	—	—	—	—	—	2	—	—	—	—	—	—	—
Fatty-acid-derived	—	2	—	—	—	—	—	—	—	1	—	—	—	—	—	—	—	6
Monoterpenes	—	—	—	—	—	—	—	—	—	—	—	—	—	—	—	—	—	—
Sesquiterpenes	—	—	—	—	—	7	—	—	—	—	—	—	3	—	—	—	—	—
Triterpenes	—	—	—	—	—	—	—	—	—	1	—	—	1	—	—	—	—	—
Shikimic- and anthranilic-derived	—	5	—	6	1	—	—	8	—	2	1	—	—	—	—	—	—	—
Aliphatic-amino-acid-derived	—	1	—	—	—	—	—	—	—	—	—	—	—	—	—	—	—	—
Aromatic-amino-acid-derived	—	4	—	2	10	3	—	7	2	1	2	—	1	1	1	—	—	—
Diketopiperazines	—	—	—	2	2	5	1	1	—	1	2	1	3	—	1	2	—	—
Derived from two amino acids (nondiketopiperazine)	—	[b]	—	—	—	—	—	1	—	1	—	—	1	—	—	—	—	1
Derived from several amino acids	—	1	—	—	—	—	—	—	—	—	—	—	—	—	—	—	—	—
Miscellaneous	—	—	—	—	—	—	—	—	—	—	—	—	—	—	—	—	—	—

Subgenus *Biverticillatum*

Metabolite type	*P. duclauxii*	*P. claviforme*	*P. mineoluteum*	*P. diversum*	*P. funiculosum*	*P. purpurogenum*	*P. verruculosum*	*P. aculeatum*	*P. islandicum*	*P. brunneum*	*P. variabile*	*P. primulinum*	*P. piceum*	*P. rugulosum*
Tri- and tetraketides	—	2	—	—	2	1	—	—	—	—	—	—	—	—
Pentaketides	—	1	—	—	—	—	—	—	—	—	1	—	—	—
Hexaketides	—	—	—	—	1	2	—	—	—	—	—	—	—	—
Heptaketides	3	—	—	—	—	—	—	—	—	—	—	1	—	—
Octaketides	—	—	—	1	1	2	—	—	22	2	1	—	—	5
Nonaketides	—	—	—	—	—	—	—	—	—	—	—	—	—	—
Decaketides	—	—	—	—	—	—	—	—	—	—	—	—	—	—
Fatty-acid-derived	—	—	1	—	1	4	—	—	—	—	—	—	—	—
Monoterpenes	—	—	—	—	—	—	—	—	—	—	—	—	—	—
Sesquiterpenes	—	—	—	2	—	—	—	—	—	—	—	—	—	—
Triterpenes	—	2	—	—	3	3	—	—	1	—	—	—	—	—
Shikimic- and anthranilic-derived	—	—	—	—	—	—	—	—	—	—	—	5	—	—
Aliphatic-amino-acid-derived	—	—	—	—	—	1	—	—	1	—	—	—	—	—
Aromatic-amino-acid-derived	1	—	—	—	—	2	—	—	3	—	—	—	—	5
Diketopiperazines	—	—	—	—	—	—	2	—	—	—	—	—	1	—
Derived from two amino acids (nondiketopiperazine)	—	—	—	—	—	—	—	—	—	—	—	—	—	—
Derived from several amino acids	—	1	—	—	1	1	—	—	1	—	—	—	—	—
Miscellaneous	—	—	—	—	1	—	—	3	1	—	—	—	—	1

[a] From Pitt (1979a). Species are bracketed to indicate affiliations in distinct series.
[b] Penicillins.

important metabolites, but other antibiotics such as griseofulvin (*113*) and antitumor drugs such as mycophenolic acid (*52*) are also commercially important. Other compounds such as ochratoxin A (*72*), citrinin (*70*), aflatoxin B_1 (*201*), penitrem A (*317*), patulin (*46*), penicillic acid (*49*), xanthomegnin (*101*), and agroclavine (*313*) are toxins with diverse target systems in animals. Gliotoxin (*331*) is structurally analogous to sporidesmin, an important hepatotoxin produced by *Pithomyces chartarum*, and it therefore seems likely that analogous enzymes involved in the synthesis of the diketopiperazine with a sulfide bridge, and with the aromatic amino acid side chain cyclized to form a pyrrole, occur across the generic boundary.

In the subgenus *Aspergilloides*, there are no metabolites derived from aromatic amino acids or their precursors, except for the gliotoxins. These Phe–Ser diketopiperazines are produced only by isolates of *P. spinulosum* and *P. turbatum* in the *Aspergilloides*, but also by *P. fellutanum* (subgenus *Furcatum*). The simple conclusion is that the enzymes responsible for the biosynthesis of a Phe–Ser anhydride are distributed fairly widely in the *Penicillia*. Provided that the identification of the producing fungi was sound, this biosynthetic potential may be saying something of importance regarding phylogenetic relationships.

Tables Ia and Ib and Table II emphasize the special position of *P. chrysogenum* within the *Penicillia* with respect to β-lactam formation, though this facility is not exclusive to *P. chrysogenum*, being evident also in *Cephalosporium*, *Aspergillus*, and *Streptomyces*. The linking of two amino acids without the formation of a diketopiperazine otherwise seems to be a *Penicillium* facility only with respect to the cyclopiamines (*359*, *360*) and asperphenamate (*358*). Both types are produced by pairs of closely related species, *P. aurantiogriseum*–*P. griseofulvum* and *P. arenicola*–*P. brevicompactum*, respectively. The question arises as to whether the commonality of particular metabolite biosynthesis should be subjugated in favor of assigning subdivisions of the section *Penicillium* on the basis of colony growth rates and the texture of the stipe wall (Pitt, 1979a). The taxonomic dilemma is to reconcile the need for a good working key to species identification with the philosophical objective of seeking the most fundamental phylogenetic affiliations. While a biochemical approach exploiting the diversity of secondary metabolites in the genus might seem to offer a useful taxonomic dimension, perusal of the tables does not readily allow either taxonomic consolidation or prediction of where particular biosynthetic potential may be found.

For example, penitrem A (*317*) is produced by *P. crustosum* and other species within the subgenus *Penicillium* only when the fungi are grown as sporulating mycelia in stationary liquid culture. These species do not sporulate in submerged culture, nor does penitrem biosynthesis occur. A connection between the two is implied. Eventually, penitrem production in submerged fermentation was obtained using a strain of *P. janczewskii*, which happens to respond to addition of calcium chloride to the medium by profuse sporulation in submerged culture (Mantle *et al.*, 1984). It would have been wild speculation to have predicted that a suitable species could be found across subgeneric boundaries.

Nevertheless, assistance in predicting species that are likely to contain hidden biosynthetic potential is highly desirable in biotechnological exploitation of *Penicillia*. Otherwise, reliance is placed on the old tedious empirical approach of isolating fungi from the wild, identifying them as near as possible (not always easy with *Penicillia* on account of the variability of wild types), and growing representatives of morphologically different types on liquid media. Simple tests for compounds the chromatographic mobilities of which are well known, and for which diagnostic reagents or bioassays are available, will allow detection of known compounds and of some unknowns. However, since about half the *Penicillia* have not been reported to produce secondary metabolites, the prospects for finding new compounds among these species, or even among the well-worked species, is exceptionally laborious without recourse to the use of biosynthetic probes. Although an expensive approach, the relatively few building blocks from which fungal metabolites are formed and which are frequently recognizable as such in the structure offer considerable probing potential by feeding ^{14}C-labeled acetate, mevalonate, shikimate, anthranilate, or several of the key amino acids to replicated cultures. Autoradiography of thin-layer chromatograms of culture extract, or complementary scintillation counting of fractions from other separation techniques, allows detection of compounds of interest. The new *Penicillium* metabolites nigrifortine (*339*) and auranthine (*283*) were both initially shown to be of interest by autoradiography following biosynthetic probing.

It must also be recognized that interesting metabolites are not usually formed while the organism is in an active replicatory phase. Since all metabolites have carbon skeletons, it is unlikely that such metabolites will be elaborated if utilizable carbon sources are exhausted. Frequently, growth limitation by nitrogen source or phosphate components of the nutritional environment, or by removal of catabolite repression, leads to diversion from primary to secondary processes. Some process of cellular differentiation, such as occurs during sporulation, may be necessary before secondary metabolism can commence. These considerations, based largely on the finding that some metabolic regulatory mechanisms that are well established for primary metabolism also operate for secondary processes, are important in designing efficient commercial overproduction of metabolites as well as in choosing experimental conditions under which hidden biosynthetic potential is to be explored.

It is notable that a number of biologically active fungal metabolites contain chlorine, e.g., griseofulvin (*113*), ochratoxin A (*72*), penitrem A (*317*), and islanditoxin (*361*). While the dechloro analogues possess less biological activity, there is probably insufficient reason to expect that chlorine *per se* confers biological activity. Chlorine is always substituted on a ring, which usually remains aromatic. The mechanism of halogenation is unclear, but the enzyme(s) involved may catalyze the formation of the bromo analogue if bromide is substituted for chloride in the medium (Mantle *et al.*, 1983).

Penicillium spp. that are anamorphic states of the Ascomycete genera *Eupenicillium* and *Talaromyces* are notable in that no amino-acid-derived metabolites have been reported (Table II). Isoprenoid metabolites are also rare. Three species with *Eupenicillium* perfect states (*P. gladioli*, *P. vanbeymae*, and *P. molle*) seem to be similarly versatile with respect to producing small polyketides. The extent to which the literature may underestimate the biosynthetic potential of *Penicillium* anamorphic states of *Eupenicillium* and *Talaromyces* is evident, since only 25% of both groups of the *Penicillium* spp. have been reported to produce secondary metabolites.

A feature of some polyketides is their biosynthetic interrelationships expressed within and among a small group of *Penicillium* spp. so as to form a metabolic grid. This is well expressed in the *P. griseofulvum* group and is

Table II. Distribution Frequency of Secondary Metabolites among *Penicillia* That Are Anamorphic States of *Eupenicillium* and *Talaromyces*

	Eupenicillium									*Talaromyces*			
Metabolite type	*P. phoenicium*	*P. hirayamae*	*P. ochrosalmoneum*	*P. dodgei*	*P. lapidosum*	*P. gladioli*	*P. vanbeymae*	*P. molle*	*P. nilense*	*P. dangeardii*	*P. emmonsii*	*P. kloeckeri*	*P. lehmanii*
Tri- and tetraketides	1	—	—	—	1	3	4	2	—	—	8	—	—
Pentaketides	—	—	—	—	—	—	—	—	—	1	—	—	—
Hexaketides	—	—	—	—	—	—	—	—	—	—	—	2	—
Heptaketides	—	—	—	1	1	—	—	—	—	—	1	—	—
Octaketides	—	2	—	—	—	—	—	—	—	—	3	3	—
Nonaketides	—	—	1	—	—	—	—	—	—	—	—	—	—
Fatty-acid-derived	—	—	—	2	—	—	—	—	—	—	—	—	2
Triterpenes	—	—	—	—	—	—	—	—	—	—	—	2	—
Aromatic-amino-acid-derived	—	—	—	—	—	—	—	—	1	1	—	—	—
Miscellaneous	—	1	—	1	—	—	—	—	—	—	—	—	—

possible only on account of the flexible specificities of the enzymes present in a member organism with respect to substrates that arise within the organism by other pathways. An extension of this situation that presents relatively unexplored biotechnological potential is the use of organisms that do not produce a particular compound to produce it, or an analogous end product, when given an intermediate on the pathway.

For example, a strain of the ergot fungus *Claviceps purpurea* that does not elaborate ergot alkaloids will transform agroclavine (*313*) (a conventional intermediate in ergot alkaloid biosynthesis) into lysergic acid amide, an unusual end product (Willingale *et al.*, 1983a). Agroclavine, and some other related alkaloids, are also *Penicillium* metabolites, and it is therefore at least conceivable that some *Penicillia* may be able to effect interesting and even useful transformations of ergot alkaloids.

The novel metabolite auranthine (*283*) has a benzodiazepine moiety similar to the antidepressant drug diazepam. This is matched among microbial metabolites only by cyclopenin, produced by a closely related *Penicillium* sp. Such metabolic potential might be a fruitful area for exploring structure–activity relationships in drugs the target of which in man is the benzodiazepine receptor or other receptors of hormonal regulators.

One or more of at least 15 amino acids are involved in the biosynthesis of *Penicillium* metabolites; these amino acids are in addition to methionine, which contributes C1 groups but is not incorporated whole. Several general principles seem to operate. Anthranilate usually combines with phenylalanine, glutamic acid, or glutamine, amino acids that are all also closely involved in aromatic amino acid biosynthesis. Tryptophan derivatives usually retain the aliphatic side chain, but where this is lost, all the products are tremorgenic mycotoxins [paxilline (*316*), the penitrems (*317–322*), and the janthitrems (*323–325*)] with much of the indole–isoprenoid structure in common. The producing organisms therefore probably have several enzymes in common, including the tryptophanase that cleaves tryptophan to provide the indole for the tremorgens. The *Penicillia* also form a wide range of diketopiperazine metabolites, cyclic dipeptides especially involving aromatic amino acids and often having isoprenoid substitutions. Results of recent double-labeling experiments using precursors of the prenylated diketopiperazines roquefortine (*351*) and verruculogen (*347*) are consistent with initial formation of the cyclic dipeptide, prenylation occurring later (Laws, 1985; Mantle and Shipston, 1987).

Excluded from discussion here are the penicillins, the chemistry of which is treated in Chapter 5.

2.1. Polyketides

2.1.1. Polyketide Lactones

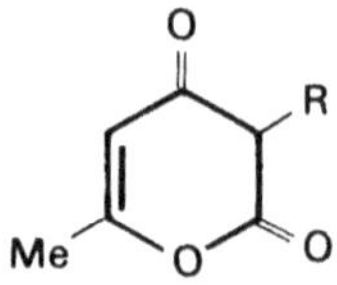

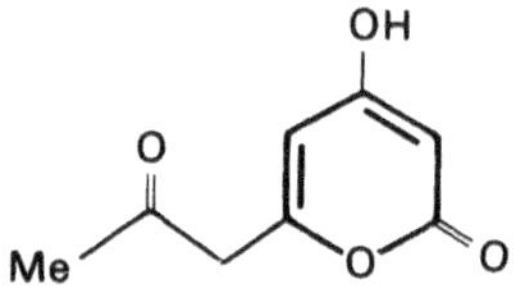

R = H: Triacetic acid lactone (*1*)

Tetraacetic acid lactone (*3*)

R = Me: 3-Methyl triacetic acid lactone (*2*)

P. emmonsii (Acker *et al.*, 1966; Bentley and Zwitkowits, 1967), *P. griseofulvum* (A. I. Scott *et al.*, 1971)

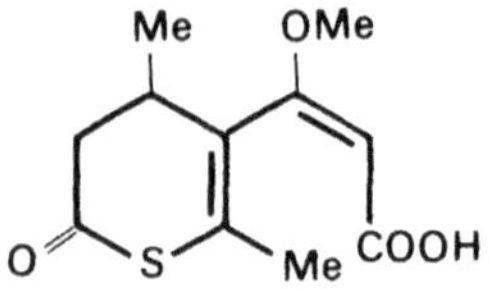

Citreothiolactone (*4*)
P. citreonigrum (Shizuri *et al.*, 1983, 1984a)

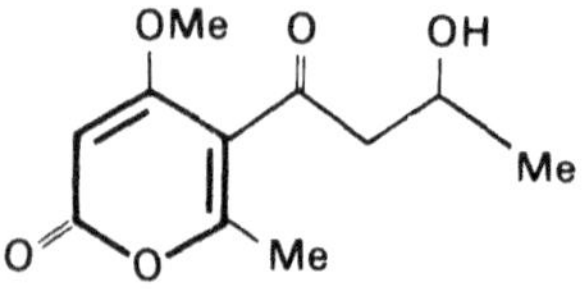

Pyrenocine B (*5*)
P. citreonigrum (Shizuri *et al.*, 1984a)

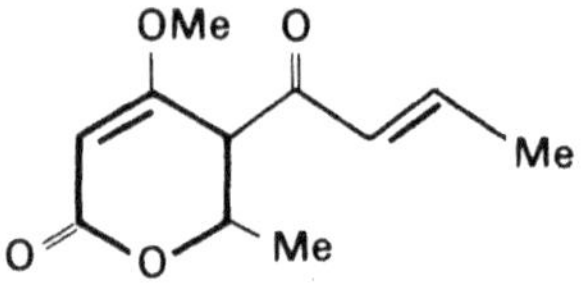

Citreopyrone (*6*)
P. citreonigrum (Niwa *et al.*, 1980; Shizuri *et al.*, 1984a)

2.1.2. Tetraketides

Folding Pattern

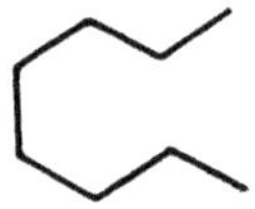

2.1.2a. Polyketide Chain Intact

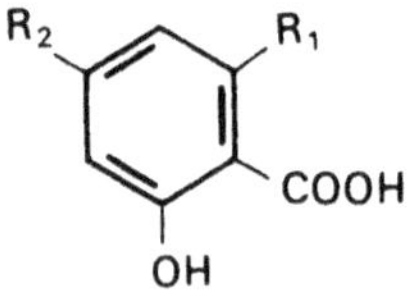

R_2 = H, R_1 = Me: 6-Methyl salicylic acid (*7*)
P. griseofulvum (Dimroth *et al.*, 1970)

R_1 = CHO: 6-Formyl salicylic acid (*8*)
P. griseofulvum (Bassett and Tanenbaum, 1958)

R_1 = COOH: 3-Hydroxyphthalic acid (*9*)
P. griseofulvum (Bassett and Tanenbaum, 1958), *P. islandicum* (Gatenbeck, 1957)

$R_2 = OH$, $R_1 = Me$: Orsellinic acid (*10*)
P. griseofulvum, *P. vanbeymae*, *P. aurantiogriseum*, *P. madriti*, *P. spinulosum* (Pettersson, 1965b)

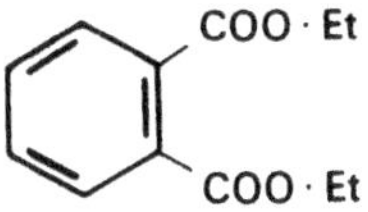

Diethylphthalate (*11*)
P. funiculosum (Qureshi *et al.*, 1980)

2.1.2b. Folded Polyketide Chain Modified by Loss of a Terminal Carbon or by Addition of One or More Carbons or by Both

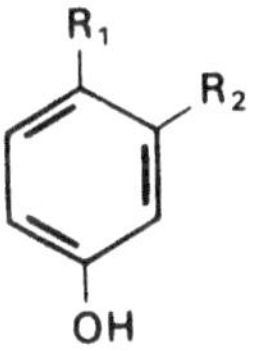

$R_1 = H$, $R_2 = CH_2OH$: *m*-Hydroxybenzyl alcohol (*12*)
P. griseofulvum (Rebstock, 1964)
$R_1 = H$, $R_2 = COOH$: *m*-Hydroxybenzoic acid (*13*)
P. griseofulvum (Simonart and Wiaux, 1959)
$R_1 = OH$, $R_2 = CH_2OH$: Gentisyl alcohol (*14*)
P. griseofulvum (Murphy and Lynen, 1975)
$R_1 = OH$, $R_2 = COOH$: Gentisic acid (*15*)
P. griseofulvum (Gatenbeck and Lonnroth, 1962)
$R_1 = H$, $R_2 = CHO$: *m*-Hydroxybenzaldehyde (*16*)
$R_1 = OH$, $R_2 = CHO$: Gentisaldehyde (*17*)
P. griseofulvum (A. I. Scott and Beadling, 1974)

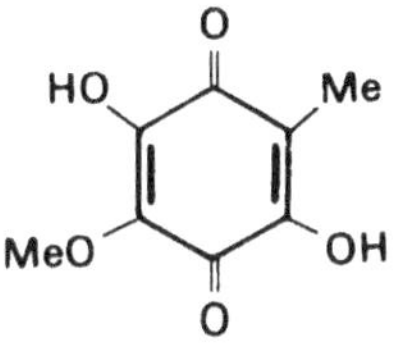

Spinulosin (*18*)
P. spinulosum (Pettersson, 1965a)

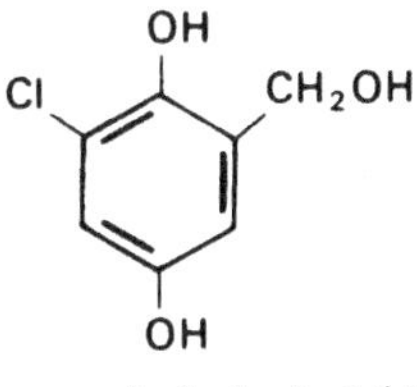

Chlorogentisyl alcohol (*19*)
P. arenicola (McCorkindale *et al.*, 1972), *P. molle* (Turner and Aldridge, 1983)

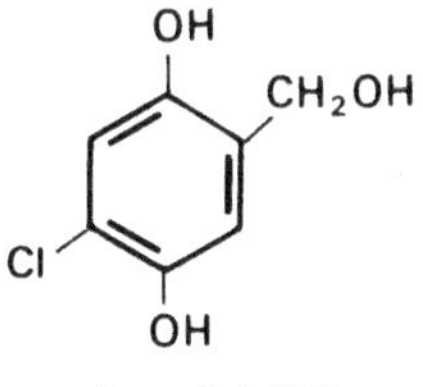

Amudol (*20*)
P. aurantiogriseum (Kamal *et al.*, 1970b)

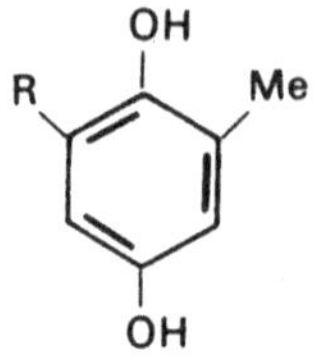

R = H: 2,5-Dihydroxytoluene (*21*)
P. puberulum (Turner and Aldridge, 1983), *P. griseofulvum* (Murphy *et al.*, 1974)
R = OMe: 1,4-Dihydroxy-2-methoxy-6-methylbenzene (*22*)
P. vanbeymae (Better and Gatenbeck, 1976)

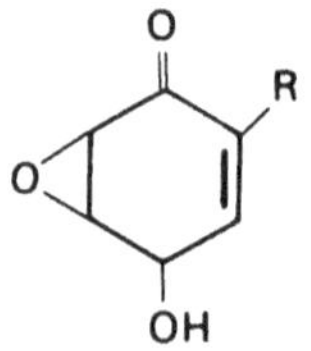

R = Me: Deoxyepoxydon (*23*)
P. griseofulvum (A. I. Scott *et al.*, 1973), *P. claviforme* (Yamamoto *et al.*, 1973)
R = CH_2OH: Isoepoxydon (*24*)
P. griseofulvum (Sekiguchi and Gaucher, 1979a,b)
Epoxydon (*25*)
P. molle (Turner and Aldridge, 1983)

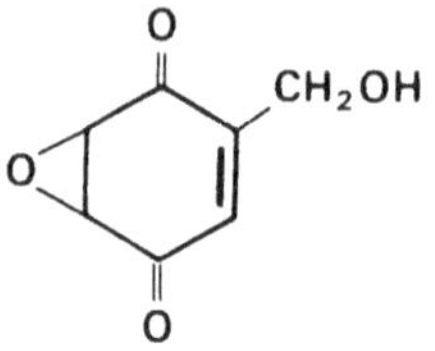

Phyllostine (*26*)
P. griseofulvum (Sekiguchi and Gaucher, 1979a,b)

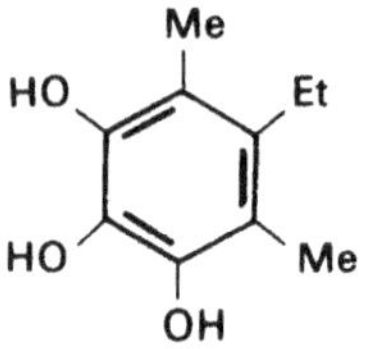

Barnol (*27*)
P. vanbeymae (Ljungcrantz and Mosbach, 1964)

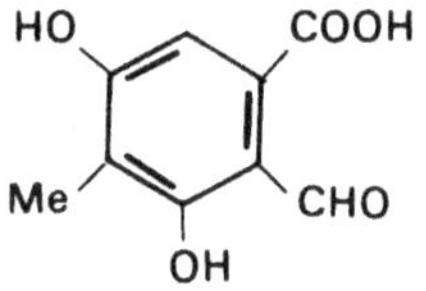

Funiculosic acid (*28*)
P. funiculosum (Qureshi *et al.*, 1980)

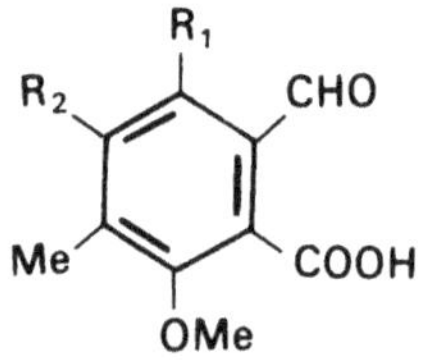

R_1 = CHO, R_2 = OH: Cyclopaldic acid (*29*)
R_1 = CH_2OH, R_2 = OH: Cyclopolic acid (*30*)
P. aurantiogriseum (Birch and Kocor, 1960)
R_1 = CHO, R_2 = H: Gladiolic acid (*31*)
R_1 = CH_2OH, R_2 = H: Dihydrogladiolic acid (*32*)
P. gladioli (Raistrick and Ross, 1952)

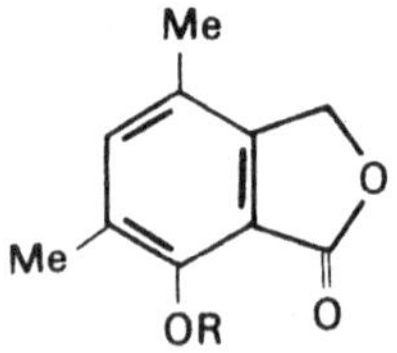

R = H: 3,5-Dimethyl-6-hydroxyphthalide (*33*)
P. gladioli (Birch and Pride, 1962)
R = Me: 3,5-Dimethyl-6-methyoxyphthalide (*34*)
Penicillium sp. (Turner and Aldridge, 1983)

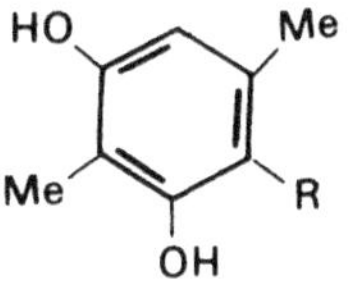

R = H: 4-Methylorcinol (*35*)
R = COOH: 3-Methylorsellinic acid (*36*)
P. spinulosum (Verachtert and Hanssens, 1975)

2.1.2c. Tetraketide Dimer

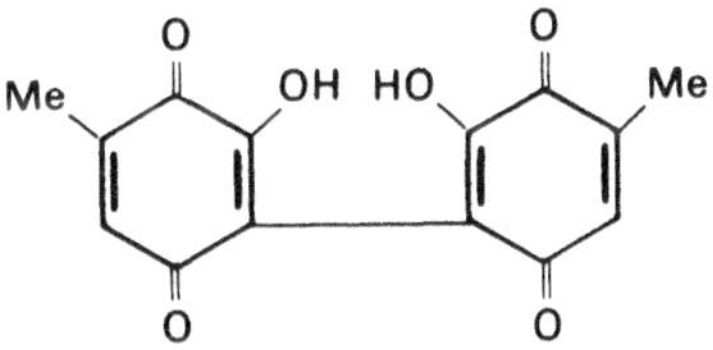

Phoenicin (*37*)
P. phoenicium, *P. purpurogenum* (Steiner *et al.*, 1974)

2.1.2d. Incorporation of an Additional Carbon from Methionine into a Seven-Membered Ring

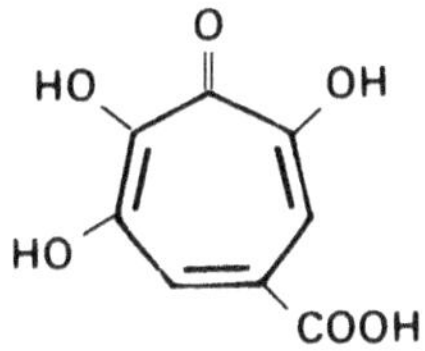

Puberulic acid (*38*)
P. puberulum, *P. aurantiogriseum* (Corbett *et al.*, 1950)

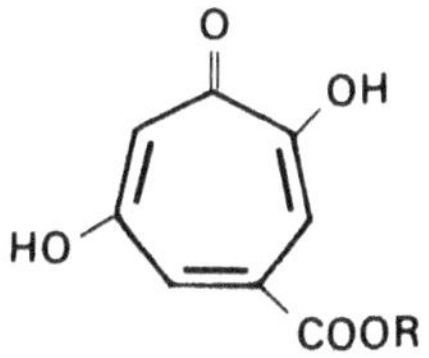

R = H: Stipatitic acid (*39*)
R = Et: Ethyl stipitatate (*40*)
P. emmonsii (Dewar, 1945; Divekar *et al.*, 1961)

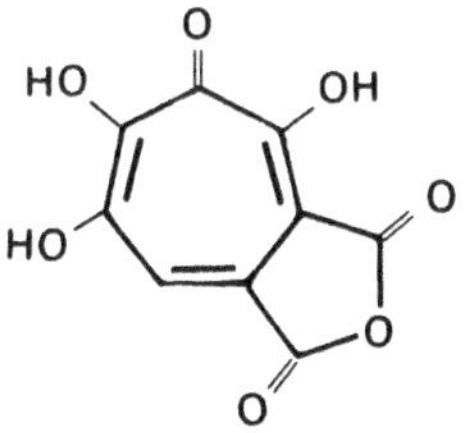

Puberulonic acid (*41*)
P. puberulum (Aulin-Erdtman, 1951)

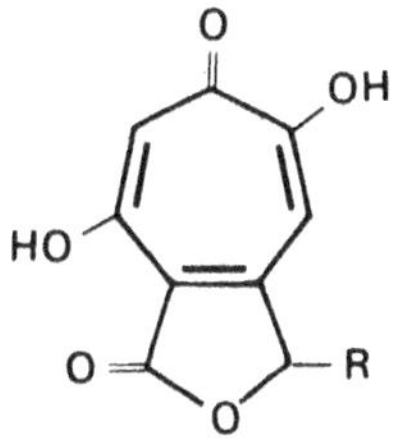

R = O: Stipitatonic acid (*42*)
R = H: Stipitalide (*43*)
R = OH: Stipitaldehydic acid (*44*)
P. emmonsii (Segal, 1959; Bryant and Light, 1974)

2.1.2e. Tetraketides Modified by Ring Scission

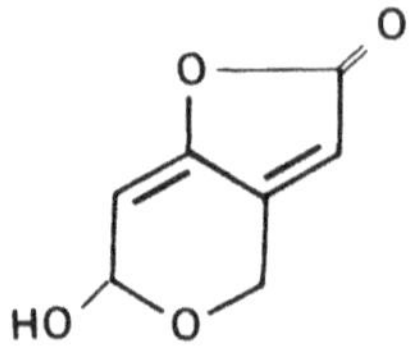

Isopatulin (*45*)
P. griseofulvum (Sekiguchi *et al.*, 1979)

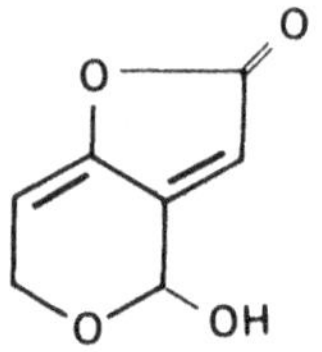

Patulin (*46*)
P. claviforme, *P. expansum*, *P. griseofulvum*, *P. melinii*, *P. novae-zeelandiae* (Abraham and Florey, 1949); *P. aurantiogriseum*, *P. granulatum*, *P. griseoroseum*, *P. jensenii*, *P. lapidosum* (P. M. Scott, 1977); *P. roquefortii* (Olivigni and Bullerman, 1978) (also Bu'Lock and Ryan, 1958; Tanenbaum and Bassett, 1959; Forrester and Gaucher, 1972a,b; A. I. Scott *et al.*, 1973)

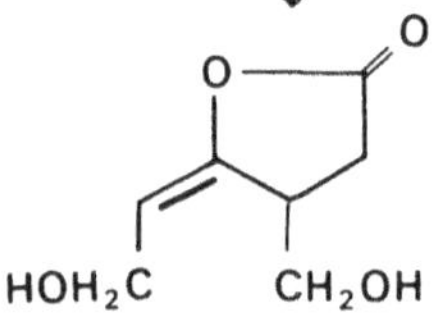

Ascladiol (*47*)
P. griseofulvum (Sekiguchi *et al.*, 1983)

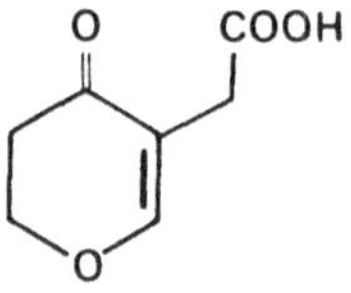

Deoxypatulinic acid (*48*)
P. griseofulvum (P. M. Scott *et al.*, 1972)

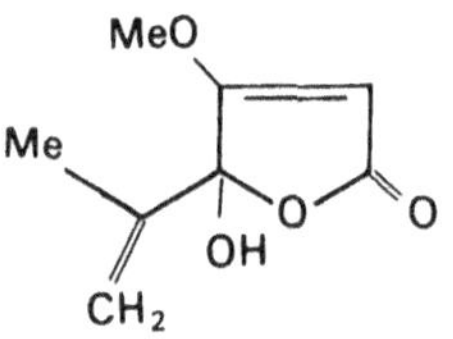

Penicillic acid (*49*)
P. aurantiogriseum, *P. vanbeymae*, *P. expansum*, *P. fennelliae*, *P. griseofulvum*, *P. madriti*, *P. puberulum*, *P. roquefortii*, *P. simplicissimum*, *P. thomii*, *P. viridicatum* (P. M. Scott, 1977) (also Bentley and Keil, 1962; Birkinshaw and Gowlland, 1962; Olivigni and Bullerman, 1978; Mosbach, 1960)

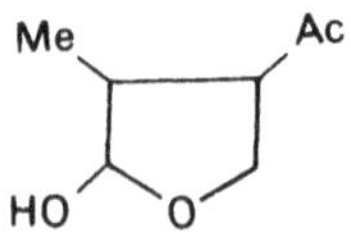

Botryodiplodin (*50*)
P. hirsutum (Fujimoto *et al.*, 1980), *P. roquefortii* (Renauld *et al.*, 1984)

2.1.2f. Tetraketide–Isoprenoid Compounds

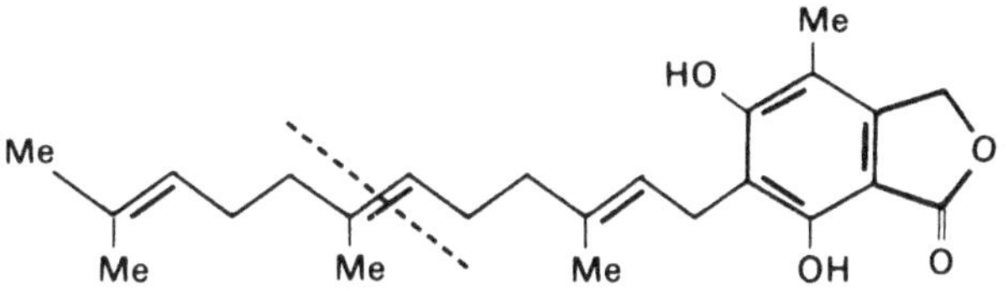

6-Farnesyl-5,7-dihydroxy-4-methylphthalide (*51*)
P. brevicompactum (Canonica *et al.*, 1971)

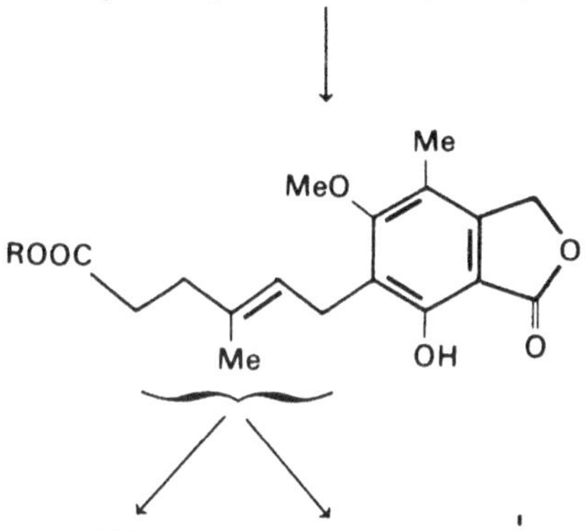

R = H: Mycophenolic acid (*52*)
R = Et: Ethyl mycophenolate (*53*)

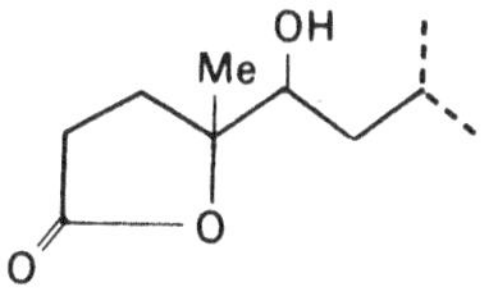

Mycophenolic acid diol lactone (*55*)

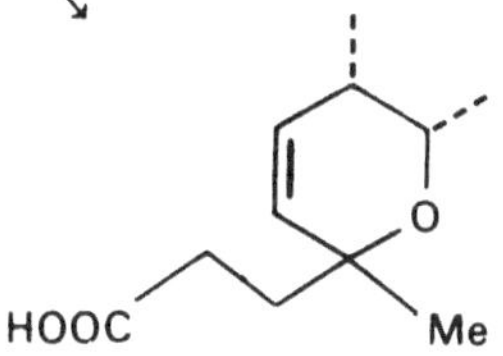

Mycochromenic acid (*54*)

P. brevicompactum (Campbell *et al.*, 1966; Bird and Campbell, 1982b; Colombo *et al.*, 1982)

For mycophenolic acid, also: *P. paxilli* (Turner and Aldridge, 1983), *P. roquefortii* (Lafont *et al.*, 1979), *P. olivicolor* (Nakajima and Nozawa, 1979)

2.1.2g. Polyketide, Possibly of Tetraketide Origin

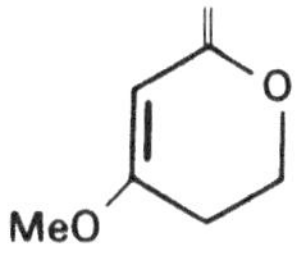

5,6-Dihydro-4-methoxy-2H-pyran-2-one (*56*)
P. italicum (Gorst-Allman and Steyn, 1983)

2.1.3. Pentaketides

Folding Pattern

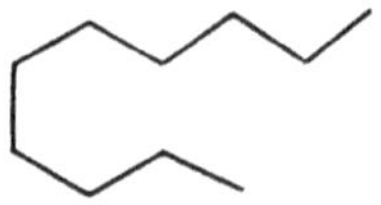

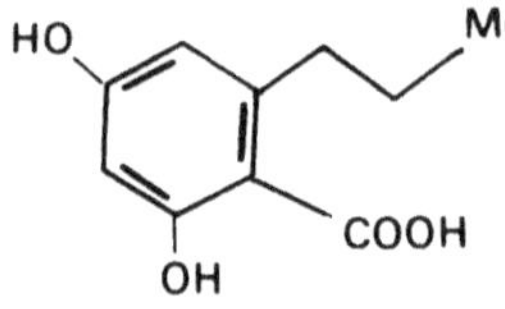

2,4-Dihydroxy-6-propyl benzoic acid (*57*)
P. brevicompactum (Godin, 1955)

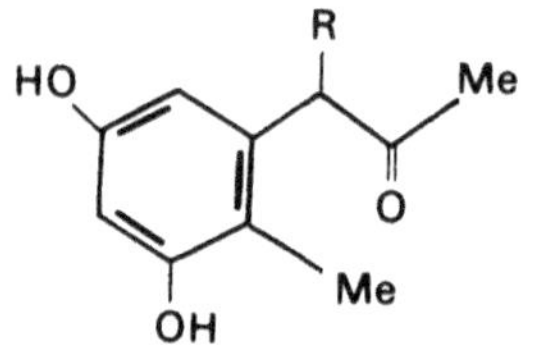

R = H: 2-Carboxy-3,5-dihydroxybenzylmethyl ketone (*58*)
R = OH: 2-Carboxy-3,5-dihydroxyphenylacetyl carbinol (*59*)
R = O: 2,4-Dihydroxy-6-pyruvylbenzoic acid (*60*)
P. brevicompactum (Oxford and Raistrick, 1933)

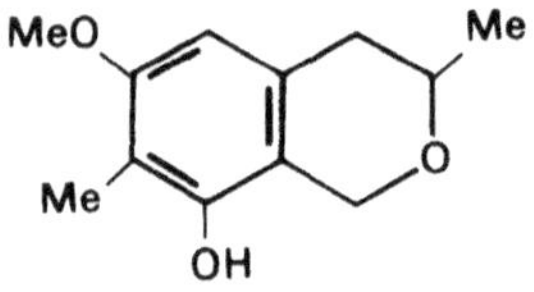

3,7-Dimethyl-8-hydroxy-6-methoxy isochroman (*61*)
P. citrinum (Cox *et al.*, 1979)

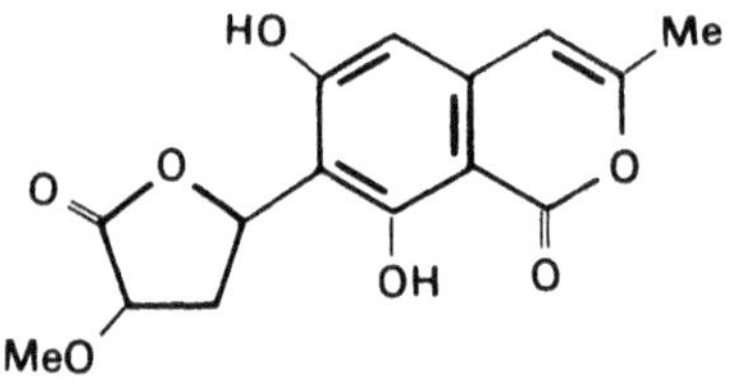

Canescin (*62*)
P. canescens (Birch *et al.*, 1969)

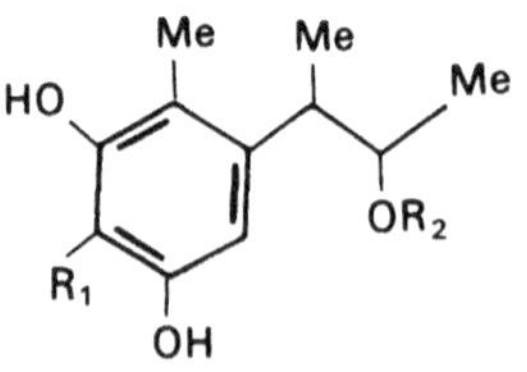

R_1 = H, R_2 = H: "Phenol A" (*63*)
R_1 = COOH, R_2 = H: 2,6-Dihydroxy-4-(3-hydroxy-2-butyl)-*m*-toluic acid (*64*)
R_1 = H, R_2 = CHO: 5-(2-Formyloxy-1-methylpropyl)-4-methyl-resorcinol (*65*)
P. citrinum (Curtis *et al.*, 1968)

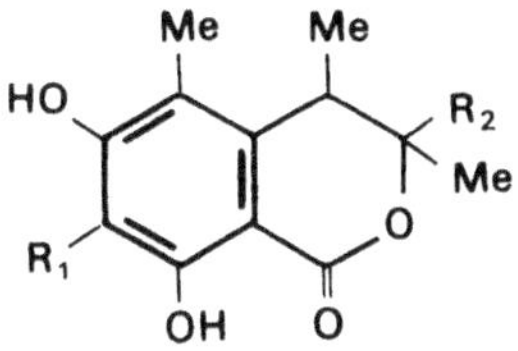

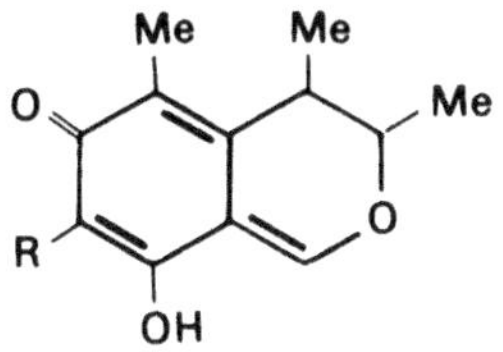

R_1 = COOH, R_2 = H: Dihydrocitrinone (*66*)
R_1 = Me, R_2 = OH: Sclerotinin A (*67*)
R_1 = H, R_2 = H: Decarboxydihydrocitrinone (*68*)

P. citrinum (Curtis *et al.*, 1968)

R = H: Decarboxycitrinin (*69*)

R = COOH: Citrinin (*70*)

P. citrinum, P. citreonigrum, P. canescens, P. claviforme, P. expansum, P. fellutanum, P. implicatum, P. jensenii, P. lividum, P. chrysogenum, P. velutinum (P. M. Scott, 1977); *P. viridicatum* (P. M. Scott, 1977; Damoglou *et al.*, 1984); *P. purpurescens, P. roquefortii, P. spinulosum* (Leistner and Pitt, 1977); *P. janthinellum* (Czech patent, 1982)

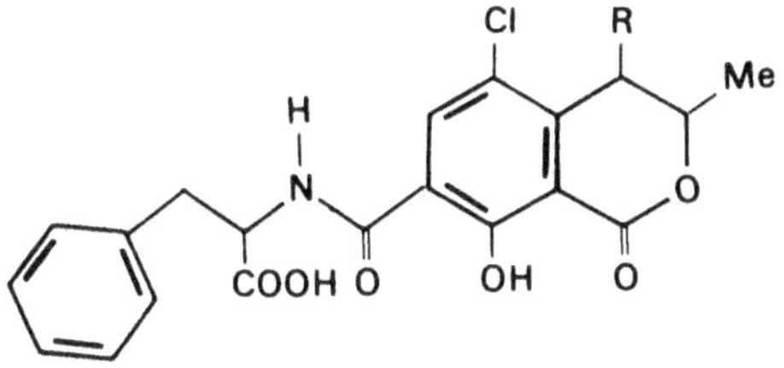

R = OH: 4-Hydroxyochratoxin A (*71*)
P. viridicatum (Hutchison *et al.*, 1971)

↑

R = H: Ochratoxin A (*72*)
P. viridicatum (van Walbeek *et al.*, 1969; van der Merwe *et al.*, 1965; Damoglou *et al.*, 1984); *P. aurantiogriseum, P. purpurescens, P. variabile, P. puberulum* (Ciegler *et al.*, 1972); *P. chrysogenum, P. verrucosum* (Leistner and Pitt, 1977)

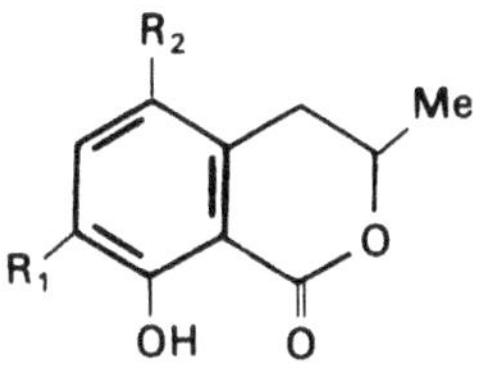

R_1 = COOH, R_2 = H: 7-Carboxy-3,4-dihydro-8-hydroxy-3-methylcoumarin (*73*)
P. viridicatum (Hutchison *et al.*, 1971)
R_1 = H, R_2 = OH: 6-Hydroxymellein (*74*)
R_1 = H, R_2 = OMe: 6-Methoxymellein (*75*)
P. thomii (Turner and Aldridge, 1983)

Folding Pattern

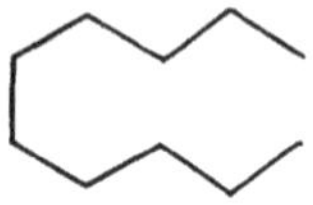

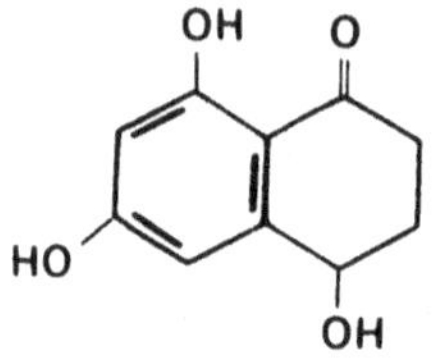

Scytalone (*76*)
P. hirsutum (Turner and Aldridge, 1983; Sankawa *et al.*, 1977)

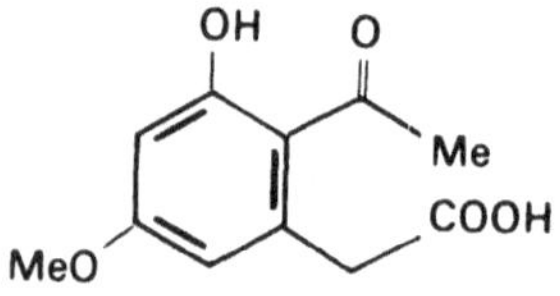

Mono-*O*-methyl curvulinic acid (*77*)
P. janczewskii, *P. novae-zeelandiae* (Turner and Aldridge, 1983)

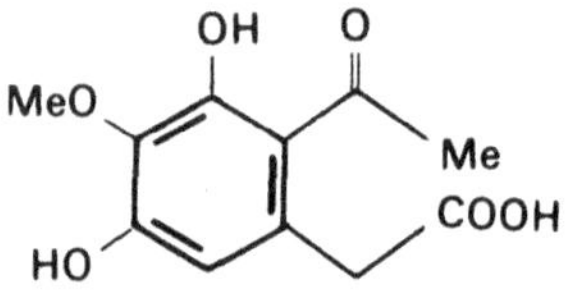

Curvulic acid (*78*)
P. janthinellum (Nakakita *et al.*, 1984a)

2.1.3a. Pentaketide Lactones

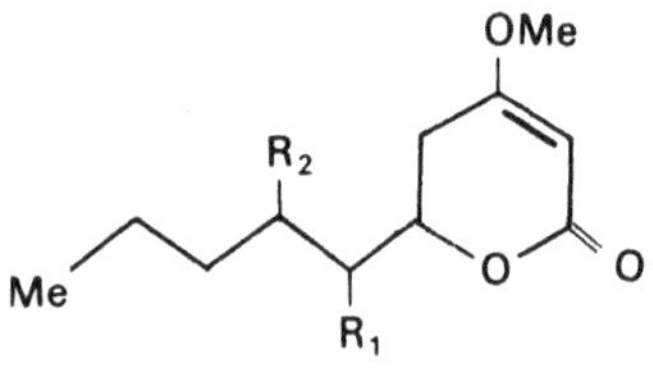

$R_1 = O$, $R_2 = H$: 5-Hydroxy-3-methoxy-6-oxo-2-decenoic acid δ-lactone (*79*)

$R_1 = OH$, $R_2 = H$: Pestalotin (*80*)
P. citreonigrum (Kimura *et al.*, 1980)

$R_1 = OH$, $R_2 = OH$: Hydroxy-pestalotin (*81*)
Penicillium sp. (McGahren *et al.*, 1973)

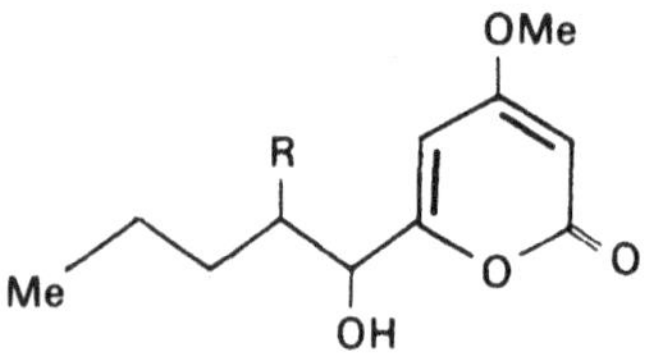

R = H: 6-(1-Hydroxypentyl)-4-methoxypyran-2-one (*82*)
P. citreonigrum (Kimura *et al.*, 1980)

R = OH: Compound LL-P880γ (*83*)
Penicillium sp. (McGahren *et al.*, 1973)

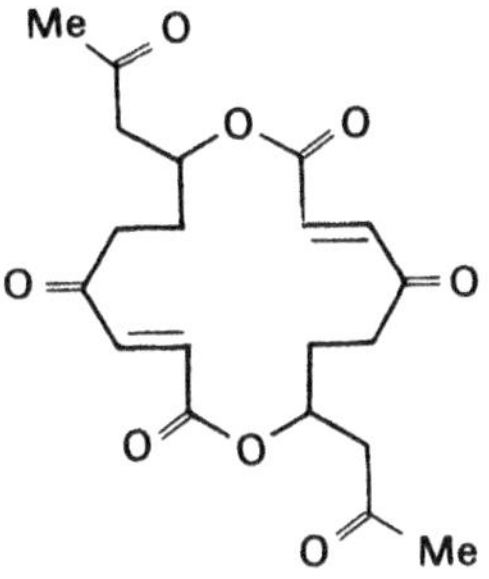

Vermiculine (*84*)
P. dangeardii (Boeckman *et al.*, 1974; Fuska and Proksa, 1983; Fuska *et al.*, 1979)

2.1.4. Hexaketides

Folding Pattern

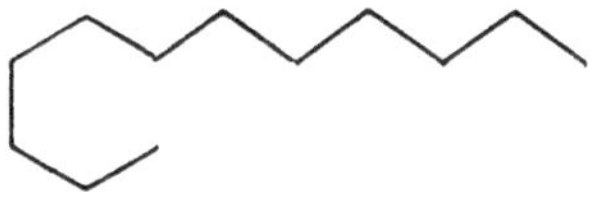

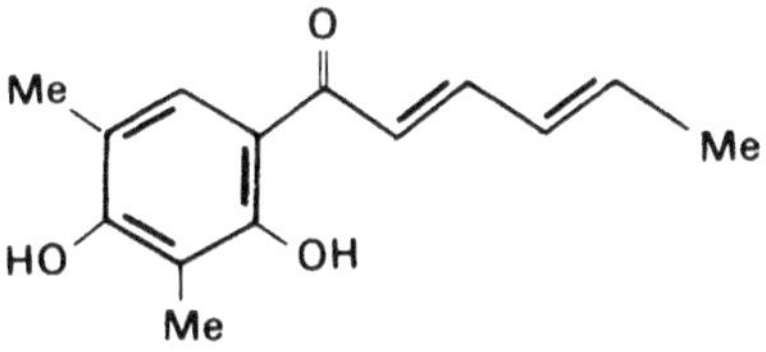

Sorbicillin (*85*)
P. chrysogenum (Cram, 1948)

Folding Pattern

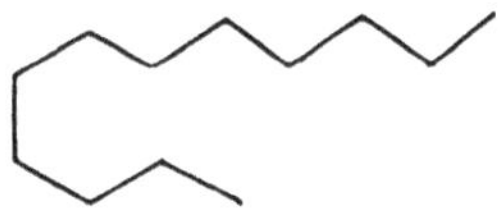

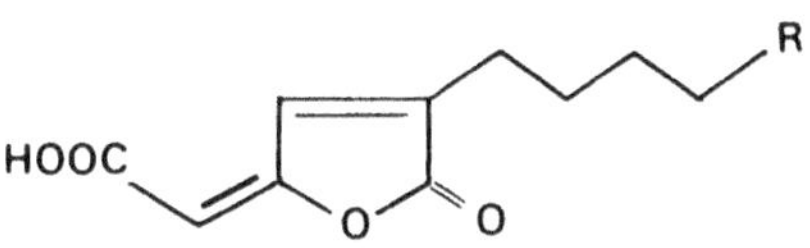

(*modified by ring scission*)
R = Me: Multicolanic acid (*86*)
R = CH_2OH: Multicolic acid (*87*)
R = COOH: Multicolosic acid (*88*)
P. sclerotiorum (Gudgeon *et al.*, 1979), *P. spinulosum* (Turner and Aldridge, 1983)

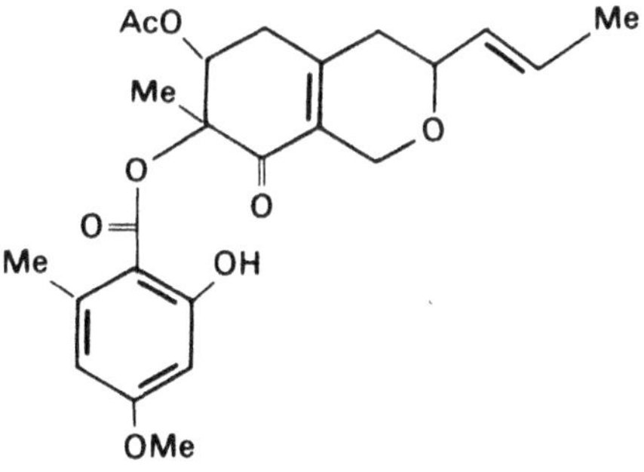

Wortmin (*89*)
P. kloeckeri (Yoshihira *et al.*, 1972)

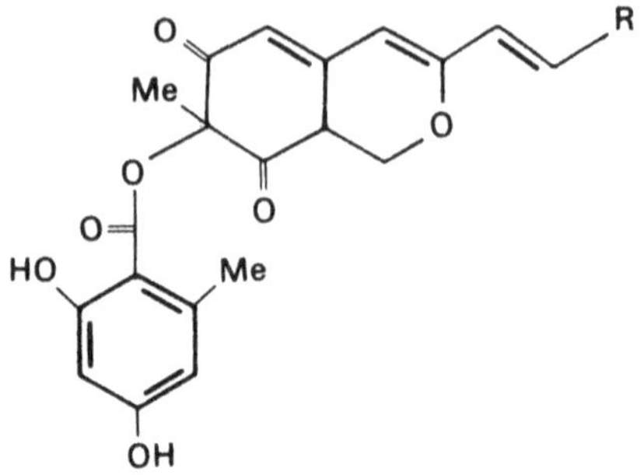

R = Me: Mitorubrin (*90*)
P. purpurogenum (Buchi *et al.*, 1965)

R = CH_2OH: Mitorubrinol (*91*)
P. purpurogenum (Buchi *et al.*, 1965), *P. kloeckeri* (Yoshihira *et al.*, 1972)

R = COOH: Mitorubrinic acid (*92*)
P. funiculosum (Locci *et al.*, 1967)

2.1.5. Heptaketides

Folding Pattern

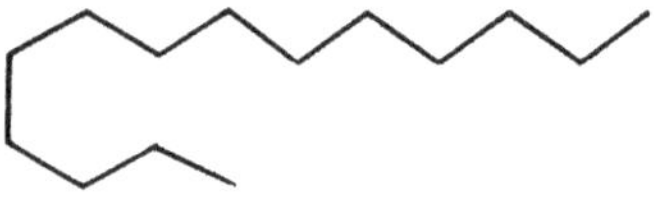

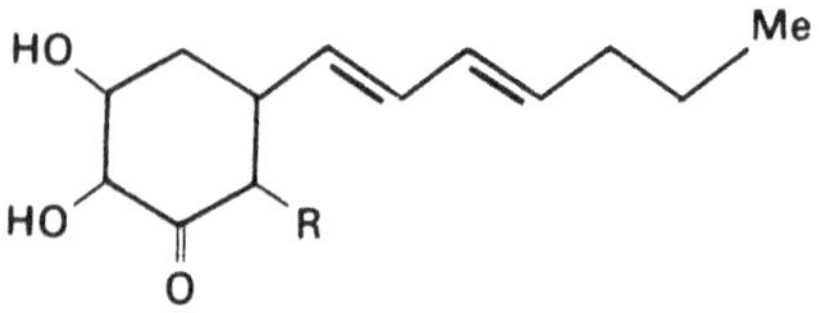

R = CHO: Frequentin (*93*)
P. glabrum (Sigg, 1963)

R = CH_2OH: Palitantin (*94*)
P. viridicatum (Chaplen and Thomas, 1960)

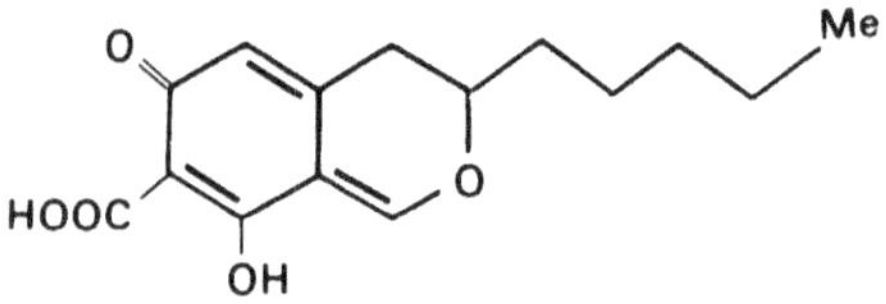

Pulvilloric acid (*95*)
P. simplicissimum (Tanenbaum and Nakajima, 1969)

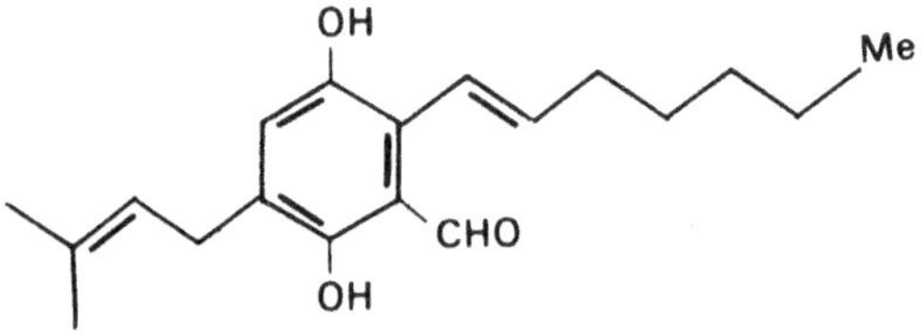

Tetrahydroauroglaucin (*96*)
P. fellutanum (Yoshihira *et al.*, 1972)

Folding Pattern

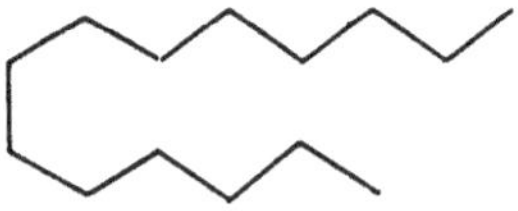

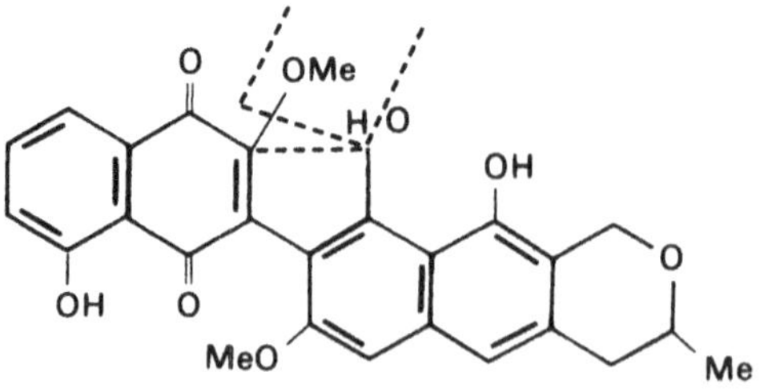

(*linked with a pentaketide moiety*)
Xanthoviridicatin D (*97*)
----Xanthoviridicatin G (*98*)
P. viridicatum (Stack *et al.*, 1979)

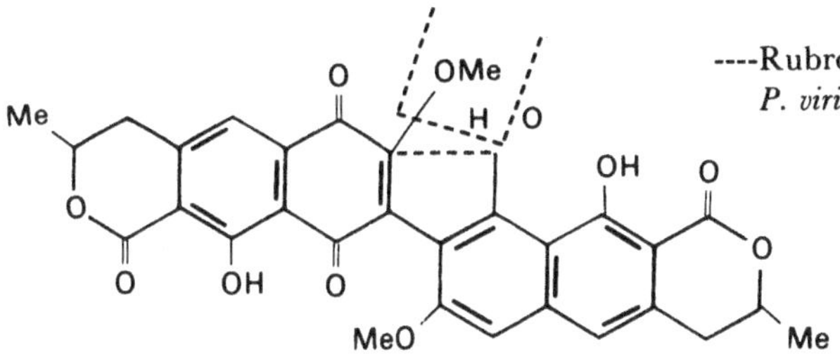

----Rubrosulfin (*99*)
P. viridicatum (Stack *et al.*, 1977)

Viomellein (*100*)
P. viridicatum (Stack *et al.*, 1977),
P. citreonigrum (Zeeck *et al.*, 1979)

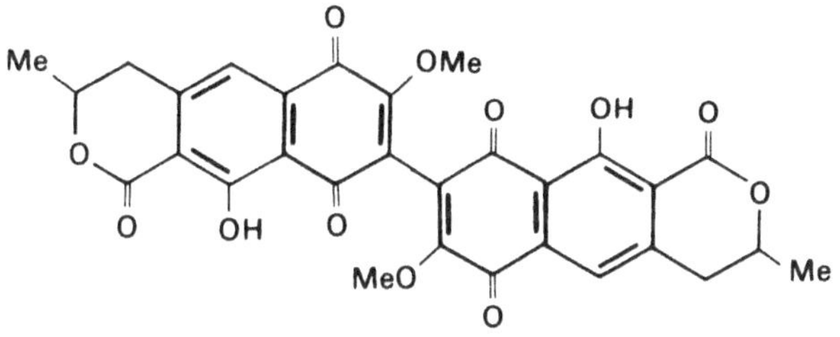

Xanthomegnin (*101*)
P. viridicatum (Stack *et al.*, 1977; Hofle and Roser, 1978),
P. aurantiogriseum (Stack and Mislivec, 1978),
P. citreonigrum (Zeeck *et al.*, 1979)

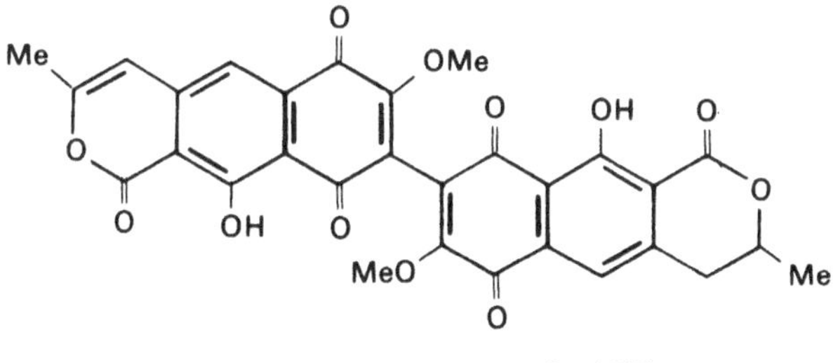

3,4-Dehydroxanthomegnin (*102*)

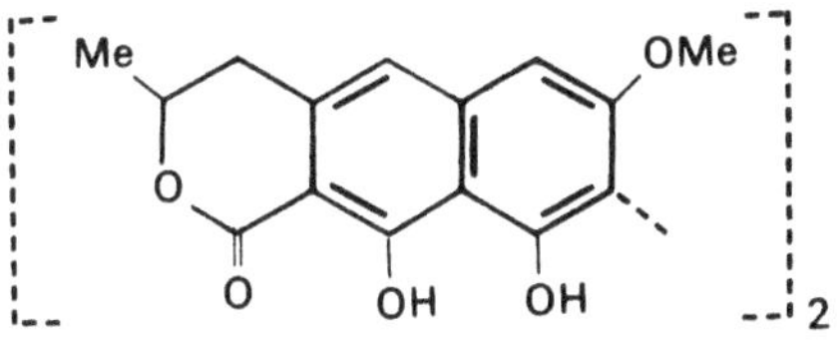

Semivioxanthin (*103*)
[Vioxanthin]$_2$ (*104*)
P. citreonigrum (Zeeck *et al.*, 1979)

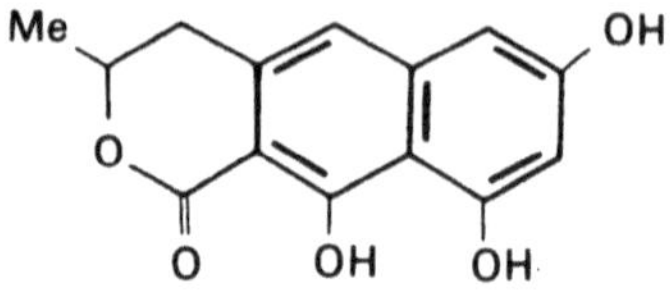

7-De-*O*-methylsemivioxanthin (*105*)
P. janthinellum (De Jesus *et al.*, 1983d)

2.1.5a. Probably Modified by Ring Scission

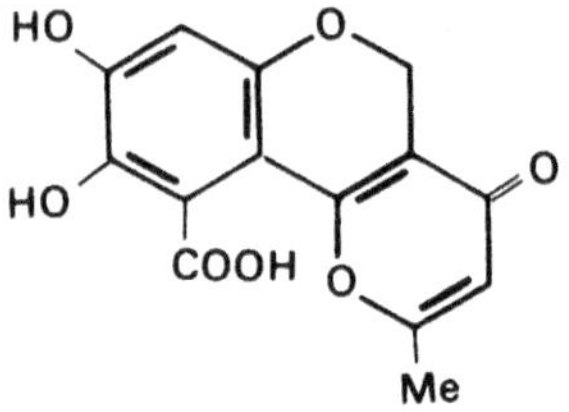

Citromycetin (*106*)
P. citrinum (Turner and Aldridge, 1983), *P. glabrum* (Gatenbeck and Mosbach, 1963)

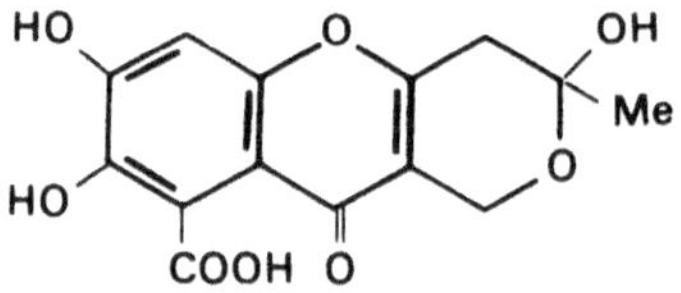

Fulvic acid (*107*) (Kurobane *et al.*, 1981)
P. griseofulvum (Dean *et al.*, 1957); *P. dodgei* (Dean *et al.*, 1957), *P. glabrum*, *P. janthinellum* (Turner and Aldridge, 1983)

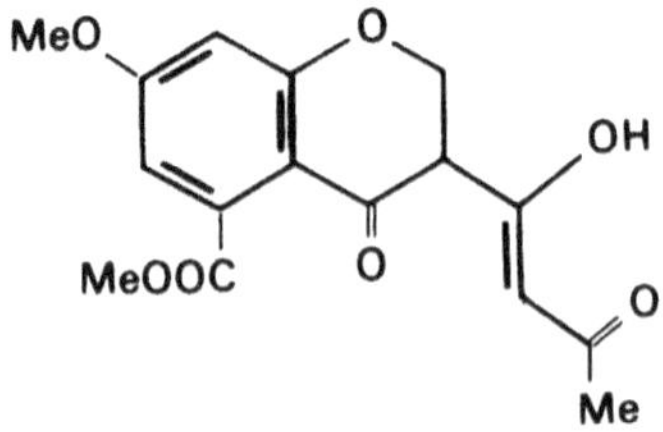

Lapidosin (*108*)
P. lapidosum (Turner, 1978)

Folding Pattern 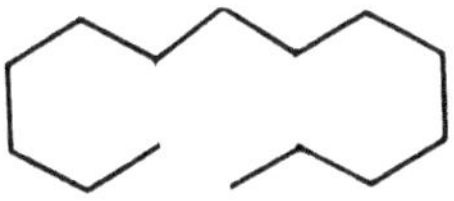

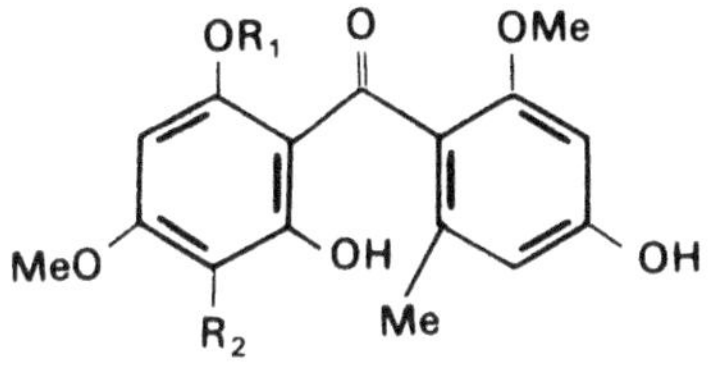

R_1 = Me, R_2 = Cl: Griseophenone A (*109*)
R_1 = H, R_2 = H: ↑↑ Griseophenone C (*110*)
R_1 = H, R_2 = Cl: ↓ Griseophenone B (*111*)
P. griseofulvum (Rhodes *et al.*, 1961)

↓

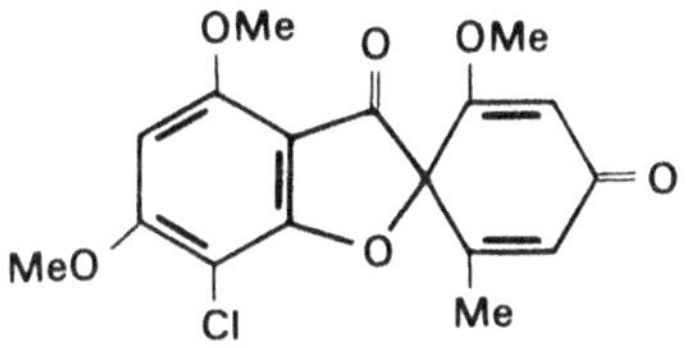

Dehydrogriseofulvin (*112*)
P. griseofulvum (McMaster *et al.*, 1960)

↓

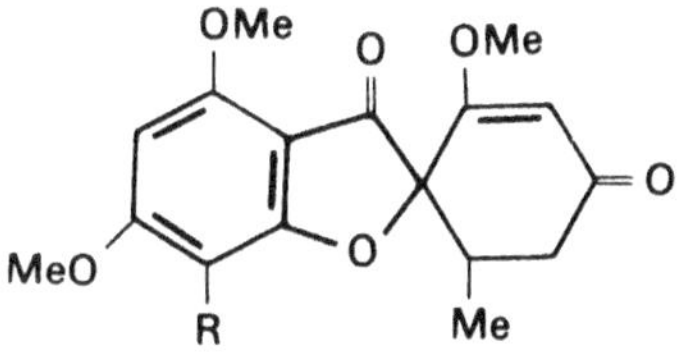

R = Cl: Griseofulvin (*113*)
P. griseofulvum (Harris *et al.*, 1976), *P. janczewskii* (MacMillan, 1954)

R = Br: Bromogriseofulvin (*114*)

R = H: Dechlorogriseofulvin (*115*)
P. griseofulvum, *P. janczewskii* (MacMillan, 1953, 1954)

R = Cl, 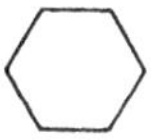: Dihydrogriseofulvin (*116*)
P. aurantiogriseum (Kamal *et al.*, 1970a)

R = H, 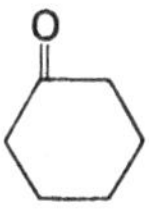: 4,6-Dimethoxy-2′-methylgrisan-3,4′,6′-trione (*117*)
P. griseofulvum (McMaster *et al.*, 1960)

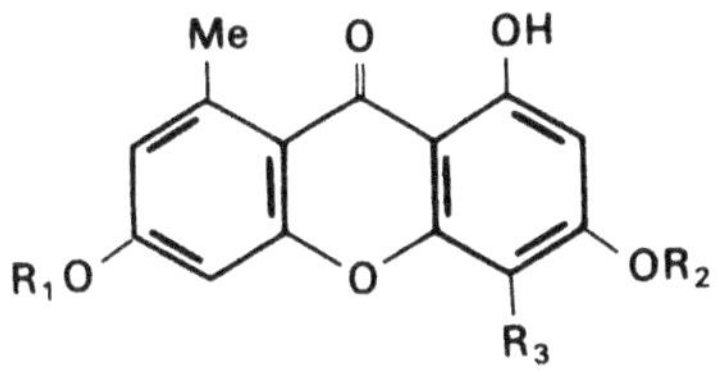

R_1	R_2	R_3	
H	Me	Cl	Griseoxanthone B (*118*)
Me	H	H	Griseoxanthone C (*119*)
H	H	H	Norlichexanthone (*120*)

P. griseofulvum (McMaster *et al.*, 1960; Broadbent *et al.*, 1975)

Folding Pattern 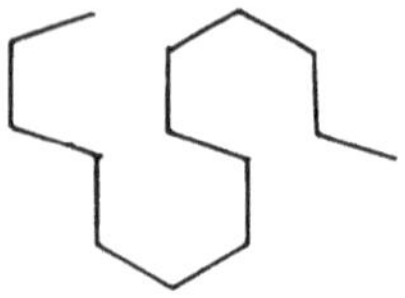 + isoprene

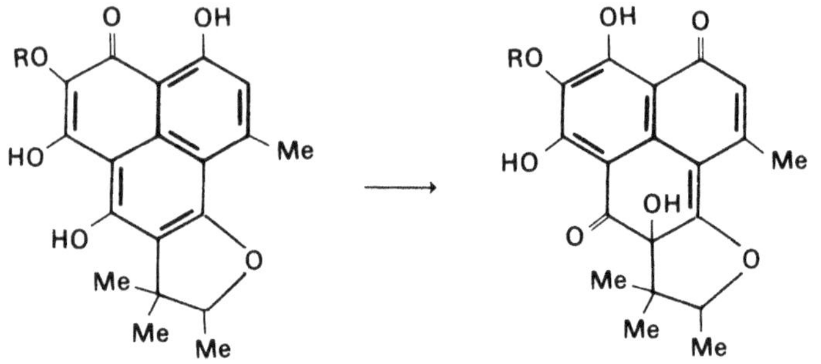

R = H: Atrovenetin (*121*)
P. herquei (Narasimhachari *et al.*, 1963), *P. melinii* (Thomas, 1961)
R = Me: Deoxyherqueinone (*122*)
P. herquei (Kriegler and Thomas, 1971)

R = Me: Herqueinone (*123*)
Isoherqueinone (*124*)
R = H: Norherqueinone (*125*)
Isonorherqueinone (*126*)
P. herquei (Quick *et al.*, 1980), *P. primulinum* (Fujimoto *et al.*, 1983)

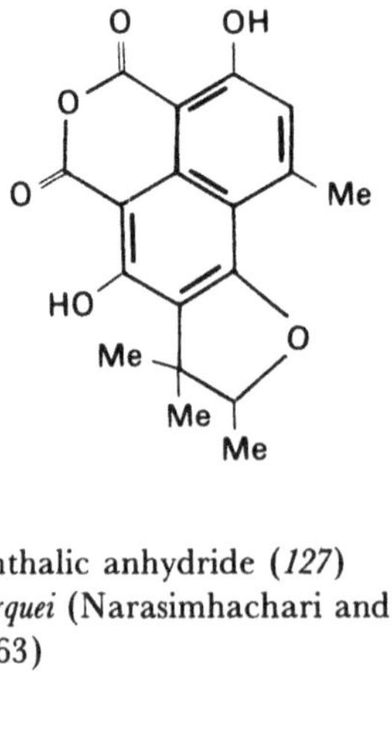

Naphthalic anhydride (*127*)
P. herquei (Narasimhachari and Vining, 1963)

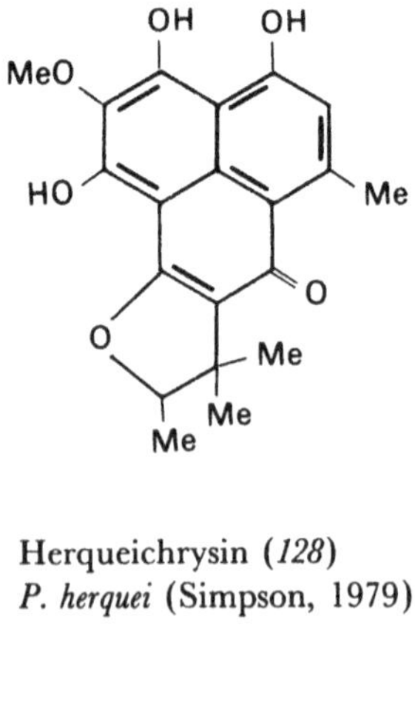

Herqueichrysin (*128*)
P. herquei (Simpson, 1979)

2.1.5b. Dimers from *P. duclauxii* (Sankawa *et al.*, 1966)

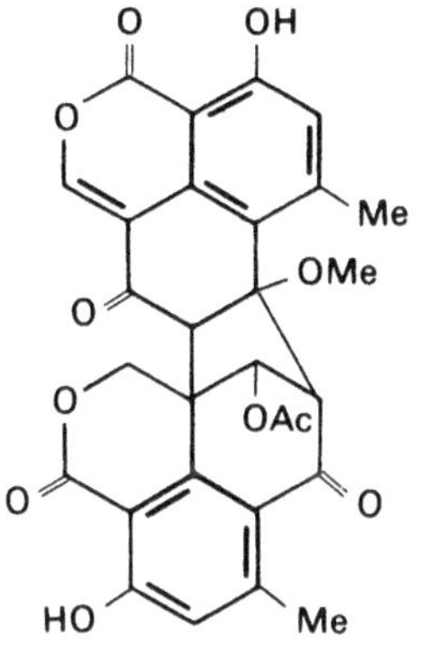

Duclauxin (*129*)
Also *P. emmonsii* (Kuhr *et al.*, 1973)

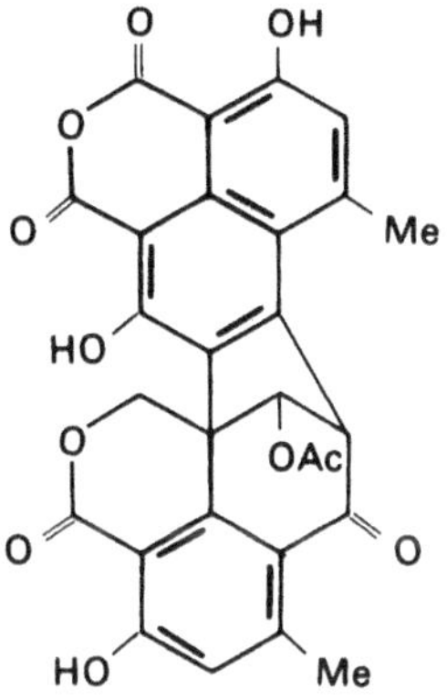

Xenoclauxin (*130*)

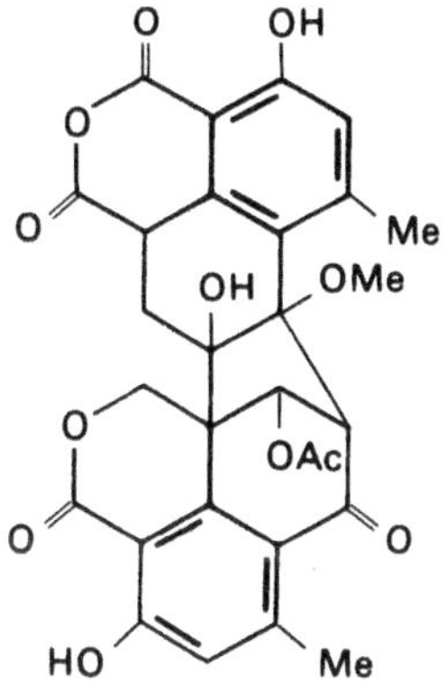

Cryptoclauxin (*131*)

2.1.6. Octaketides

Folding Pattern

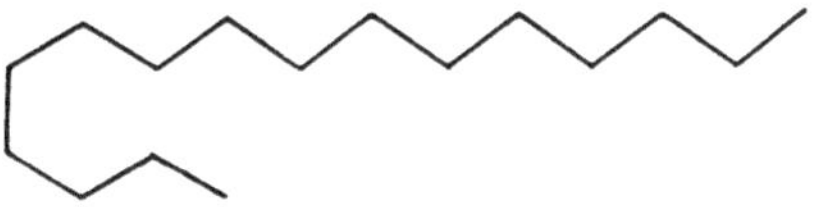

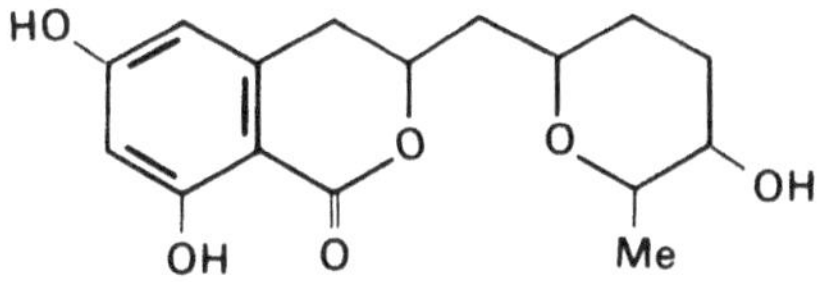

5′-Hydroxyasperentin (*132*)
Penicillium sp. (Turner and Aldridge, 1983)

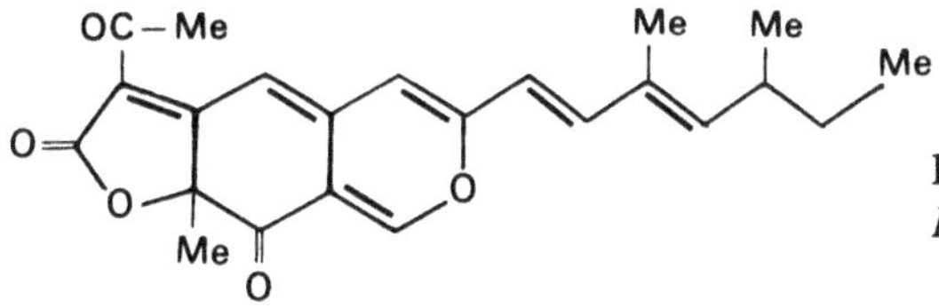

Rotiorin (*133*)
P. sclerotiorum (Holker *et al.*, 1964)

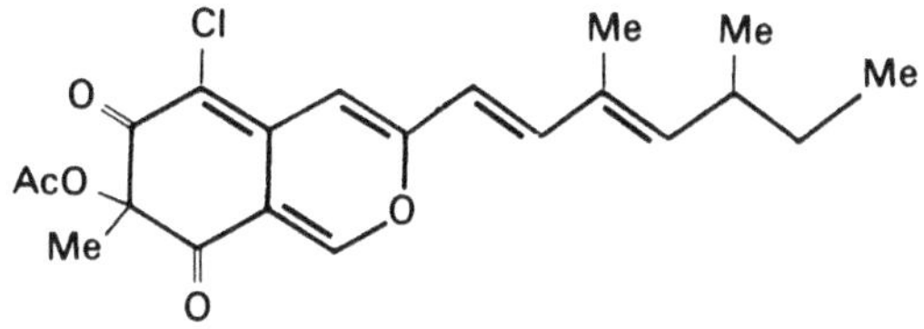

Sclerotiorin (*134*)
Episclerotiorin (*135*)
P. sclerotiorum (Holker *et al.*, 1962), *P. hirayamae* (Udagawa, 1963)

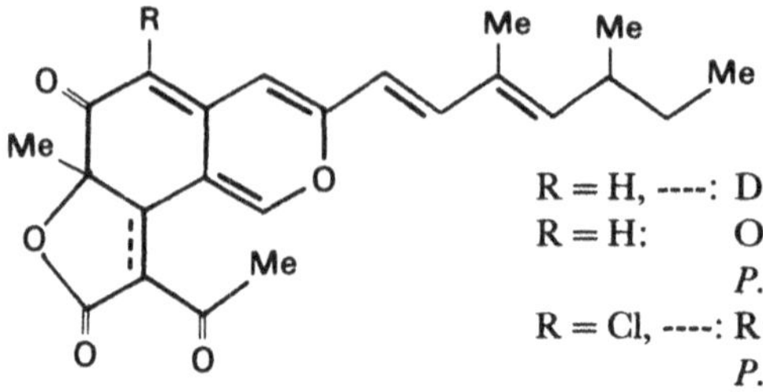

R = H, ----: Dehydrorubrorotiorin (*136*)
R = H: Ochrephilone (*137*)
P. sclerotiorum (Seto and Tanabe, 1974)
R = Cl, ----: Rubrorotiorin (*138*)
P. hirayamae (Gray and Whalley, 1971)

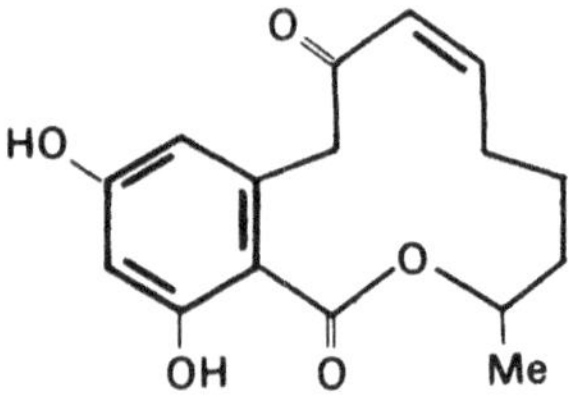

Resorcylide (*139*)
Penicillium sp. (Oyama *et al.*, 1978)

Folding Pattern

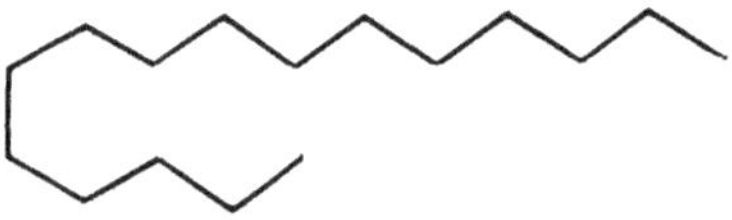

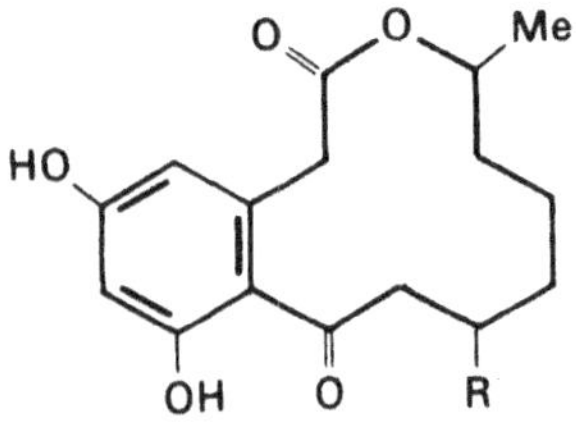

R = H: Curvularin (*140*)
P. restrictum (Raistrick and Rice, 1971), *P. citrinum* (Vesonder *et al.*, 1976)
R = OH: β-Hydroxycurvularin (*141*)
P. roseopurpureum (Turner and Aldridge, 1983)

Folding Pattern

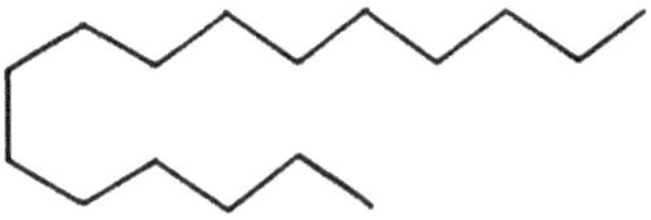

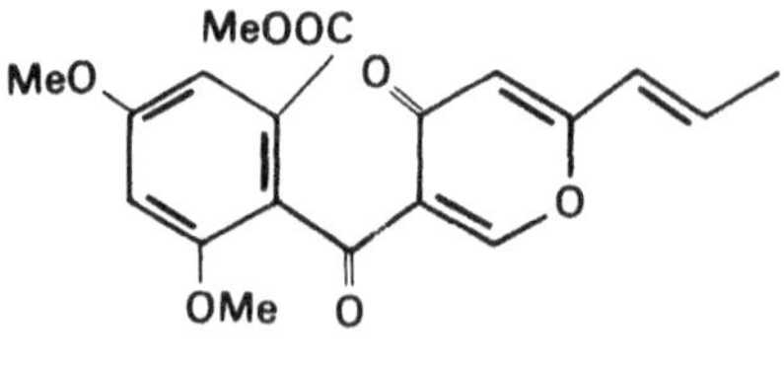

Funicone (*142*)
P. funiculosum (Merlini *et al.*, 1970)

2.1.6a. Octaketides Not Involving Aromatization

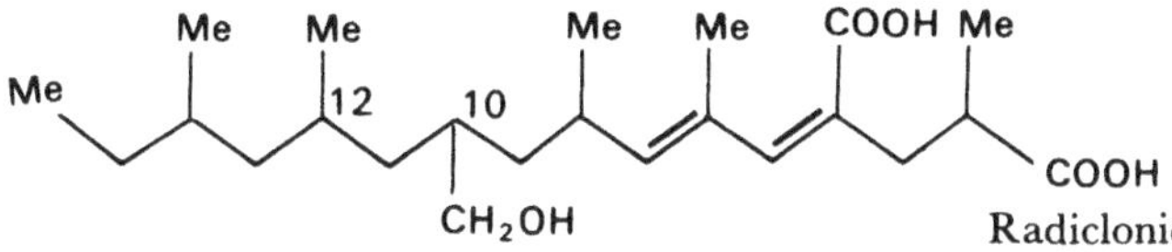

Radiclonic acid (*143*)
Penicillium sp. (Sassa *et al.*, 1973)
Also 10-CH_2OAc analogue: *O*-Acetylradiclonic acid (*144*)
12-Hydroxyradiclonic acid (*145*)
Penicillium sp. (Sassa *et al.*, 1975)

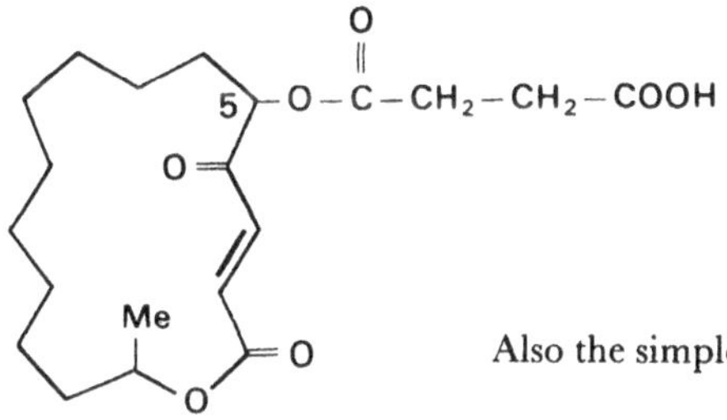

Antibiotic A26771B (*146*)
P. turbatum (Michel *et al.*, 1977)
Also the simpler 5-OH lactone analogue (*147*)
P. turbatum (Turner and Aldridge, 1983)

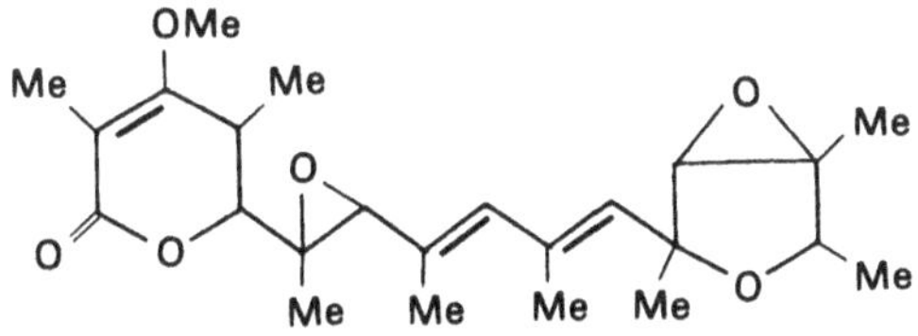

Verrucosidin (*148*)
P. aurantiogriseum (Ganguli *et. al.*, 1984)

2.1.6b. Anthraquinone Monomers

Folding Pattern

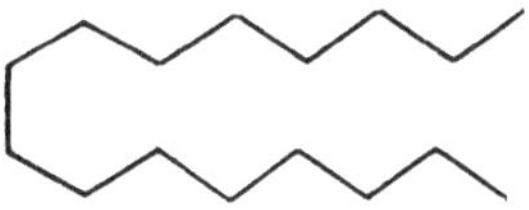

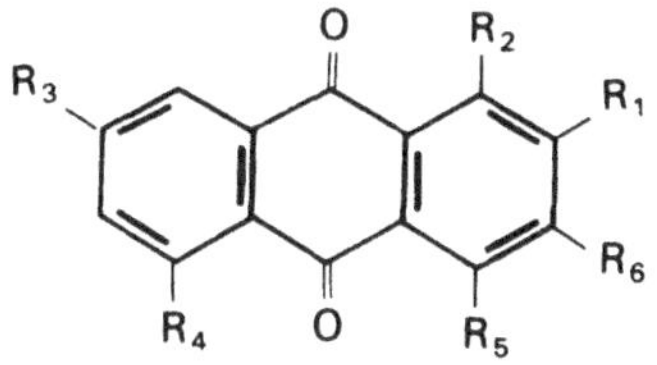

R_1	R_2	R_3	R_4	R_5	R_6	
Me	OH	H	OH	OH	H	Islandicin* (*149*) *P. islandicum* (Gatenbeck, 1962; Casey *et al.*, 1978)

* Also derived by heating a steroidal product of *P. islandicum* (pibasterol) (Ghosh *et al.*, 1978).

R_1	R_2	R_3	R_4	R_5	R_6	
Me	H	H	OH	OH	H	Chrysophanol (*150*) *P. islandicum* (B. H. Howard and Raistrick, 1950)
Me	OH	OH	H	OH	H	1,4,7,8-Tetrahydroxy-2-methyl anthraquinone (*151*) *P. islandicum* (Gatenbeck, 1959b)
Me	H	OH	OH	OH	COOH	Endocrocin (*152*) *P. islandicum* (Gatenbeck, 1959a), *P. rugulosum* (Tatsuno *et al.*, 1975)
Me	H	OH	OH	OH	H	Emodin (*153*) *P. rugulosum* (Tatsuno *et al.*, 1975), *P. emmonsii* (van Eijk, 1973)
Me	OH	OH	OH	OH	H	Catenarin (*154*) *P. emmonsii* (van Eijk, 1973)
Me	OH	OMe	OH	OH	H	Erythroglaucin (*155*) *P. emmonsii* (van Eijk, 1973)
COOH	H	OH	OH	OH	H	Emodic acid (*156*) *P. aurantiogriseum* (Anslow *et al.*, 1940)
CH_2OH	H	OH	OH	OH	H	Hydroxyemodin (*157*) *P. aurantiogriseum*, *P. griseoroseum* (Anslow *et al.*, 1940)
CH_2OH	H	OH	OMe	OH	H	Hydroxyemodin-5-methylether (*158*) *P. glabrum* (Mahmoodian and Stickings, 1964)
Me	H	OH	OMe	OH	H	Questin (*159*) *P. glabrum* (Mahmoodian and Stickings, 1964)
CH_2OH	H	OH	OH	OMe	H	Roseopurpurin (*160*) *P. roseopurpureum* (Hind, 1940a,b; Posternak, 1940)

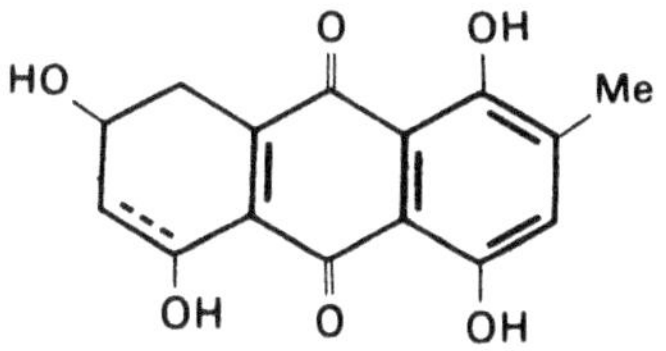

Tetrahydrocatenarin (*161*)
---: Dihydrocatenarin (*162*)
P. islandicum (Bu'Lock and Smith, 1968)

2.1.6c. *Penicillium islandicum*
Anthraquinone Dimers (Ogihara *et al.*, 1968)

Dimer	Monomer constituents
Iridioskyrin (*163*)	Islandicin × 2
Skyrin* (*164*)	Emodin × 2
Dicatenarin (*165*)	Catenarin × 2
Dianhydrorugulosin (*166*)	Chrysophanol × 2
Roseoskyrin (*167*)	Chrysophanol + islandicin
Auroskyrin (*168*)	Chrysophanol + emodin
Rhodoislandicin A (*169*)	Chrysophanol + catenarin
Rhodoislandicin B (*170*)	Emodin + islandicin
Punicoskyrin (*171*)	Islandicin + catenarin
Aurantioskyrin (*172*)	Emodin + catenarin

Also several isomers and oxo- and deoxyderivatives, notably Skyrinol (*173*), Oxyskyrin (*174*), and Oxyrugulosin (*175*) (Takeda *et al.*, 1973).

* Also found in *P. rugulosum* and *P. kloeckeri* (Breen *et al.*, 1955) and *P. brunneum* (Shibata and Udagawa, 1963).

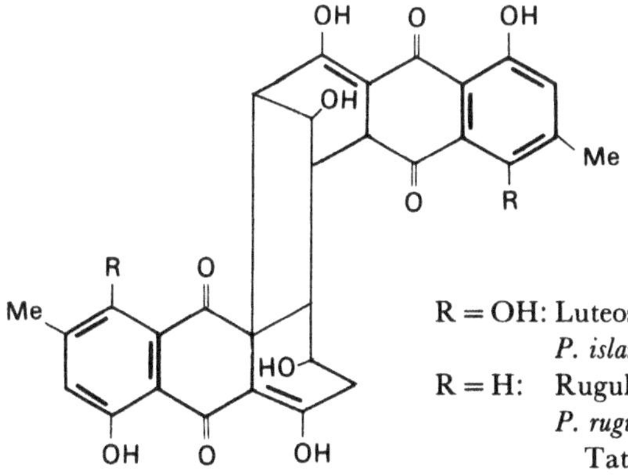

R = OH: Luteoskyrin (*176*)
P. islandicum (N. Kobayashi *et al.*, 1968)
R = H: Rugulosin (*177*)
P. rugulosum (N. Kobayashi *et al.*, 1968; Tatsuno *et al.*, 1975)

P. kloeckeri (Breen *et al.*, 1955); *P. brunneum* (Shibata and Udagawa, 1963); *P. variabile* (Yamazaki *et al.*, 1972); *P. canescens*, *P. aurantiogriseum* (Leistner and Pitt, 1977)

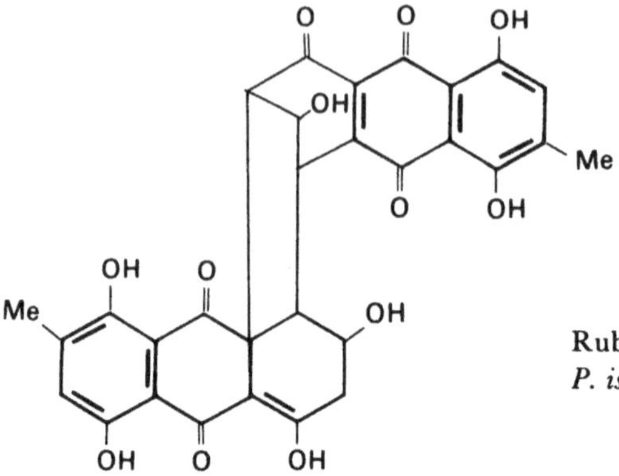

Rubroskyrin (*178*)
P. islandicum (N. Kobayashi *et al.*, 1968)

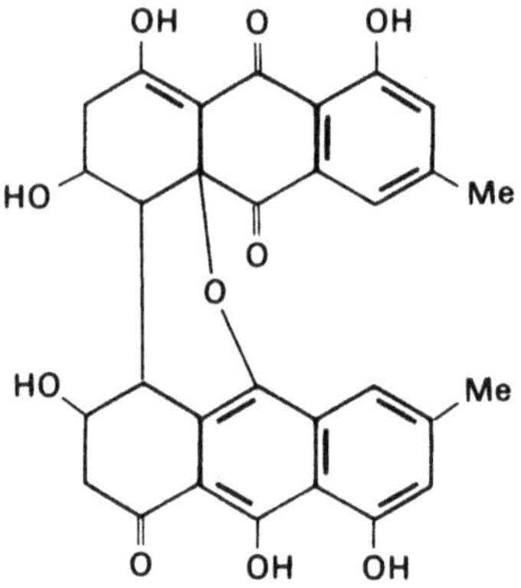

Flavoskyrin (*179*)
P. islandicum (Seo *et al.*, 1973)

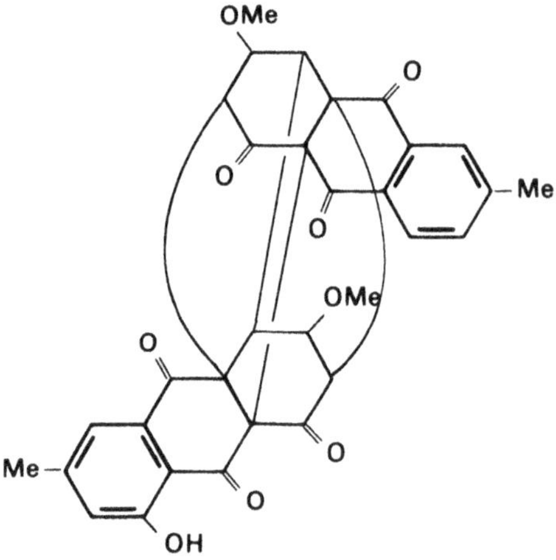

Rugulin (*180*)
P. rugulosum (Sedmera *et al.*, 1978)

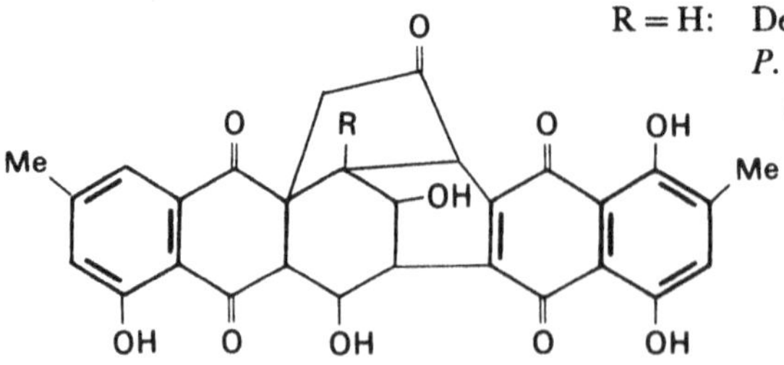

R = OH: Purpurogenone (*181*)
R = H: Deoxypurpurogenone (*182*)
P. purpurogenum
(Roberts and Thompson, 1971)

2.1.6d. Anthrone Dimer

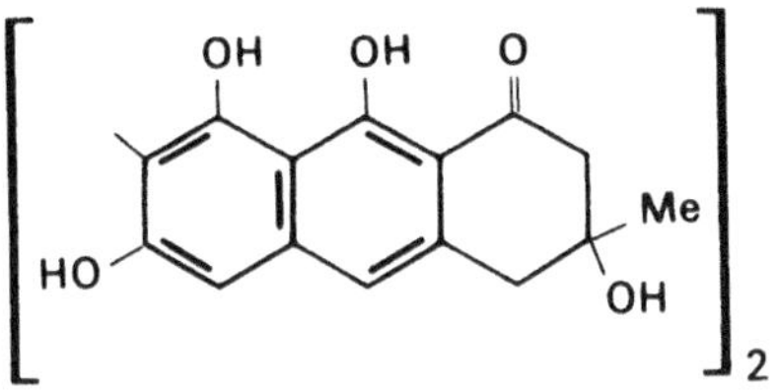

Flavomannin (*183*)
P. kloeckeri (Atherton *et al.*, 1968)

2.1.6e. Anthraquinones and Anthrones Modified by Scission

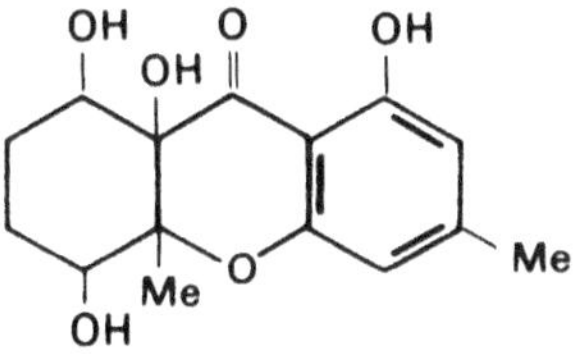

Diversonol (*184*)
P. diversum (Turner, 1978)

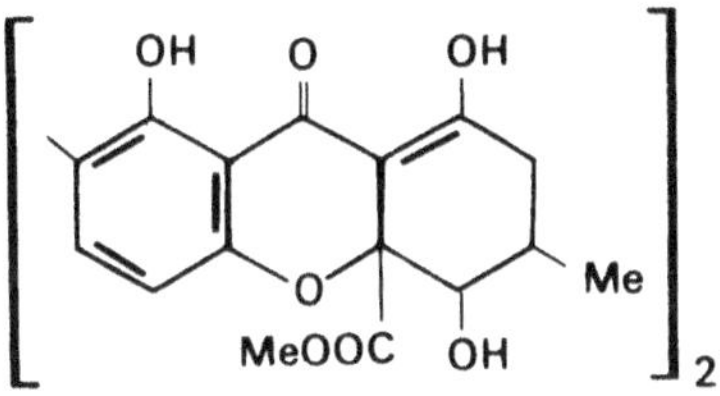

Secalonic acid D (*185*)
P. oxalicum (Franck, 1980), *P. atramentosum* (C. C. Howard *et al.*, 1976)

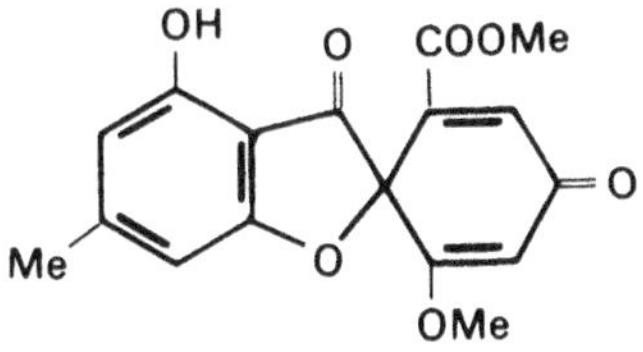

Dechlorogeodin (*186*)
P. glabrum (Mahmoodian and Stickings, 1964)

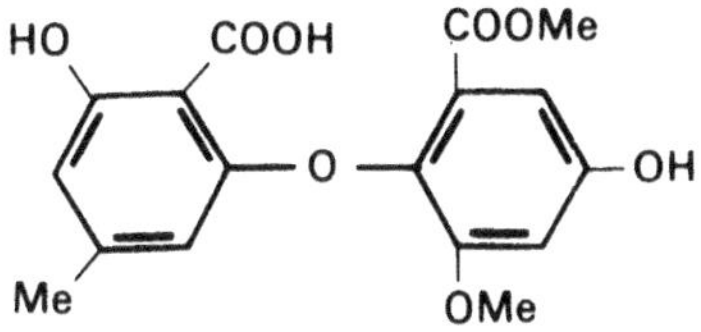

Asterric acid (*187*)
P. citrinum, *P. glabrum* (Turner and Aldridge, 1983)

2.1.7. Nonaketides

Folding Pattern

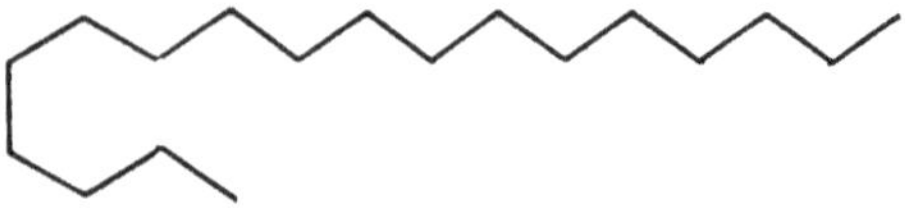

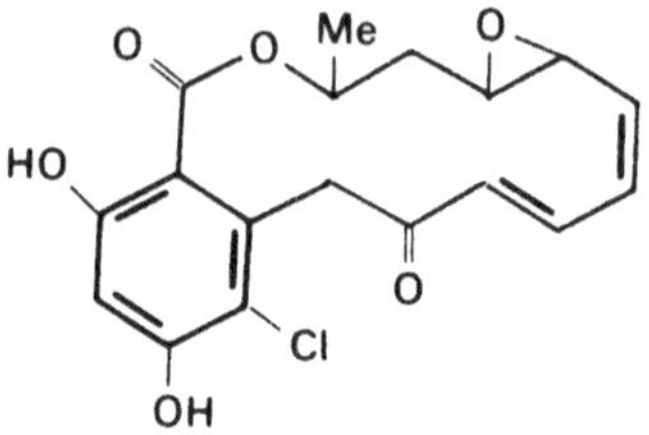

Monorden (*188*)
P. residanum (Nozawa and Nakajima, 1979; McCapra *et al.*, 1964)

Folding Pattern

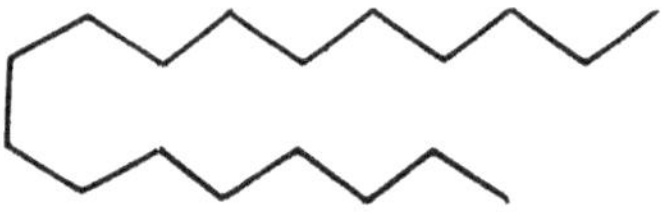

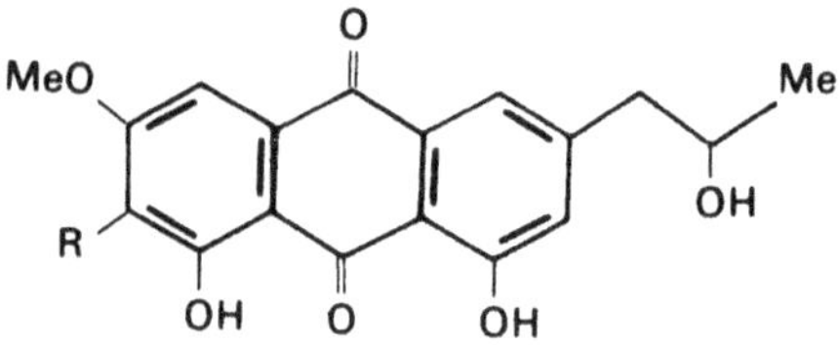

R = H: Nalgiovensin (*189*)
R = Cl: Nalgiolaxin (*190*)
P. jensenii (Birch and Stapleford, 1967)

Folding Pattern

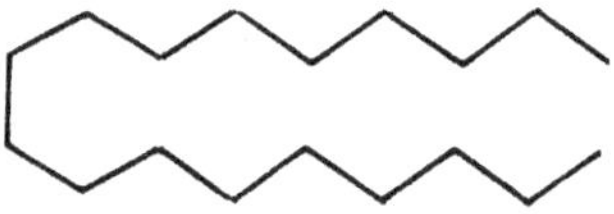

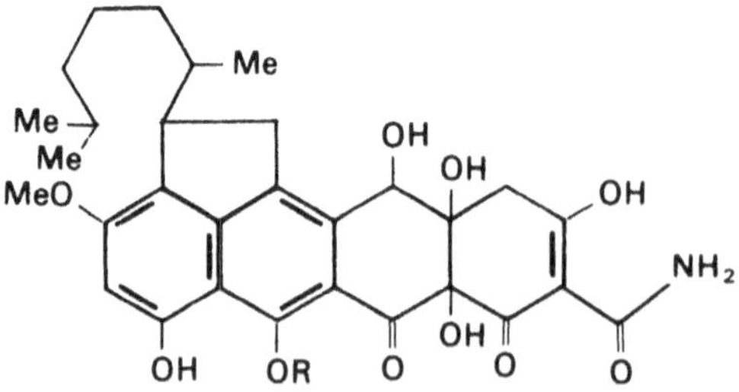

R = H: Viridicatumtoxin (*191*)
P. expansum (De Jesus *et al.*, 1982)
R = glucose: 11-*O*-β-Glucoside (*192*)
P. chrysogenum (Raju *et al.*, 1982)

Folding Pattern

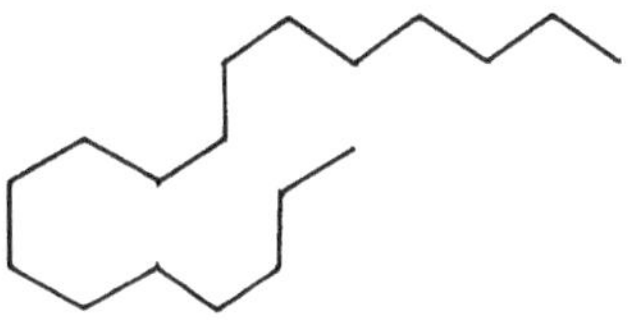

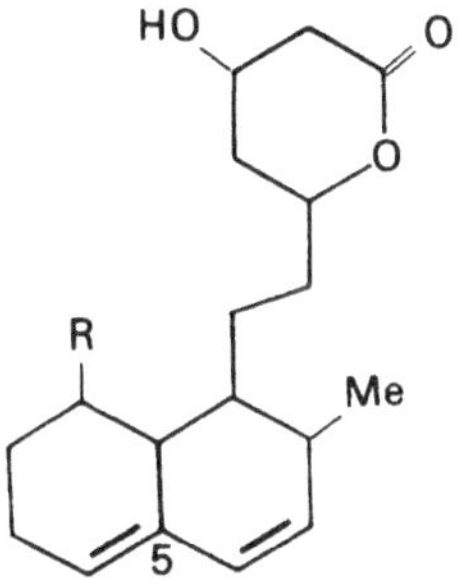

R = OH: ML236A (*193*)
R = $OCOCH(CH_3)CH_2CH_3$: Compactin (*194*)
R = $OCOCH(CH_3)CH_2CH_3$, 5,6-dihydro: Dihydrocompactin (*195*)
R = H: ML236C (*196*)

P. citrinum (Endo *et al.*, 1976; Tony Lam *et al.*, 1981), *P. brevicompactum* (Brown *et al.*, 1976)

2.1.7a. Nonaketides Not Involving Aromatization

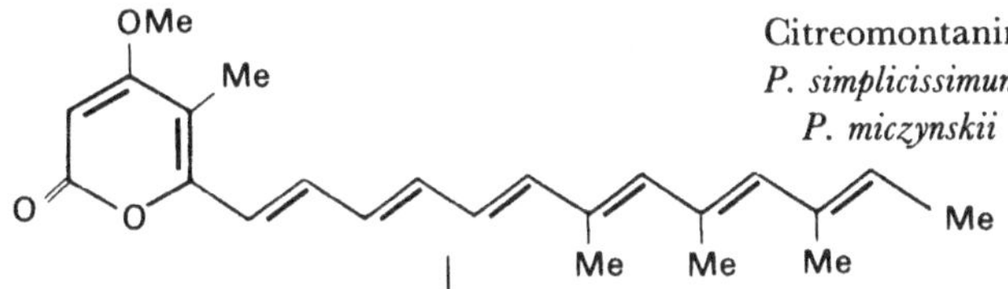

Citreomontanine (*197*)
P. simplicissimum (Rebuffat *et al.*, 1980), *P. miczynskii* (Rebuffat *et al.*, 1981)

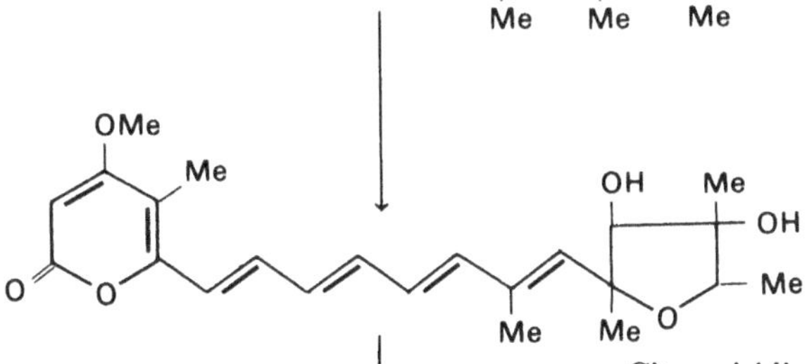

Citreoviridin (*198*)
P. citreonigrum (Sakabe *et al.*, 1977), *P. fellutanum* (Cole *et al.*, 1981), *P. miczynskii* (Leistner and Pitt, 1977), *P. simplicissimum* (Nagel *et al.*, 1972; Steyn *et al.*, 1982), *P. ochrosalmoneum* (Turner and Aldridge, 1983)

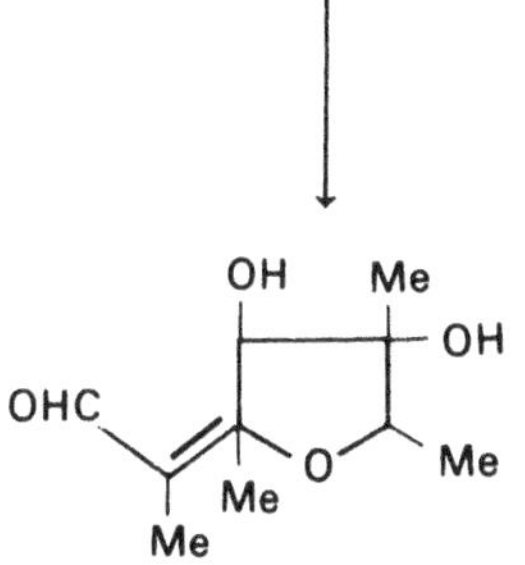

Citreoviral (*199*)
P. citreonigrum (Shizuri *et al.*, 1984b)

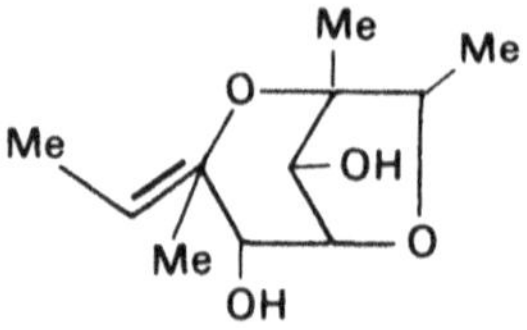

Citreoviridinol (*200*)
P. citreonigrum (Niwa *et al.*, 1981)

2.1.8. Decaketides

Folding Pattern

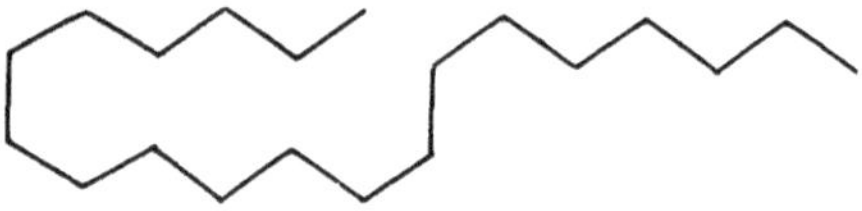

Modified by scission

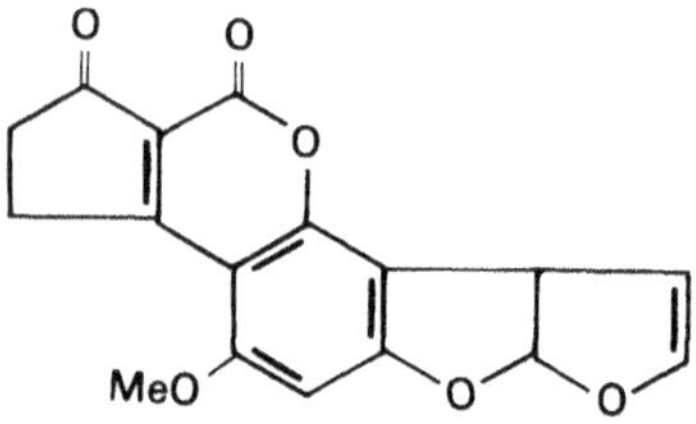

Aflatoxin B_1 (*201*)
P. puberulum (Hodges *et al.*, 1964; Biollaz *et al.*, 1968; Turner and Aldridge, 1983)

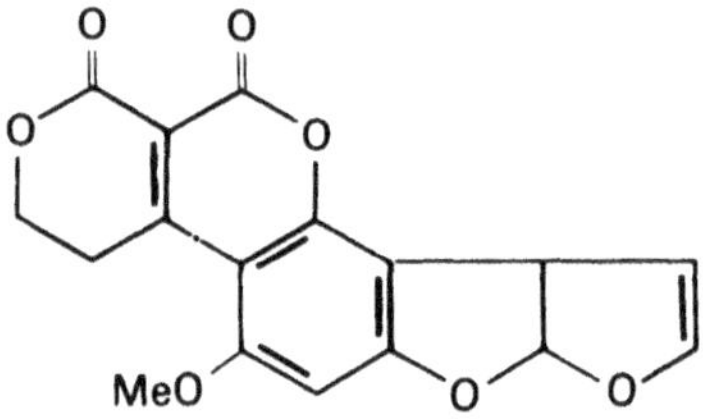

Aflatoxin G_1 (*202*)
P. puberulum (Hodges *et al.*, 1964; Biollaz *et al.*, 1968; Turner and Aldridge, 1983)

2.2. Isoprenoids

2.2.1. Monoterpene

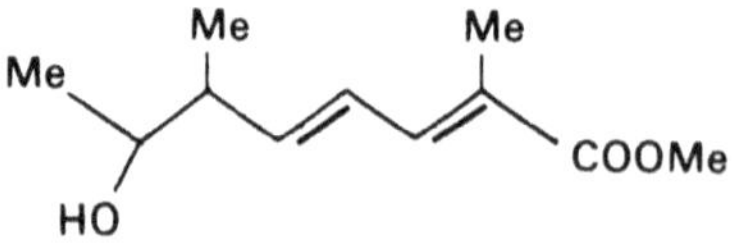

Citreodiol (*203*)
Epicitreodiol (*204*)
P. citreonigrum (Shizuri *et al.*, 1984b)

2.2.2. Sesquiterpenes

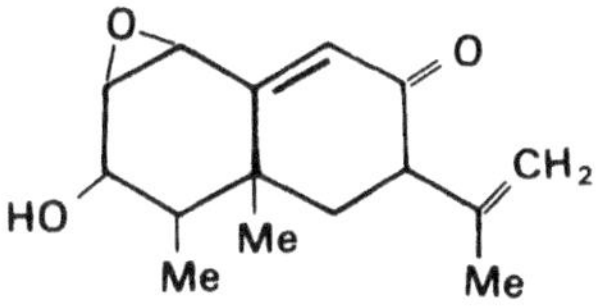

Eremefortine B (*205*)

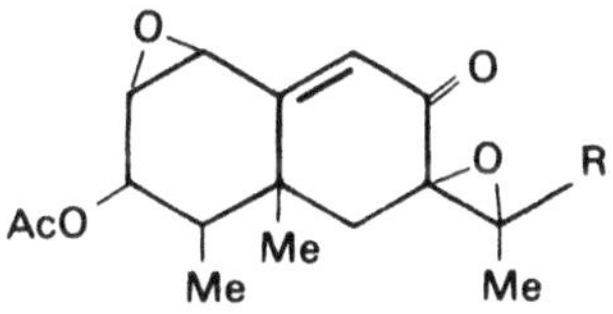

R = CHO: PR toxin (*206*)
R = Me: Eremefortine A (*207*)
$R = CH_2OH$: Eremefortine C (*208*)
$R = CO \cdot NH_2$: Eremefortine E (*209*)

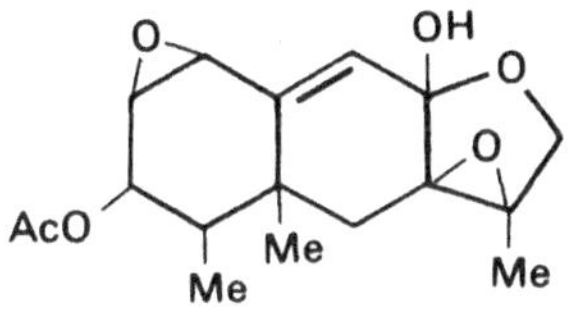

Eremefortine D (*210*)

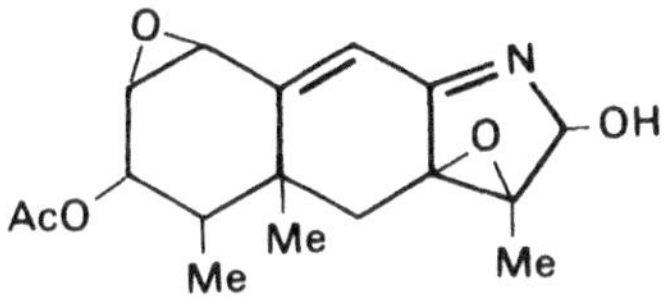

PR imine (*211*)

P. roquefortii (Wei *et al.*, 1975; Moreau *et al.*, 1980; Chalmers *et al.*, 1981)

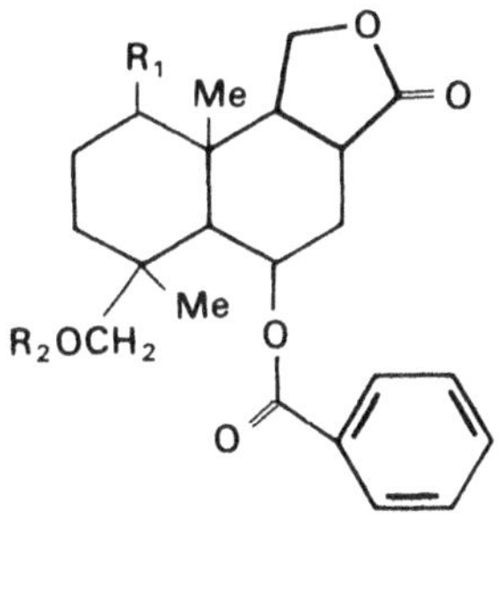

$R_1 = OH, R_2 = Ac$: Pebrolide (*212*)
$R_1 = H, R_2 = Ac$: 1-Deoxypebrolide (*213*)
$R_1 = OH, R_2 = H$: Deacetylpebrolide (*214*)
P. brevicompactum (McCorkindale *et al.*, 1981)

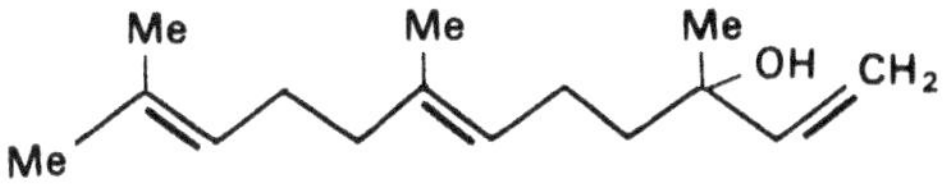

Nerolidol (*215*)

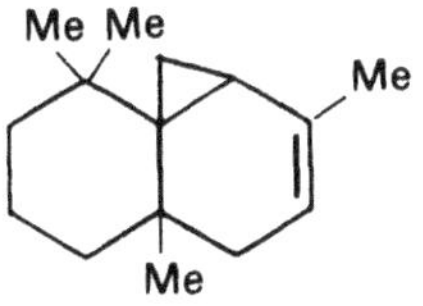

Thujopsene (*216*)

P. decumbens (Collins, 1976)

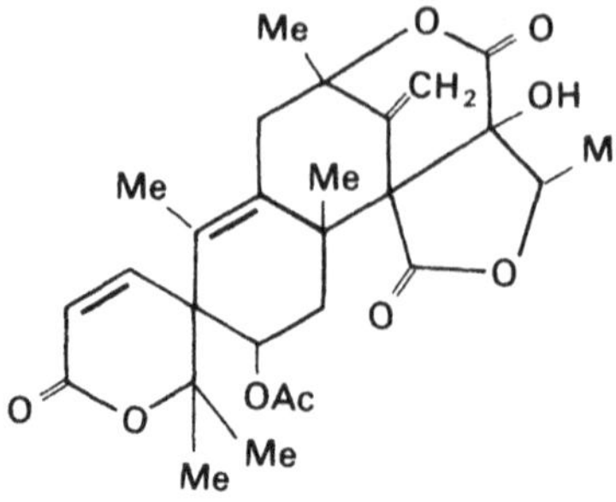

Isoaustin (*217*)

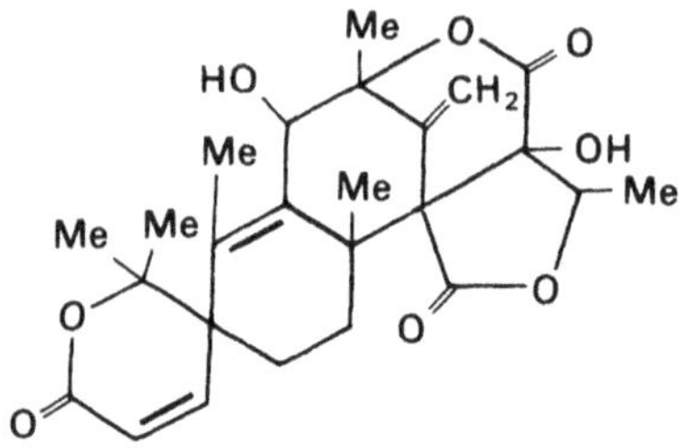

Deacetylaustin (*218*)

P. diversum (Simpson *et al.*, 1982)

2.2.3. Triterpenes

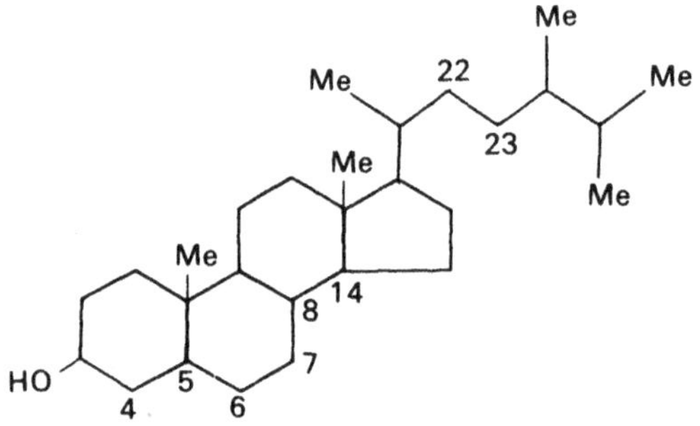

Δ5:	Cholesterol (*219*) *P. funiculosum* (Chen and Haskins, 1963)
Δ5, Δ7, Δ22:	Ergosterol (*220*) *P. brevicompactum* (Bird and Campbell, 1982b)
Δ7:	Fungisterol (*221*) *P. claviforme* (Halsall and Sayer, 1959; Weete and Laseter, 1974)
Δ6, Δ22, 5,8-OH:	Ergosta-6,22-diene-3β,5α,8α triol (*222*) *P. purpurogenum* (White *et al.*, 1973)
Δ6, Δ22, 5,8 dioxo:	Ergosterol peroxide (*223*) *P. purpurogenum* (Adam *et al.*, 1967; Bates *et al.*, 1976)
Δ4, Δ6, Δ8–14, Δ22:	Ergosta 4,6,8(14),22-tetraen-3-one (*224*) *P. citrinum*, *P. griseofulvum*, *P. islandicum* (Price and Worth, 1974); *P. purpurogenum* (White and Taylor, 1970; White *et al.*, 1973)

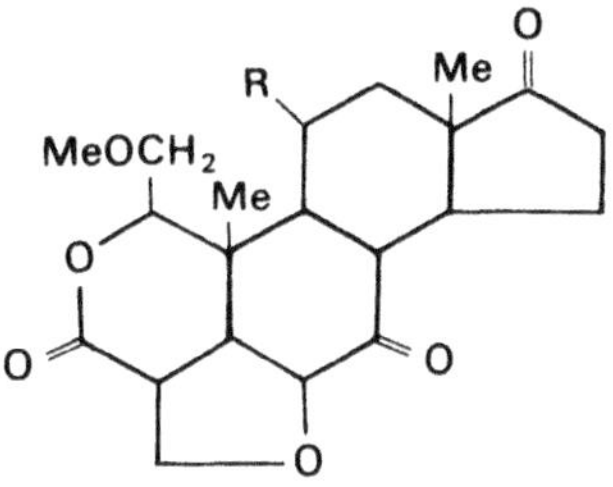

R = H: Deacetoxywortmannin (*225*)
R = AcO: Wortmannin (*226*)
P. kloeckeri (MacMillan *et al.*, 1972; Simpson *et al.*, 1979), *P. funiculosum* (Haefliger and Hauser, 1973)

2.3. Fatty-Acid-Derived Metabolites

$HOOC(CH_2)_6\,CH=CHCOOH$ 2-Decenedioic acid (*227*)
$CH_3(CH_2)_{16}\,CH_2OH$ Octadecanol (*228*)
P. chrysogenum (Cram and Tishler, 1948; Angeletti *et al.*, 1952)

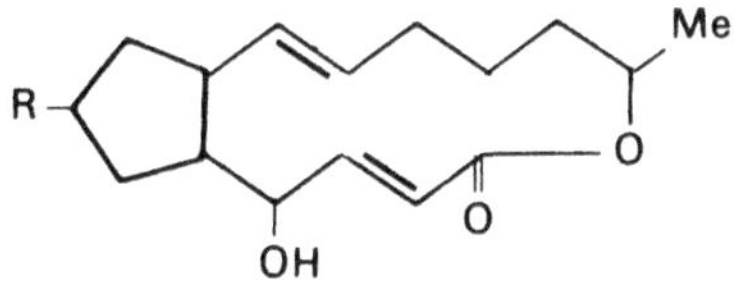

R = OH: Brefeldin A (*229*)
P. decumbens (Singleton *et al.*, 1958), *P. dodgei* (Sigg, 1964), *P. cyaneum* (Betina *et al.*, 1962; Bu'Lock and Clay, 1969), *P. simplicissimum* (Betina *et al.*, 1966), *P. janthinellum* (Turner and Aldridge, 1983)
R = H: Brefeldin C (*230*)
P. dodgei (Sunagawa *et al.*, 1979)

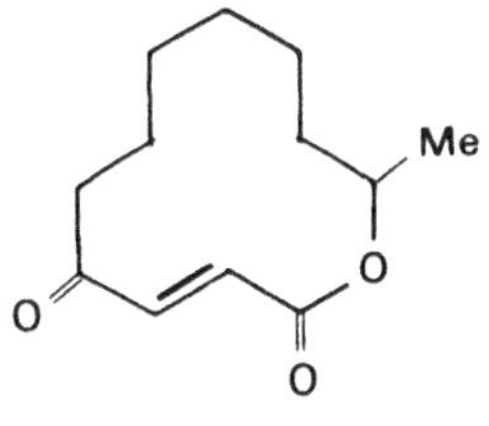

Patulolide A (*231*)
P. griseofulvum (Sekiguchi *et al.*, 1985)

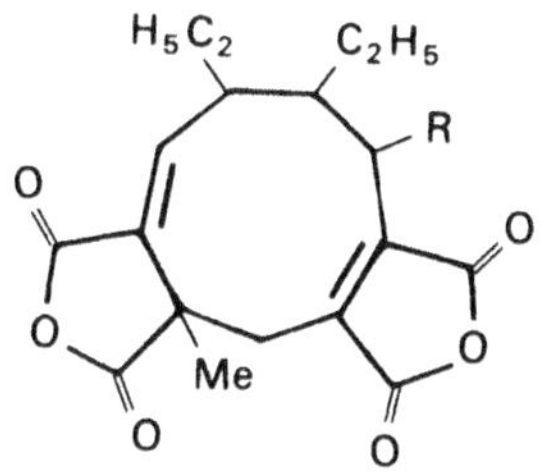

R = OH: Glauconic acid (*232*)
R = H: Glaucanic acid (*233*)
P. purpurogenum (Wijkman, 1931; Barton and Sutherland, 1965; Bloomer *et al.*, 1968)

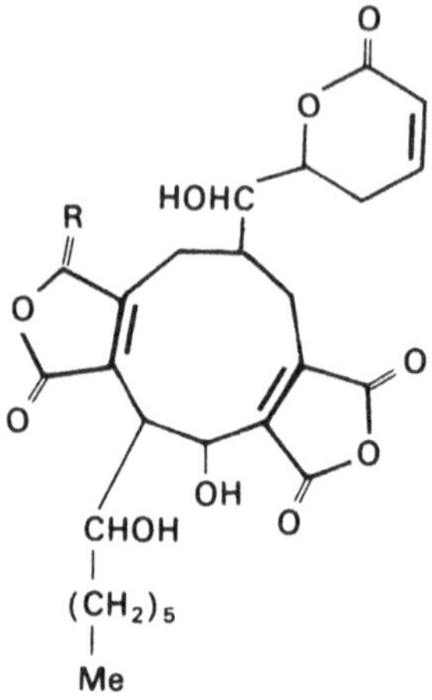

R = H, OH: Rubratoxin A (*234*)
R = O: Rubratoxin B (*235*)
P. purpurogenum (Moss *et al.*, 1969)

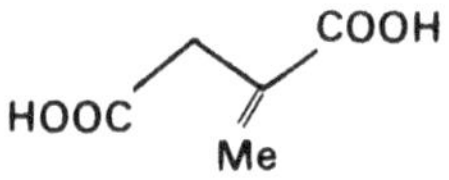

Itaconic acid (*236*)
P. fellutanum (Lybing and Reio, 1958; Bentley and Thiessen, 1957)

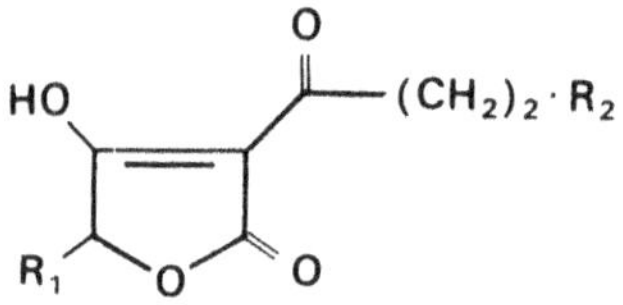

R_1	R_2	
CH_3	$CH(OH)\cdot CH_2\cdot CH_3$	Terrestric acid (*237*) *P. jensenii* (Birkinshaw and Raistrick, 1936)
$CH_2\cdot COOH$	CH_3	Carlosic acid* (*238*)
$CH_2\cdot COOH$	CH_2OH	Carlic acid* (*239*)
CH_3	CH_2OH	Carolic acid*,† (*240*)
CH_2	CH_2OH	Dehydrocarolic acid*,† (*241*)
CH_3	$COOH$	Carolinic acid* (*242*)
$CH_2\cdot COOH$	$(CH_2)_2\cdot CH_3$	Viridicatic acid* (*243*)

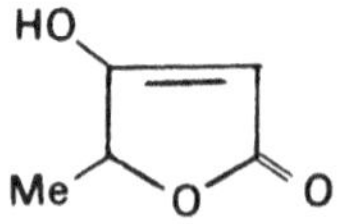

γ-Methyl tetronic acid* (*244*)

**P. fellutanum*, †*P. spinulosum* (Bracken and Raistrick, 1947; Birkinshaw and Samant, 1960; Bentley *et al.*, 1962).

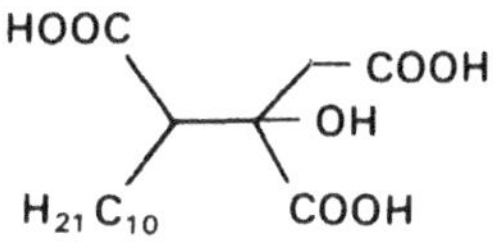

Decylcitric acid (*245*)
P. lehmanii (Gatenbeck and Mahlen, 1968)

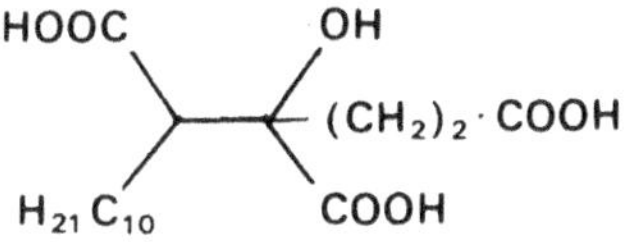

4-Hydroxy-4,5-dicarboxy pentadecanoic acid (*246*)
P. lehmanii (Tabuchi *et al.*, 1977)

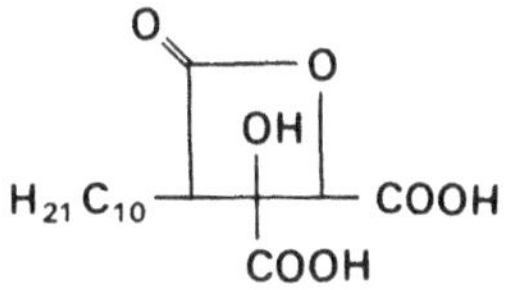

Minioluteic acid (*248*)
P. minioluteum (Birkinshaw and Raistrick, 1934)

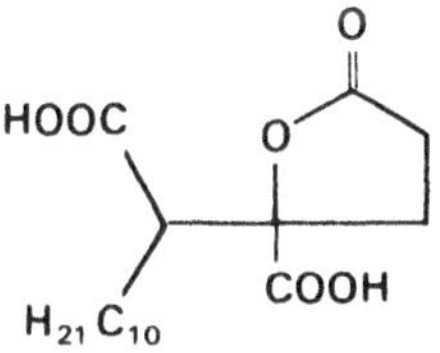

Spiculisporic acid (*247*)
P. funiculosum (Gatenbeck and Mahlen, 1964; Bodo *et al.*, 1976)

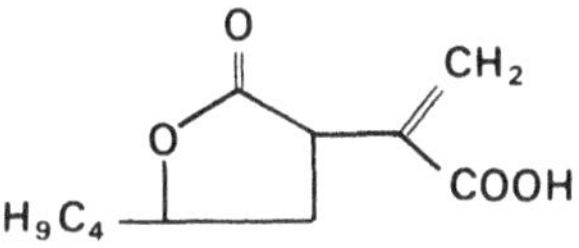

Canadensic acid (*249*)
P. arenicola (Turner, 1971)

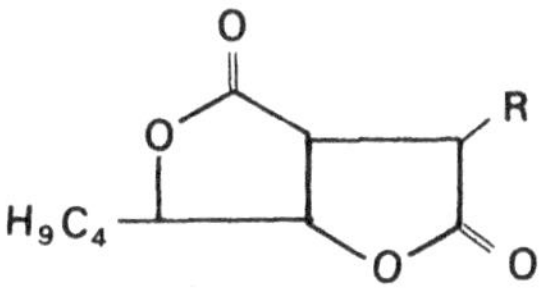

R = CH_2: Canadensolide (*250*)
R = Me: Dihydrocanadensolide (*251*)
P. arenicola (McCorkindale *et al.*, 1968; Birch *et al.*, 1968)

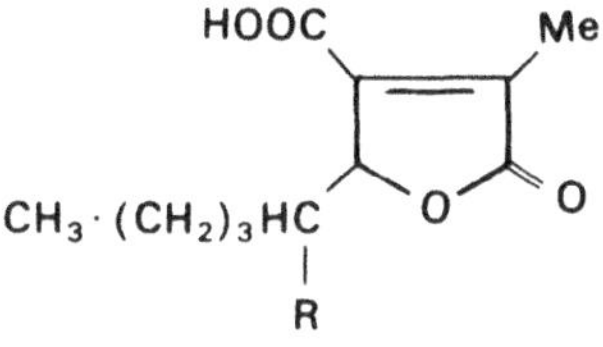

R = H: Isocanadensic acid (*252*)
R = OH: Hydroxyisocanadensic acid (*253*)
R = H, no heterocyclic double bond: Dihydroisocanadensic acid (*254*)
P. arenicola (Turner, 1971)

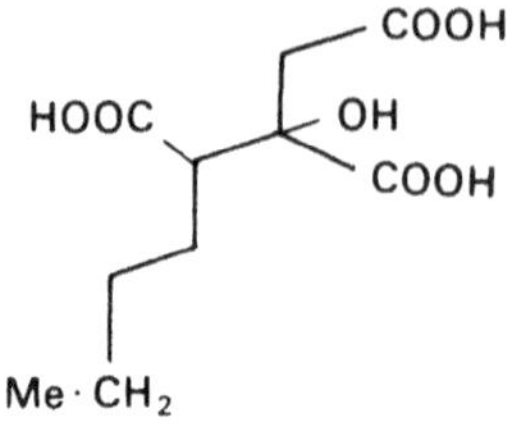

Butylcitric acid (*255*)

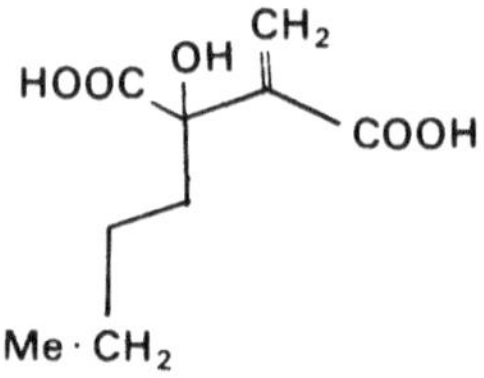

Butylhydroxyitaconic acid (*256*)

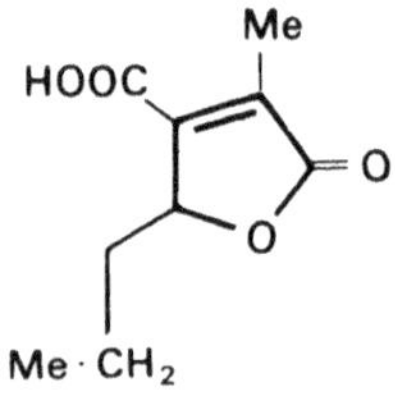

Decumbic acid (*257*)

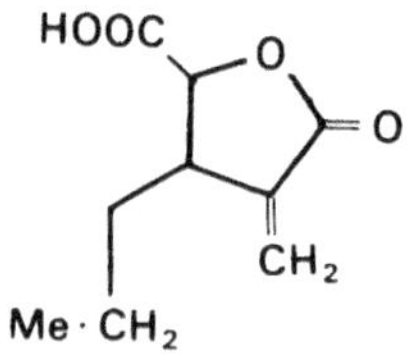

Ethisic acid (*258*)

P. decumbens (McCorkindale *et al.*, 1978b)

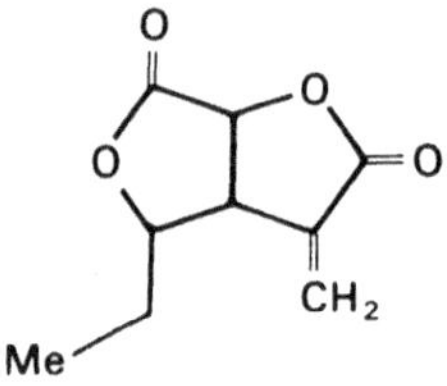

Ethisolide (*259*)
P. decumbens (Aldridge and Turner, 1971)

2.4. Metabolites Derived from End Products or Intermediates of Amino Acid Biosynthesis

2.4.1. Shikimate

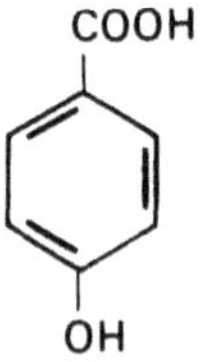

p-Hydroxybenzoic acid (*260*)
P. griseofulvum (Bassett and Tanenbaum, 1958)

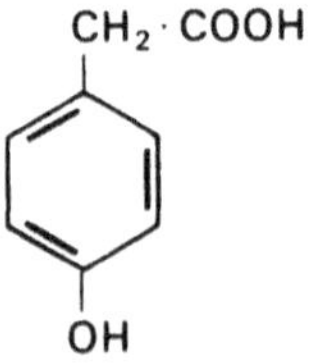

p-Hydroxyphenylacetic acid (*261*)
P. viridicatum (Turner and Aldridge, 1983)

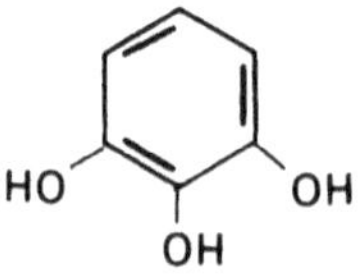

Pyrogallol (*262*)
P. griseofulvum (Bassett and Tanenbaum, 1958)

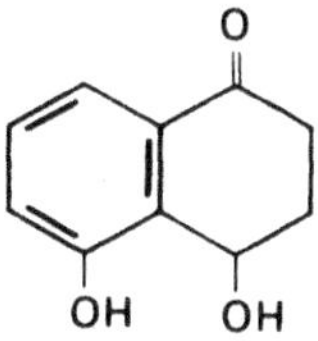

Sclerone (*263*)
Isosclerone (*264*)

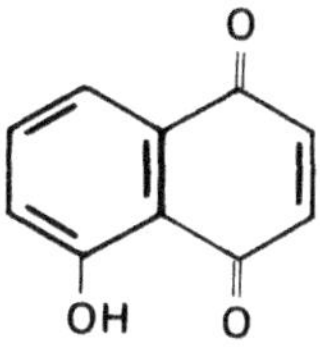

Juglone (*265*)

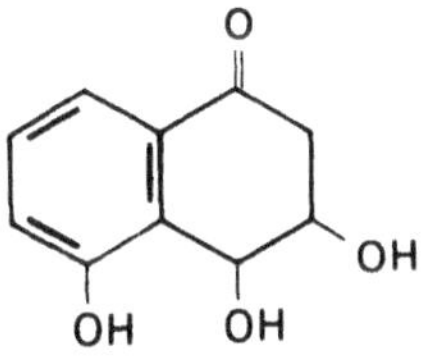

3,4,5-Trihydroxy-1-tetralone (*266*)

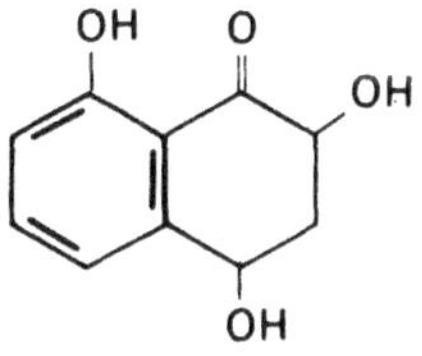

2,4,8-Trihydroxy-1-tetralone (*267*)

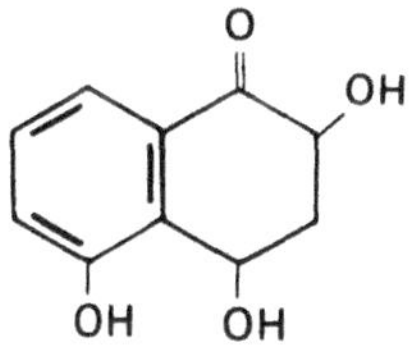

2,4,5-Trihydroxy-1-tetralone (*268*)

P. primulinum (Fujimoto *et al.*, 1983)

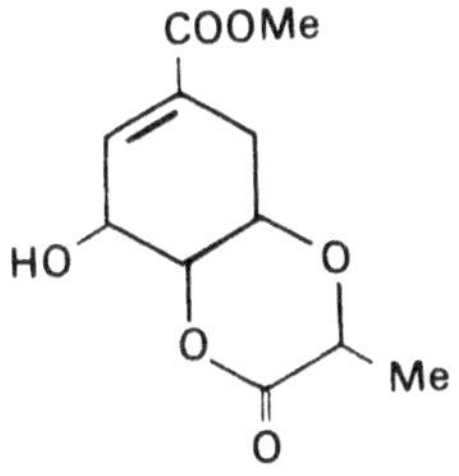

Methyl 5-lactyl-shikimate lactone (*269*)
Penicillium sp. (Isogai *et al.*, 1985)

2.4.2. Anthranilate

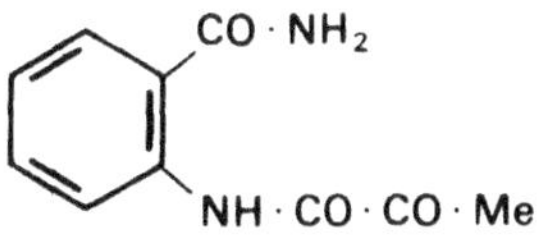

2-Pyruvoylaminobenzamide (*270*)
P. chrysogenum (Suter and Turner, 1967)

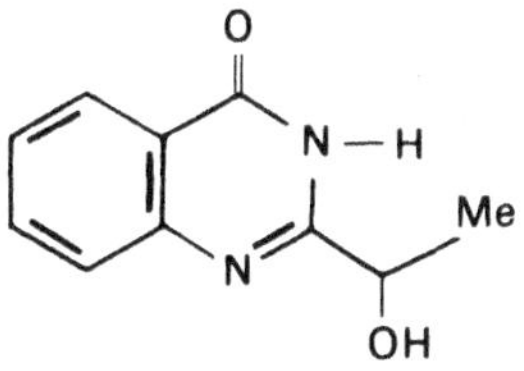

Chrysogine (*271*)
P. chrysogenum (Hikino *et al.*, 1973)

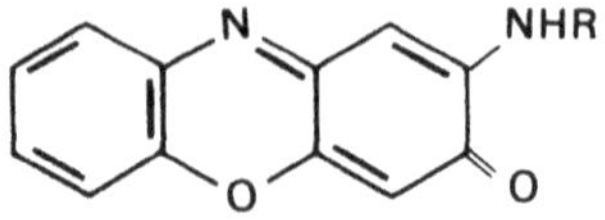

R = H: Questiomycin A (*272*)
R = Ac: *N*-Acetyl-2-amino-3H-phenoxazin-3-one (*273*)
P. chrysogenum (Bar *et al.*, 1971a; Pfeifer *et al.*, 1972)

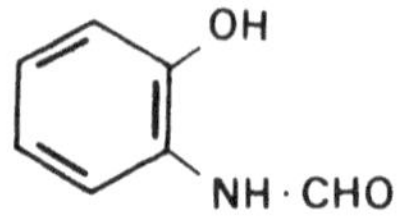

N-Formyl-2-aminophenol (*274*)
P. chrysogenum (Bar *et al.*, 1971b)

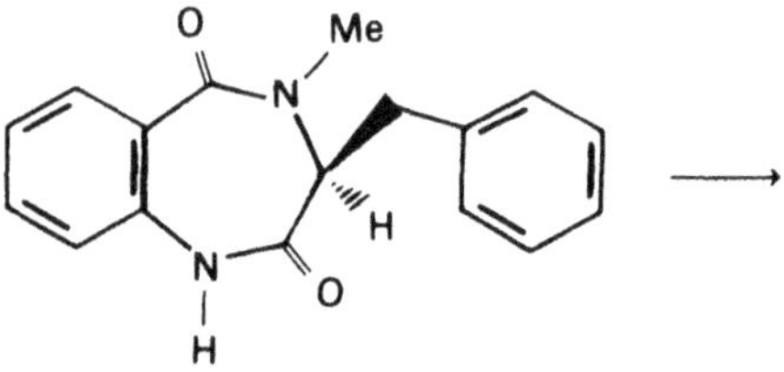

Cyclopeptine (*275*)

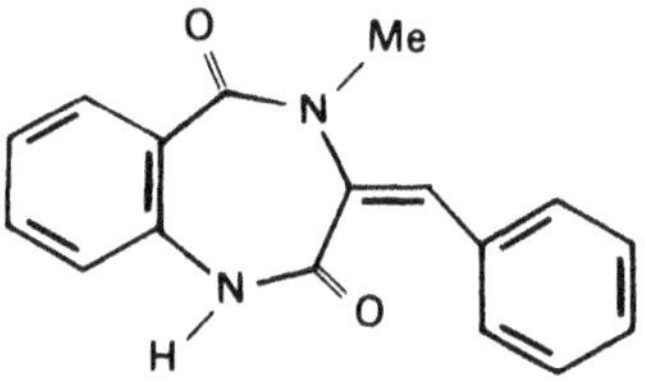

Dehydrocyclopeptine (*276*)

P. aurantiogriseum (Framm *et al.*, 1973)

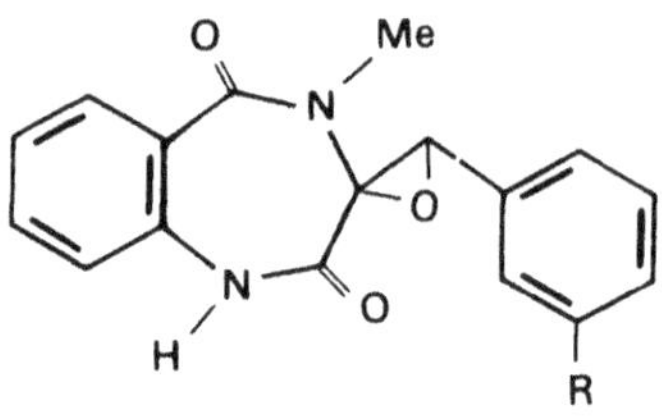

R = H: Cyclopenin (*277*)
R = OH: Cyclopenol (*278*)
P. viridicatum, *P. aurantiogriseum* (Mohammed and Luckner, 1963), *P. crustosum* (Laws, 1985)

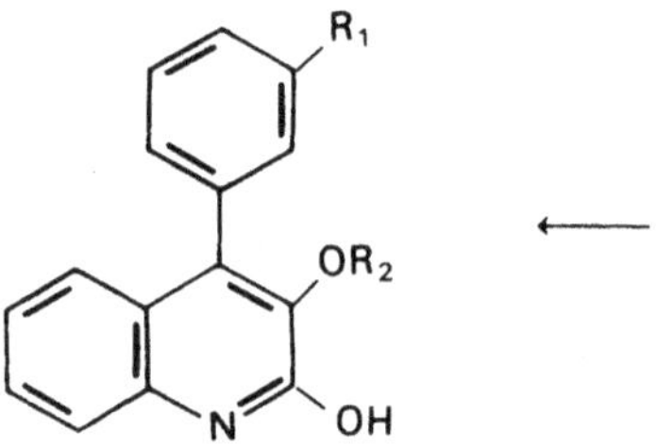

$R_1 = R_2 = H$: Viridicatin (*279*)
$R_1 = OH$, $R_2 = H$: Viridicatol (*280*)
P. viridicatum, *P. aurantiogriseum* (Bracken *et al.*, 1954; Cunningham and Freeman, 1953; Birkinshaw *et al.*, 1963a; Luckner, 1967); *P. puberulum* (Austin and Meyers, 1964)
$R_1 = H$, $R_2 = Me$: 3-*O*-Methyl viridicatin (*281*)
P. viridicatum, *P. aurantiogriseum* (Birkinshaw *et al.*, 1963a)

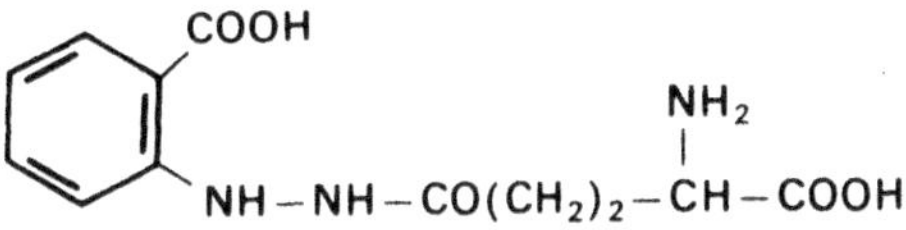

Anthglutin (*282*)
P. oxalicum (Minato, 1979)

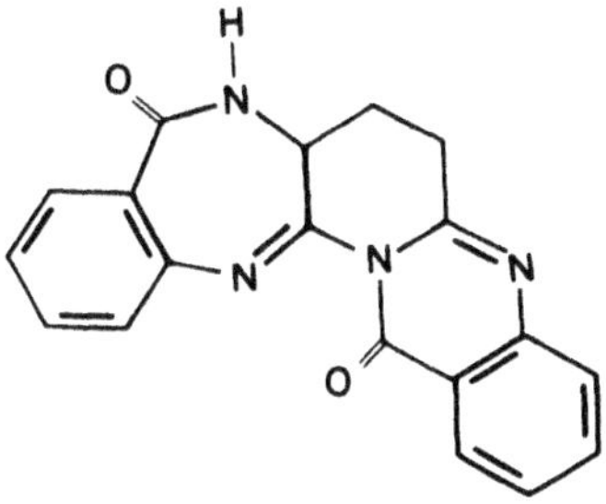

Auranthine (*283*)
P. aurantiogriseum (Yeulet *et al.*, 1986)

2.4.3. Single Aliphatic Amino Acids

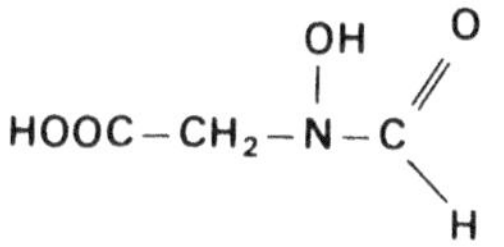

Hadacidin (from Gly) (*284*)
P. glabrum, *P. purpurescens* (Gray *et al.*, 1964); *P. thomii* (Stevens and Emery, 1966; Kaczka *et al.*, 1962)

$O_2N-CH_2-CH_2-COOH$

β-Nitropropionic acid (from Asp) (*285*)
P. melinii (Raistrick and Stossl, 1958)

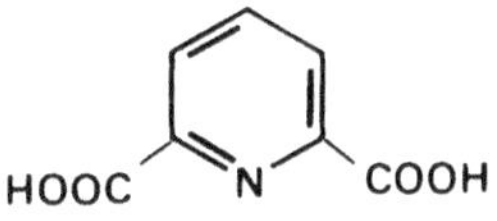

Dipicolinic acid (from Asp) (*286*)
P. citreonigrum (Ooyama *et al.*, 1960; Kalle and Deo, 1983)

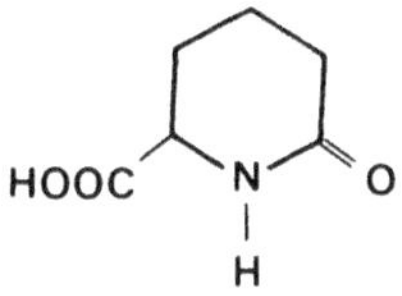

6-Oxo-piperidine-2-carboxylic acid (from Lys) (*287*)
P. chrysogenum (Brundidge *et al.*, 1980)

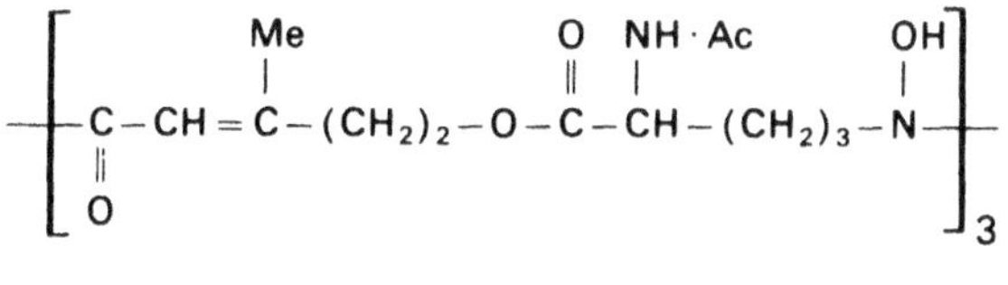

Triacetylfusarinine (from Orn) (*288*)
Penicillium sp. (Hossain *et al.*, 1980; Moore and Emery, 1976)

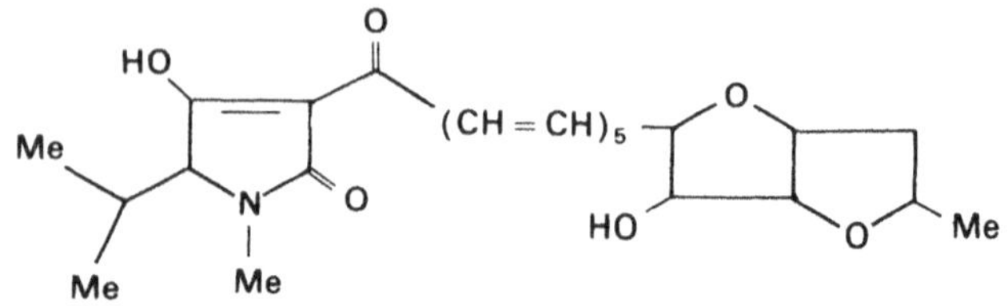

Erythroskyrin (from Val) (*289*)
P. islandicum (Shibata *et al.*, 1966)

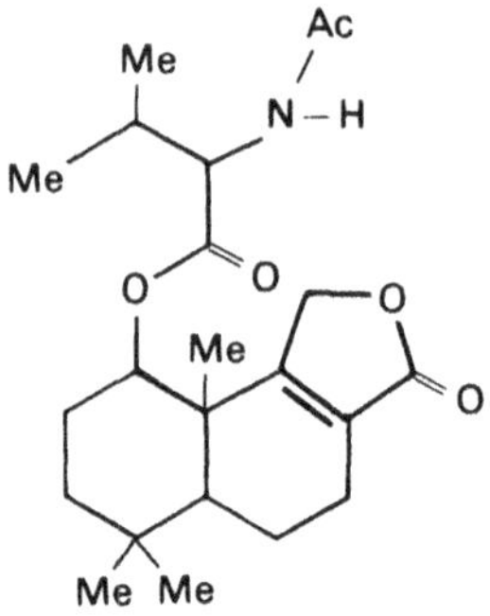

Purpuride (from Val) (*290*)
P. purpurogenum (T. J. King *et al.*, 1973)

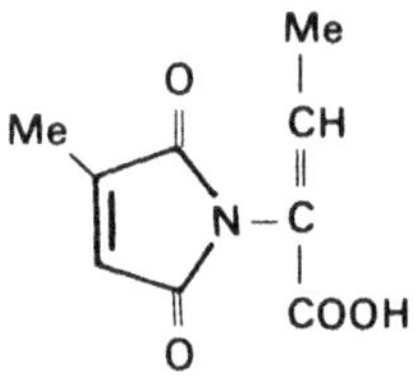

Pencolide (from Thr) (*291*)
P. sclerotiorum (Birkinshaw *et al.*, 1963b)

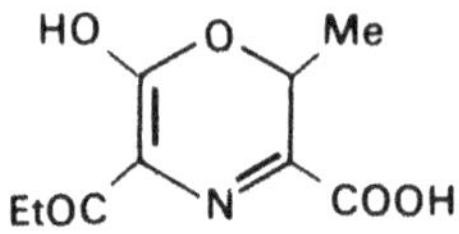

Oxathine carboxylic acid (from Gly) (*292*)
P. thomii (Japanese patent, 1984)

2.4.4. Single Aromatic Amino Acids

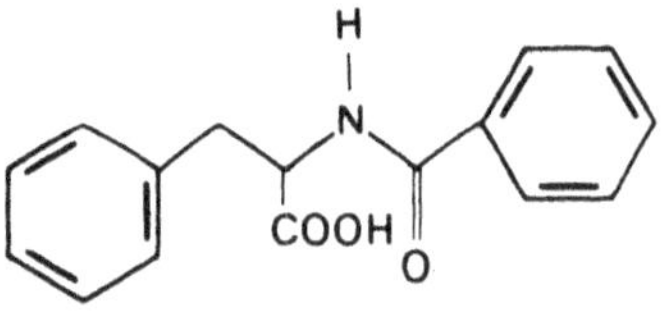

N-Benzoylphenylalanine (from Phe) (*293*)
P. brevicompactum (Doerfler *et al.*, 1981)

(from Tyr)

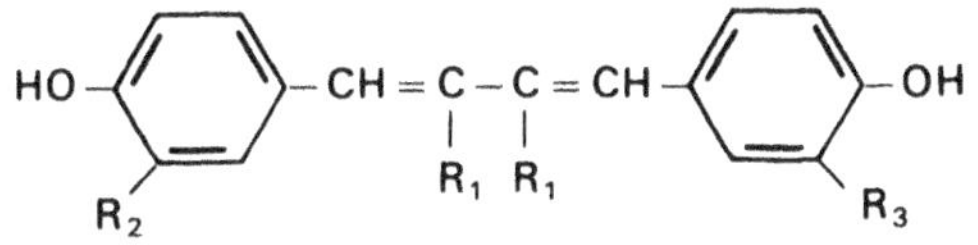

$R_1 = -\overset{\oplus}{N} \equiv \overset{\ominus}{C}$, $R_2 = R_3 = H$: Xanthocillin X (*294*)
$R_1 = -C \equiv N$, $R_2 = OH$, $R_3 = H$: Xanthocillin Y_1 (*295*)
$R_1 = -C \equiv N$, $R_2 = R_3 = OH$: Xanthocillin Y_2 (*296*)
P. chrysogenum (Achenbach *et al.*, 1972; Hagedorn *et al.*, 1960), *P. nilense* (Vesonder, 1979)
Substitution of –NC group with –NH · CHO (*297*)
P. chrysogenum (mutant) (Pfeifer *et al.*, 1972)

(from Trp)

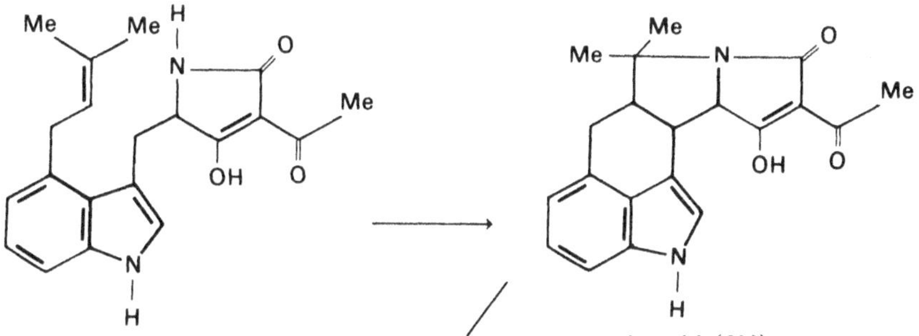

β-Cyclopiazonic acid (*298*)
P. aurantiogriseum (Holzapfel *et al.*, 1970)

Cyclopiazonic acid (*299*)
P. aurantiogriseum, *P. verrucosum*, *P. griseofulvum*, *P. viridicatum*, *P. puberulum*, *P. crustosum* (Holzapfel, 1968; Leistner and Pitt, 1977; De Jesus *et al.*, 1981; Hermansen *et al.*, 1984)

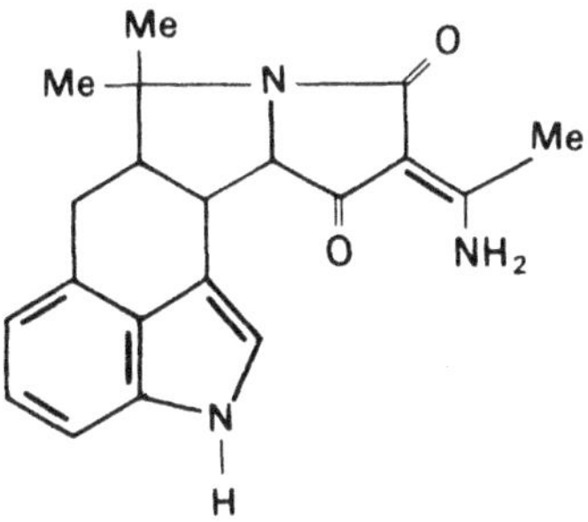

Cyclopiazonic acid imine (*300*)
P. aurantiogriseum (Holzapfel *et al.*, 1970)

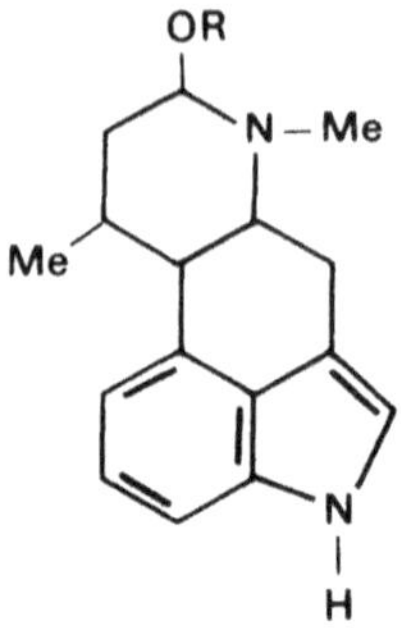

R = Ac: Roquefortine A (*301*)
R = H: Roquefortine B (*302*)
P. roquefortii (Ohmomo *et al.*, 1975), *P. crustosum* (Cole *et al.*, 1983)

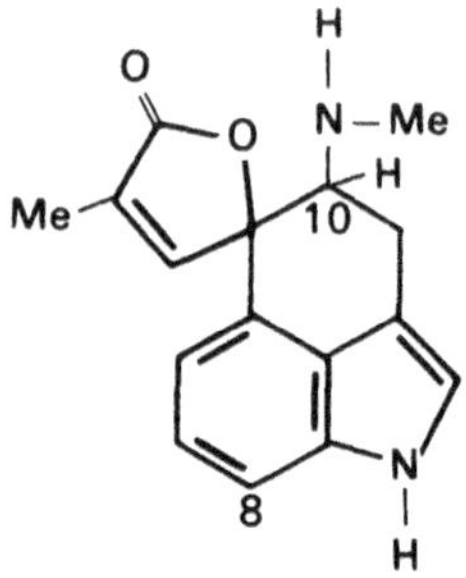

Rugulovasin(s) (derived by permutations of isomerization at position 10 and Cl-substitution at position 8) (*303*)
P. rugulosum (Yamatodani *et al.*, 1970; Ohmomo *et al.*, 1977); *P. islandicum* (Cole *et al.*, 1967a,b); *P. purpurogenum*, *P. camembertii* (Dorner *et al.*, 1980)

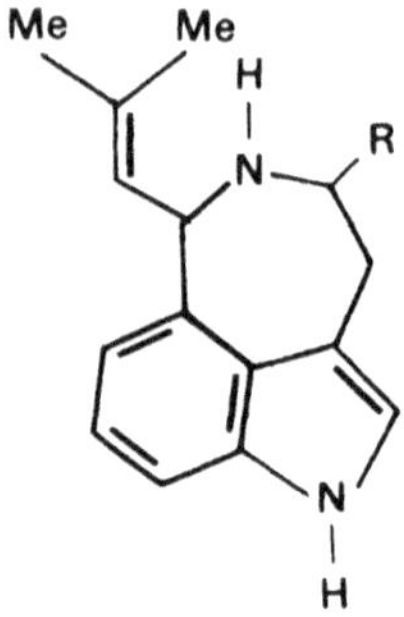

R = H: Aurantioclavine (*304*)
R = COOH: Clavicipitic acid (*305*)
P. aurantiogriseum (Kozlovskii *et al.*, 1981b; G. S. King *et al.*, 1977)

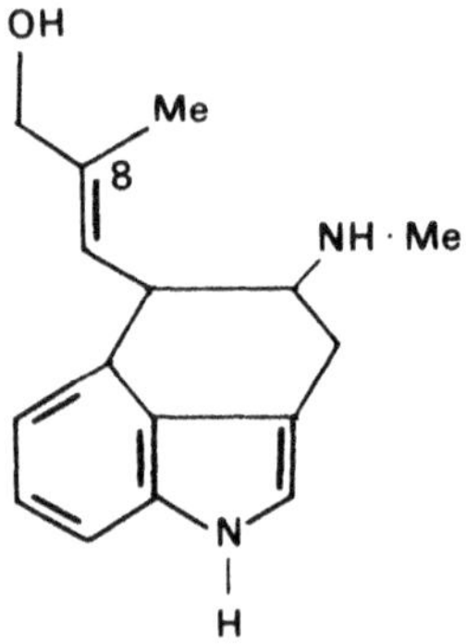

Chanoclavine (*306*)
Isochanoclavine (*307*) (epimeric at C8)
P. rugulosum (Abe *et al.*, 1969)

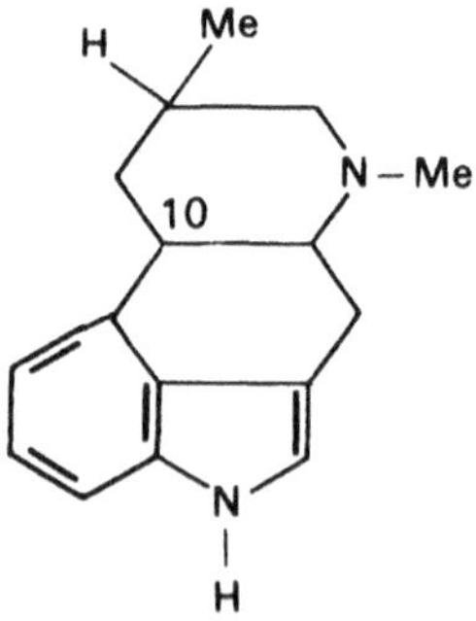

Costaclavine (*308*)
Epicostaclavine (*309*) (epimeric at C10)

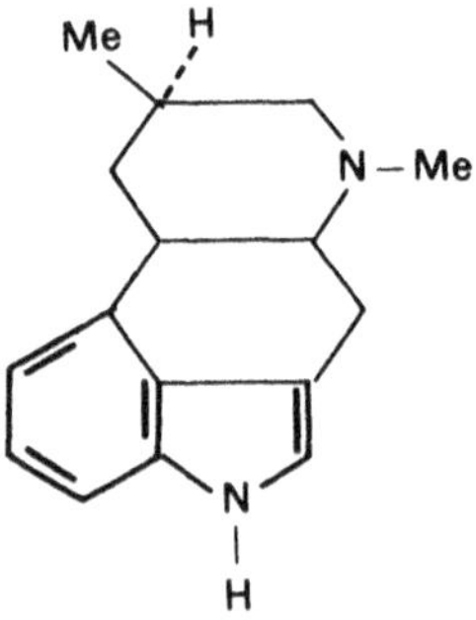

Festuclavine (*310*)
P. crustosum (Cole *et al.*, 1983)

P. citrinum (Kozlovskii *et al.*, 1981c)

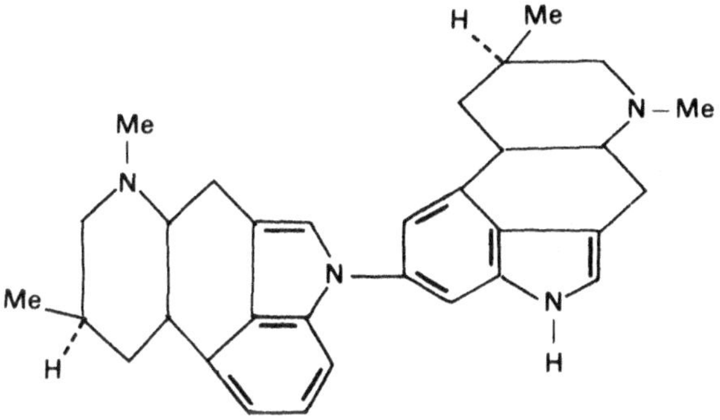

Cividiclavine (dimer) (*311*)
Pyroclavine (monomer) (*312*)
P. citrinum (Vining *et al.*, 1982)

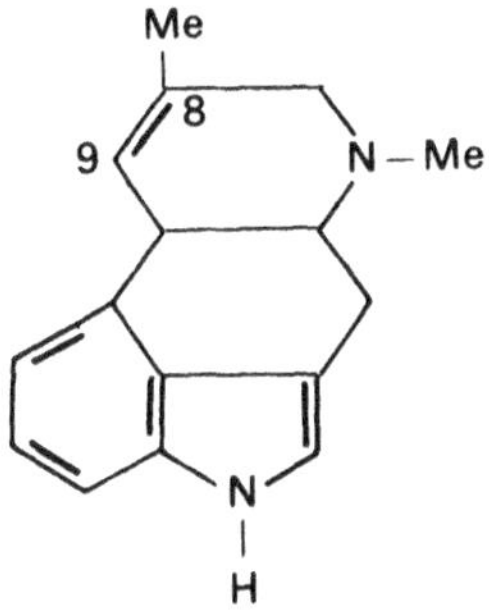

Agroclavine (*313*)
P. corylophilum (Kozlovskii *et al.*, 1982), *P. roquefortii* (Abe *et al.*, 1967)
8,9-Epoxyagroclavine (*314*)
P. corylophilum (Kozlovskii *et al.*, 1982)

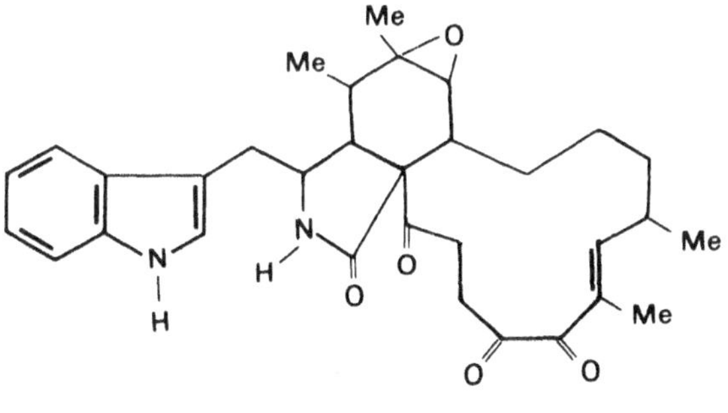

Chaetoglobosin C (*315*)
P. aurantiogriseum (Probst and Tamm, 1981)

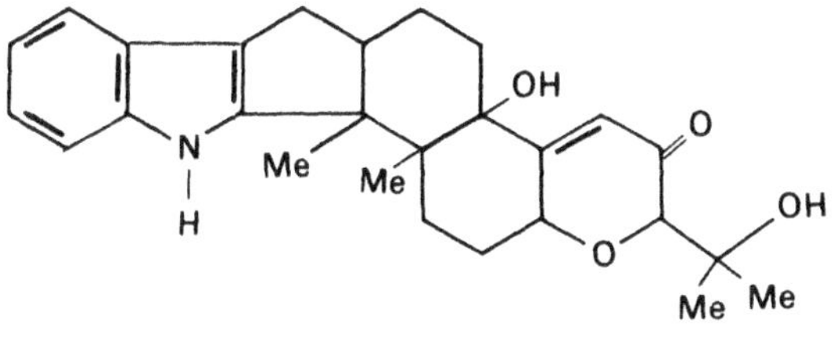

Paxilline (*316*)
P. paxilli (Cole *et al.*, 1974; Springer *et al.*, 1975)

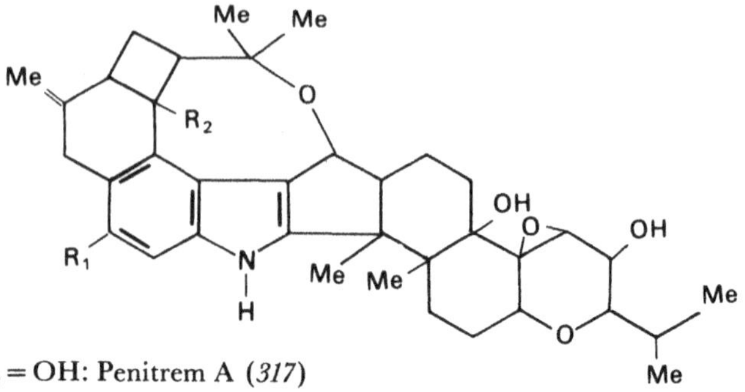

$R_1 = Cl, R_2 = OH$: Penitrem A (*317*)
$R_1 = H, R_2 = H$: Penitrem B (*318*)
$R_1 = Cl, R_2 = H$, deoxy (epoxide, with concomitant Δ): Penitrem C (*319*)
$R_1 = H, R_2 = H$, deoxy (epoxide, with concomitant Δ): Penitrem D (*320*)
$R_1 = H, R_2 = OH$: Penitrem E (*321*)
$R_1 = Cl, R_2 = H$: Penitrem F (*322*)

P. crustosum (Wilson *et al.*, 1968; De Jesus *et al.*, 1983a–c; Pitt, 1979b; Mantle *et al.*, 1983); *P. granulatum*, *P. viridicatum*, *P. aurantiogriseum* (Ciegler and Pitt, 1970); *P. puberulum* (Wagener *et al.*, 1980); *P. duclauxii* (Patterson *et al.*, 1979); *P. janczewskii* (Mantle *et al.*, 1978)

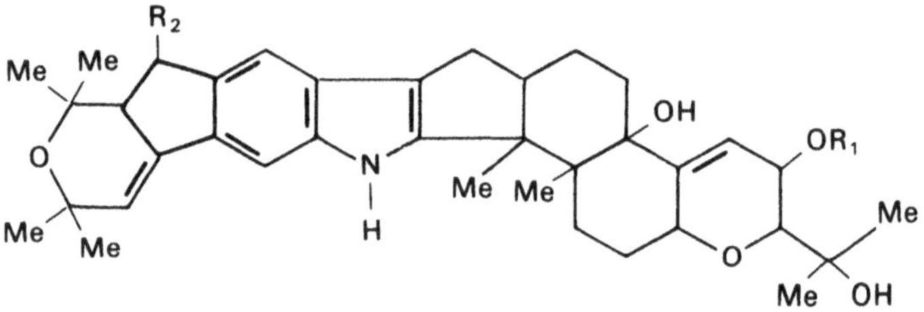

$R_1 = H, R_2 = OH$: Janthitrem E (*323*)
$R_1 = Ac, R_2 = OH$: Janthitrem F (*324*)
$R_1 = Ac, R_2 = H$: Janthitrem G (*325*)

P. janthinellum (De Jesus *et al.*, 1984)

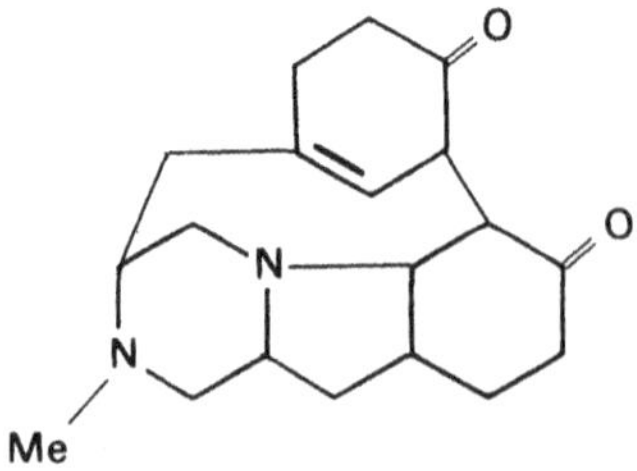

Herquline (*326*)
(possibly from aromatic amino acid and intramolecular rearrangement)
P. herquei (Furusaki *et al.*, 1980)

2.4.5. Diketopiperazines

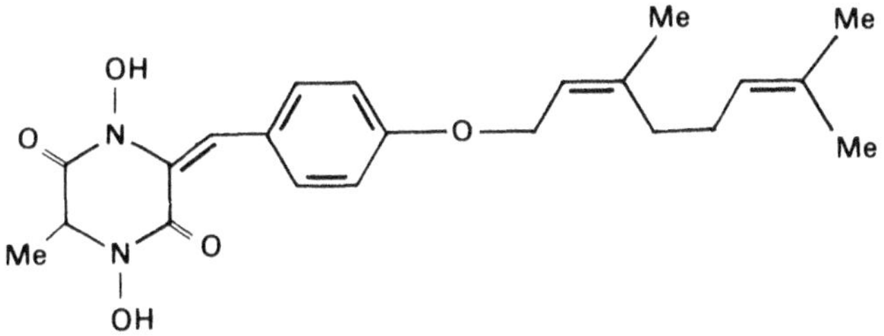

Mycelianamide (Tyr + Ala) (*327*)
P. griseofulvum (Oxford and Raistrick, 1948; Birch *et al.*, 1962)

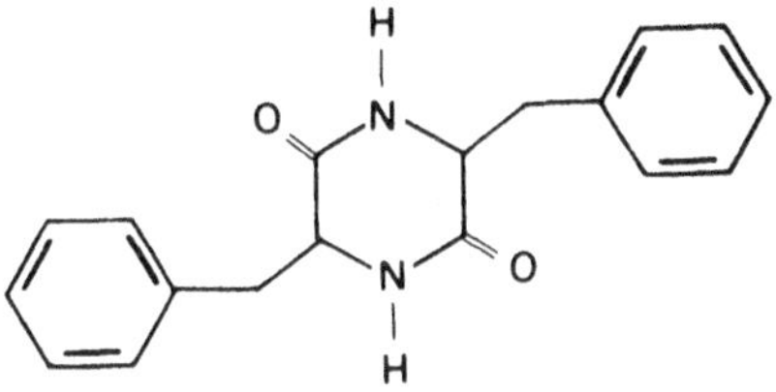

Phenylalanine anhydride (Phe + Phe) (*328*)
P. janczewskii (Birkinshaw and Mohammed, 1962)

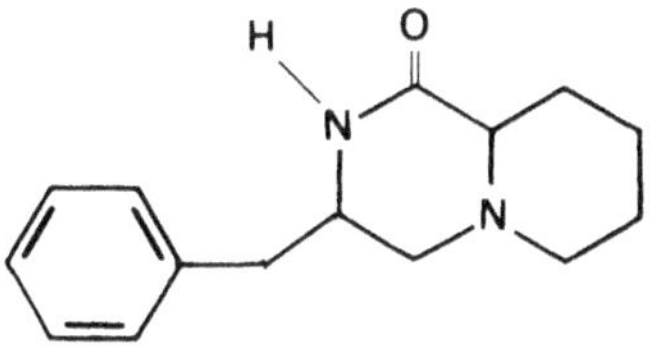

Verruculotoxin (Phe + ?, reduced) (*329*)
P. verruculosum (MacMillan *et al.*, 1976)

(Phe + Ser)

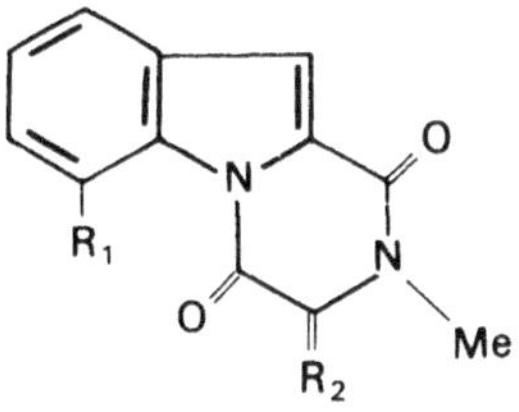

R_1 = OH, R_2 = O: Dioxopyrazinoindole-A (*336*)
R_1 = OH, R_2 = CH_2: Dioxopyrazinoindole-B (*337*)
R_1 = H, R_2 = O: Dioxopyrazinoindole-C (*338*)
P. spinulosum (Ali *et al.*, 1968)

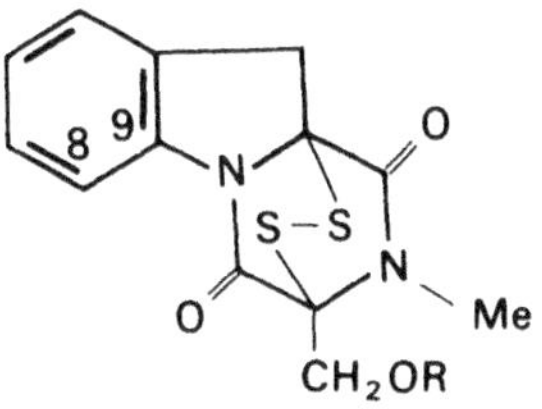

R = H: Dehydrogliotoxin (*330*)
R = H, 8,9-dihydro: Gliotoxin (*331*)
R = Ac, 8,9-dihydro: Gliotoxin acetate (*332*)
P. fellutanum (Bracken and Raistrick, 1947), *P. spinulosum* (Johnson *et al.*, 1953; Lowe *et al.*, 1966; Bose *et al.*, 1968), *P. corylophilum* (Mull *et al.*, 1945)

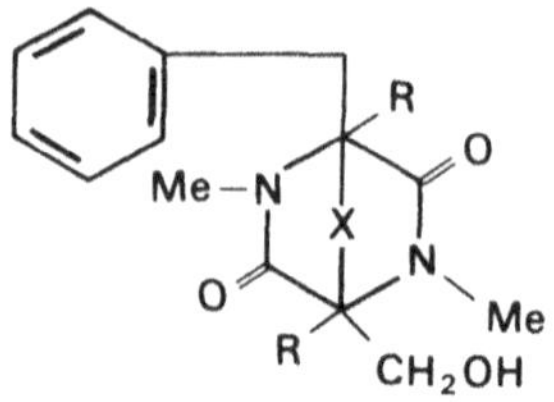

R = nil, X = -S-S-: Hyalodendrin (*333*)
R = S-Me, X = nil: Gliovictin (*334*)
R = nil, X = -S-S-S-S-: Tetrathiahyalodendrin (*335*)
P. turbatum (Michel *et al.*, 1974)

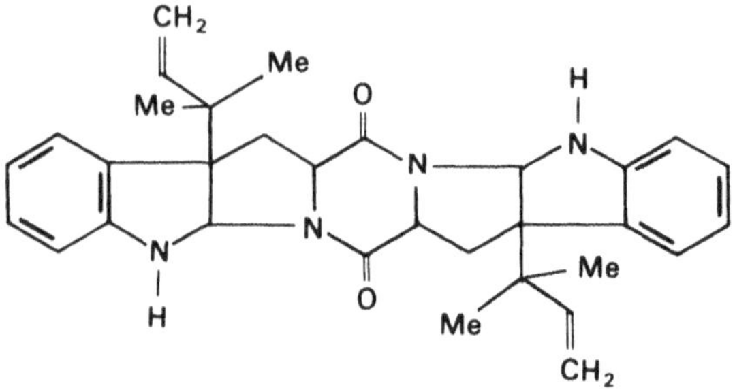

Nigrifortine (Trp + Trp) (*339*)
P. janczewskii (Laws and Mantle, 1985)

(Trp + Pro)

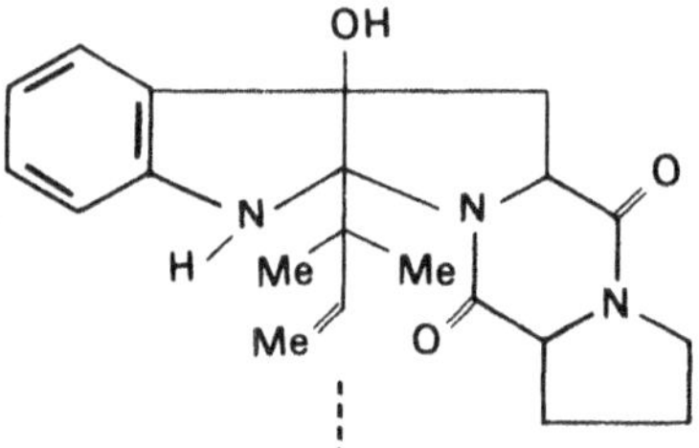

Brevianamide E (*340*)
P. brevicompactum (Birch and Wright, 1969)

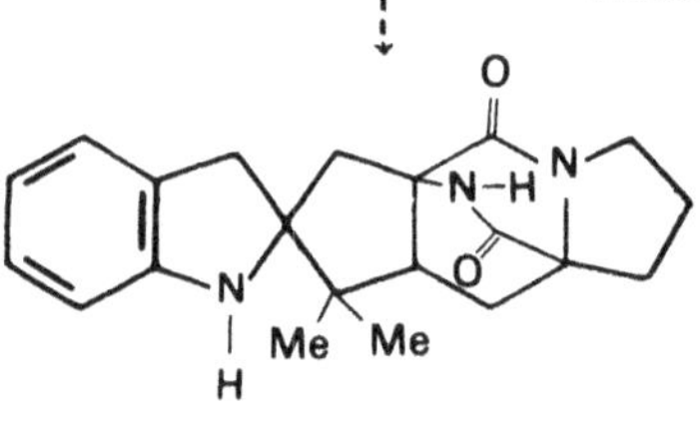

Brevianamide A (B = isomer) [*341* (*342*)]
P. olivicolor (Robbers *et al.*, 1975), *P. brevicompactum* (Birch and Wright, 1969; Bird and Campbell, 1982b), *P. viridicatum* (Wilson *et al.*, 1973)
(Brevianamides C and D are photolytic artifacts of A and B.)

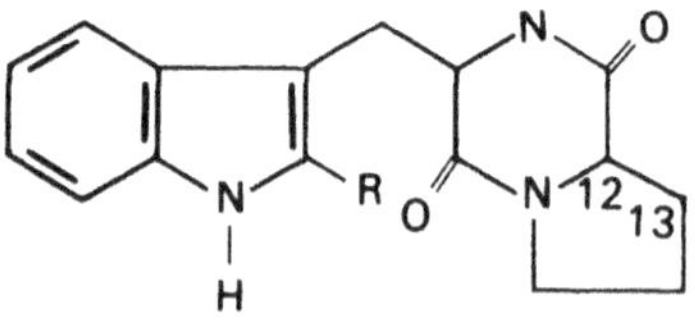

R = H: Tryptophanylproline anhydride (*343*)
P. brevicompactum (Birch and Russell, 1972)

R = 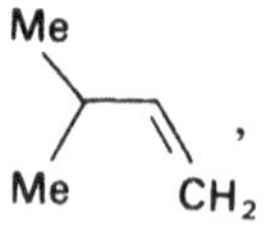, Δ12:

12,13-Dehydrodeoxybrevianamide E (*344*)

R = 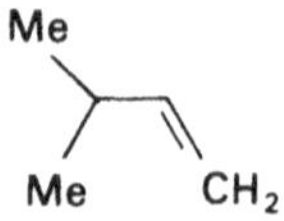

Deoxybrevianamide E (*345*)

P. italicum (Scott *et al.*, 1974)

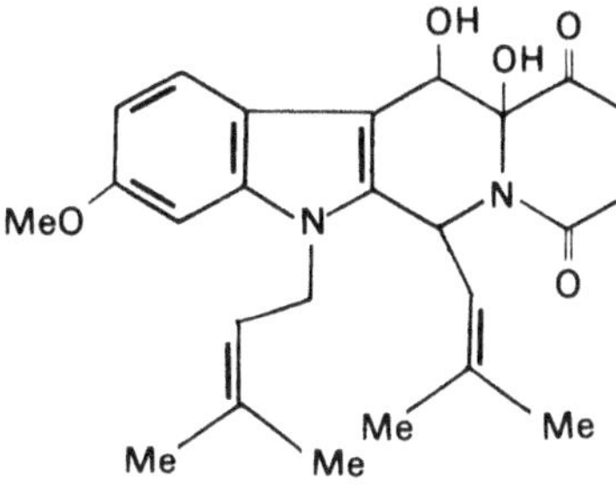

Fumitremorgin B (*346*)
P. puberulum (Dix *et al.*, 1972), *P. simplicissimum* (Gallagher and Latch, 1977)

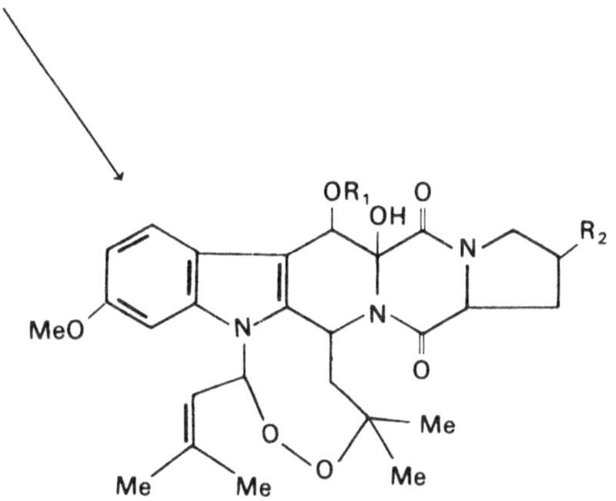

$R_1 = R_2 = H$: Verruculogen (*347*)
P. simplicissimum (Cole *et al.*, 1972; Fayos *et al.*, 1974; Pitt, 1979b; Day and Mantle, 1982; Willingale *et al.*, 1983b); *P. janthinellum* (Lanigan *et al.*, 1979); *P. piceum*, *P. janczewskii*, *P. raistrickii* (Patterson *et al.*, 1981); *P. paxilli* (Cockrum *et al.*, 1979)

$R_1 =$ 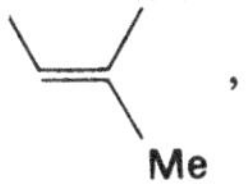

,

$R_2 = H$: Fumitremorgin A (*348*)
P. janthinellum (Lanigan *et al.*, 1979)

$R_1 = H$, $R_2 = OAc$: Acetoxyverruculogen (*349*)
P. verruculosum (Uramoto *et al.*, 1982)

(Trp + His)

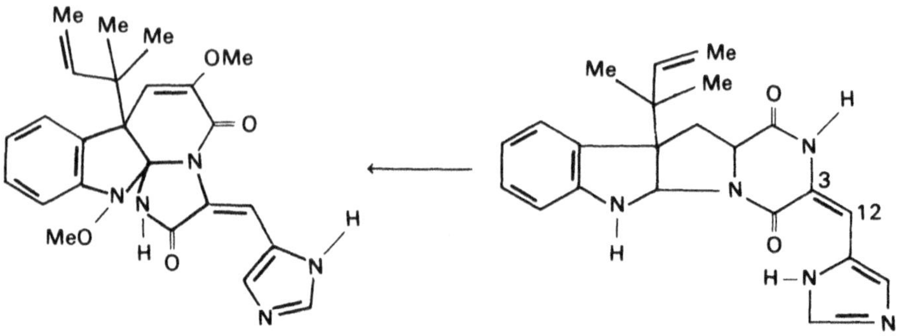

Oxaline (*350*)
P. oxalicum (Nagel *et al.*, 1976; Vleggaar and Wessels, 1980; Steyn and Vleggaar, 1983)

Roquefortine (roquefortine C) (*351*)
P. roquefortii (Scott *et al.*, 1976; Barrow *et al.*, 1979; Kozlovskii and Reshetilova, 1984; Gorst-Allman *et al.*, 1982), *P. puberulum* (Wagener *et al.*, 1980), *P. hirsutum* (Ohmomo *et al.*, 1980), *P. crustosum* (Kozlovskii *et al.*, 1981a; De Jesus *et al.*, 1983c; Mantle *et al.*, 1983; Laws, 1985), *P. verrucosum* (Kozlovskii and Reshetilova, 1984), *P. aurantiogriseum* (Vesonder *et al.*, 1980)
Also: 3,12-Dihydroroquefortine (*352*)
P. roquefortii (Ohmomo *et al.*, 1978), *P. crustosum* (Kozlovskii *et al.*, 1981a)

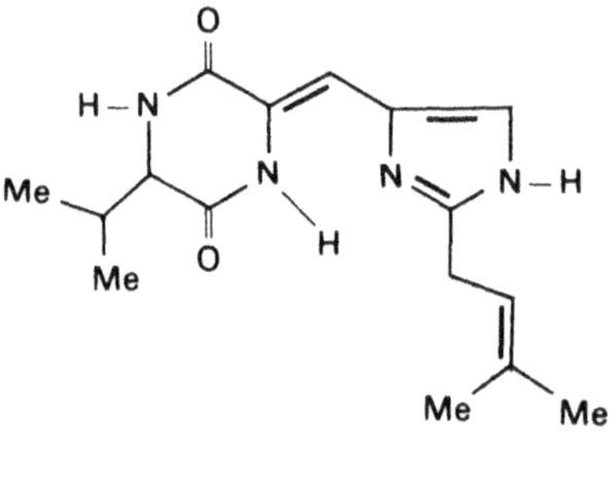

Viridamine (His + Val) (*353*)
P. viridicatum (Holzapfel and Marsh, 1977)

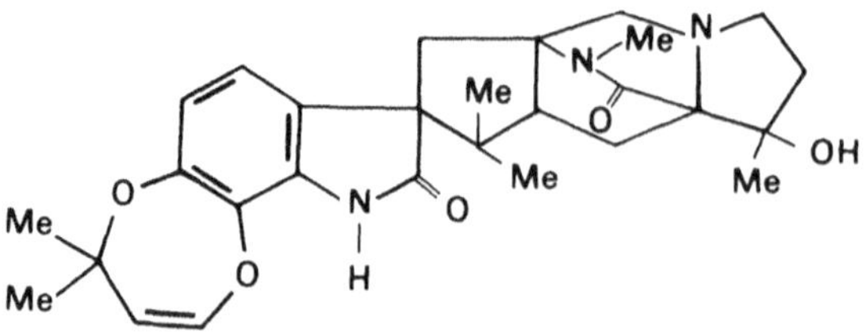

Paraherquamide (Trp + Pro, reduced) (*354*)
P. simplicissimum (Yamazaki *et al.*, 1981)

(Trp + ?Lys, reduced)

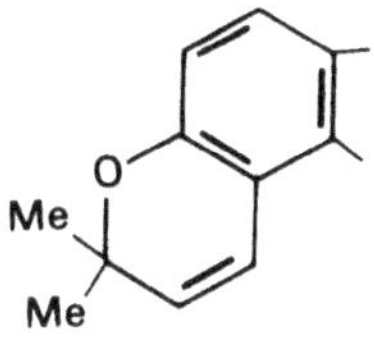

Marcfortine C (*355*)

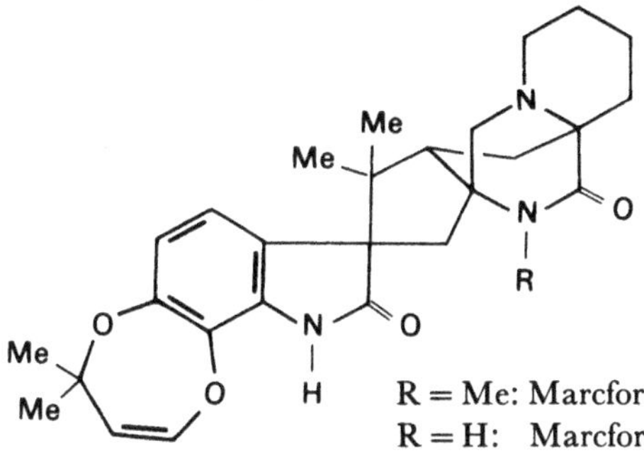

R = Me: Marcfortine A (*356*)
R = H: Marcfortine B (*357*)

P. roquefortii (Polonsky *et al.*, 1980; Prange *et al.*, 1981)

2.4.6. Two Amino Acids but without Formation of a Diketopiperazine

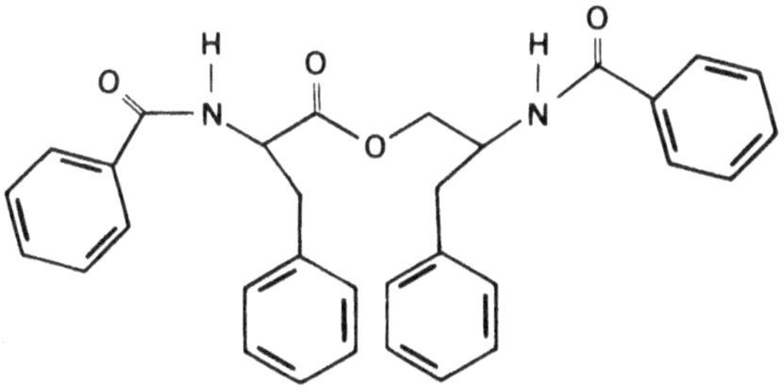

Asperphenamate (Phe + Phenylalaninol) (*358*)

P. arenicola (McCorkindale *et al.*, 1978a), *P. brevicompactum* (Bird and Campbell, 1982a,b)

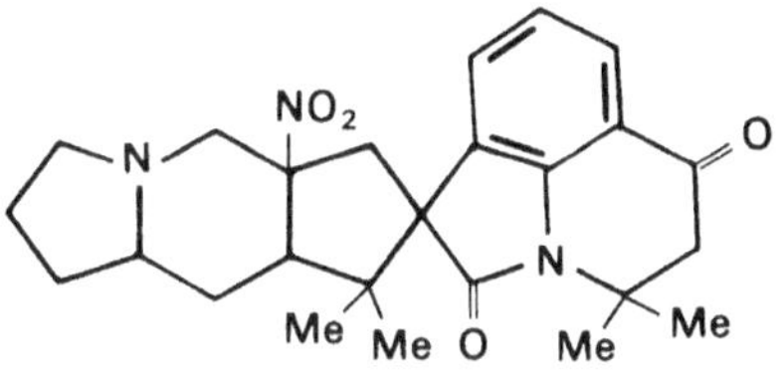

Cyclopiamines A and B (isomers) (Trp + Pro) (*359, 360*)

P. aurantiogriseum, *P. griseofulvum* (Bond *et al.*, 1979)

2.4.7. Several Amino Acids

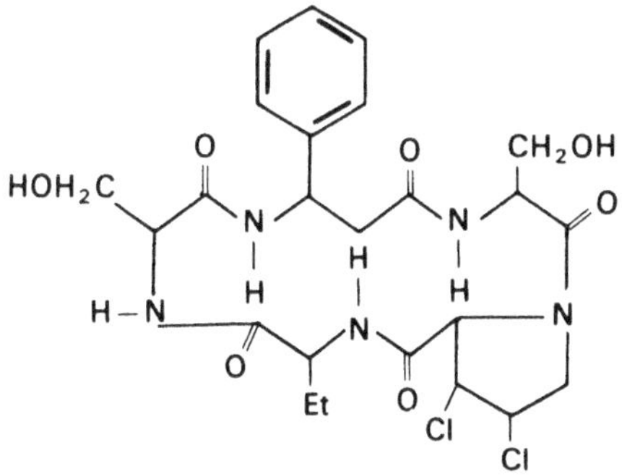

Islanditoxin (cyclochlorotine) (*361*)

P. islandicum (Marumo, 1959; Yoshioka *et al.*, 1973; Ghosh and Ramgopal, 1980)

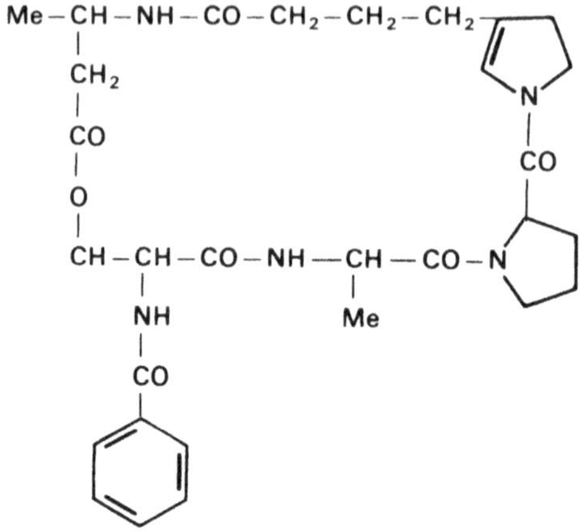

Brevigellin (*362*)
P. brevicompactum (McCorkindale and Baxter, 1981)

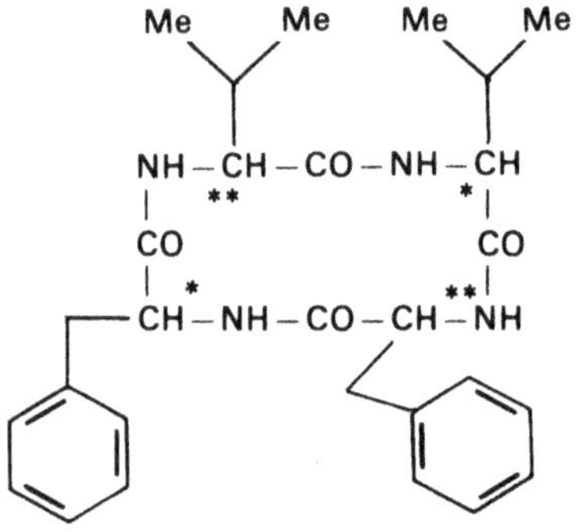

Amino acid configuration:
*L; **D.

Fungisporin (*363*)
P. chrysogenum, *P. claviforme*, *P. funiculosum*, *P. purpurogenum* (Sumiki and Miyao, 1952; Miyao, 1960; Studer, 1969)

2.5. Miscellaneous

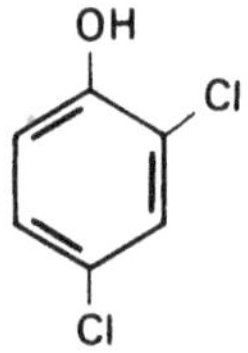

2,4-Dichlorophenol (*364*)
Penicillium sp. (Ando *et al.*, 1970)

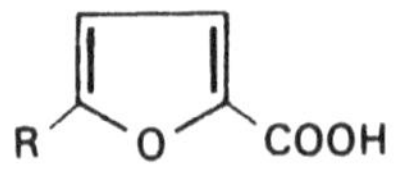

R = H: Furan-2-carboxylic acid (*365*)
P. corylophilum (Turner and Aldridge, 1983)
R = COOH: Furan-2,5-dicarboxylic acid (*366*)
P. sclerotiorum (Turner and Aldridge, 1983)

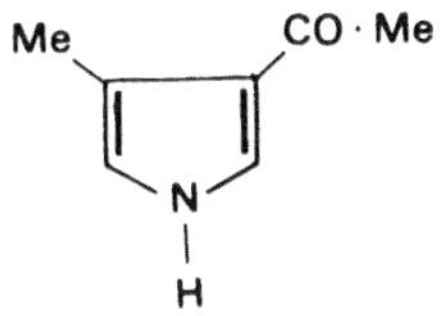

Deoxyverrucarin E (*367*)
P. hirayamae (Arndt *et al.*, 1974; Chexal *et al.*, 1980)

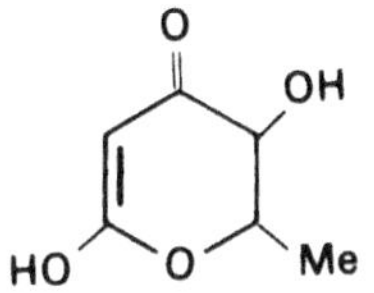

2,3-Dihydro-3,6-dihydroxy-2-methyl-4-pyrone (*368*)
P. restrictum (Raistrick and Rice, 1971)

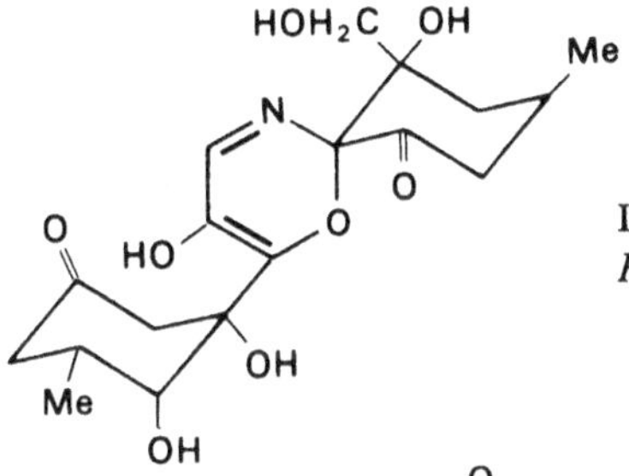

Leucogenol (*369*)
P. restrictum (Rice, 1971)

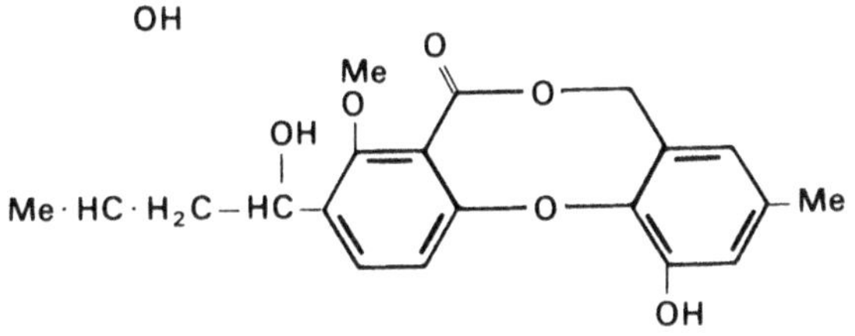

Penicillide (*370*)
Penicillium sp. (Sassa *et al.*, 1974)

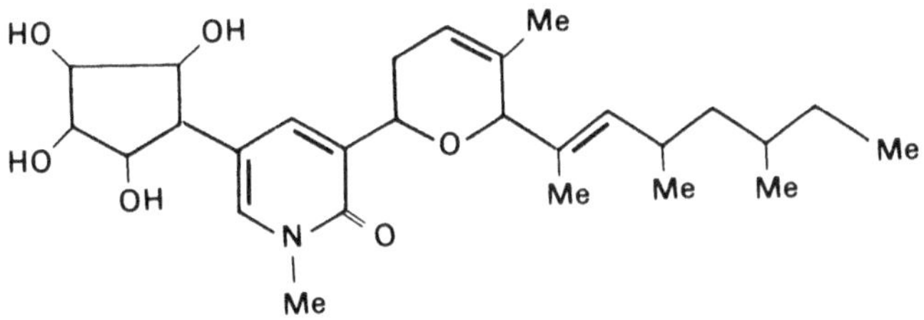

Funiculosin (*371*)
P. funiculosum (Ando *et al.*, 1978)

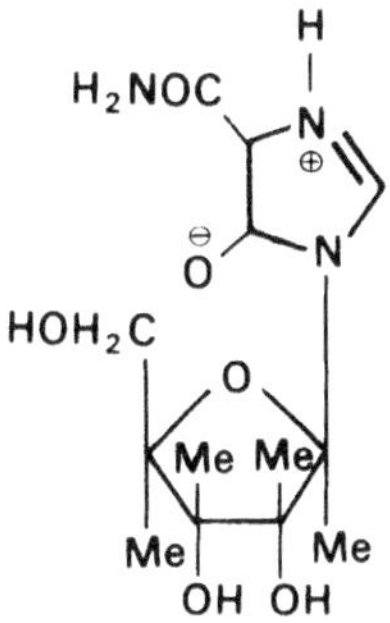

Bredenin (*372*)
P. dodgei (Mizuno *et al.*, 1974)

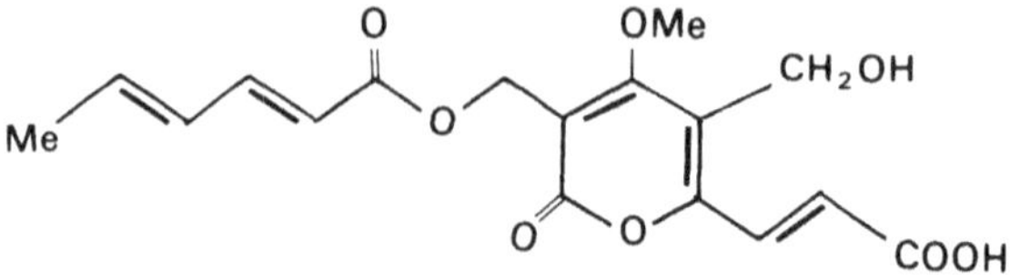

Islandic acid (*373*)
P. islandicum (Fujimoto *et al.*, 1982)

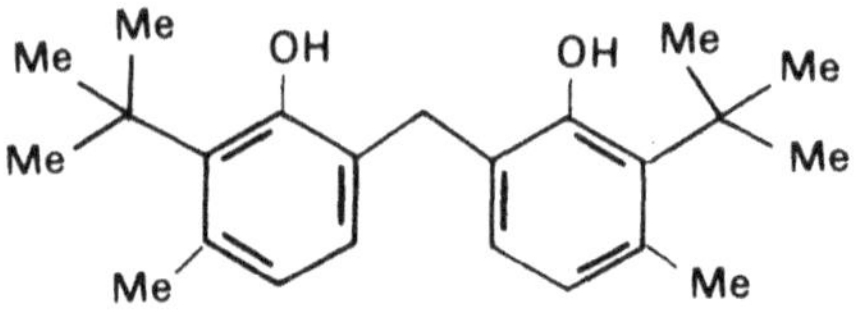

2,2′-Methylenebis(5-methyl-6-*tert*-butylphenol) (*374*)
P. janthinellum (Nakakita *et al.*, 1984b)

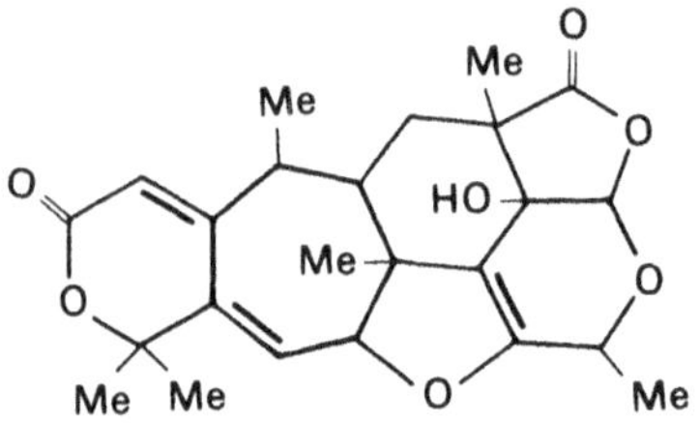

Paraherquonin (*375*)
P. simplicissimum (Okuyama *et al.*, 1983)

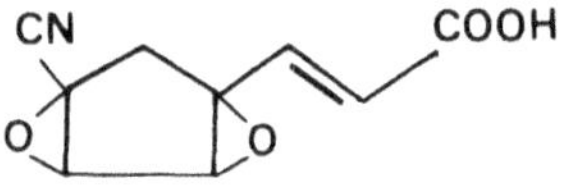

Antibiotic 2188 (*376*)
P. rugulosum (German patent, 1984)

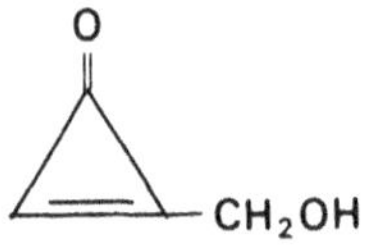

Penitricin A (*377*)

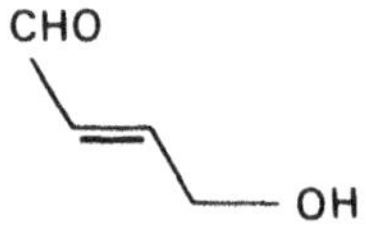

Penitricin B (*378*)

Penitricin C (*379*)

P. aculeatum (Okuda *et al.*, 1984)

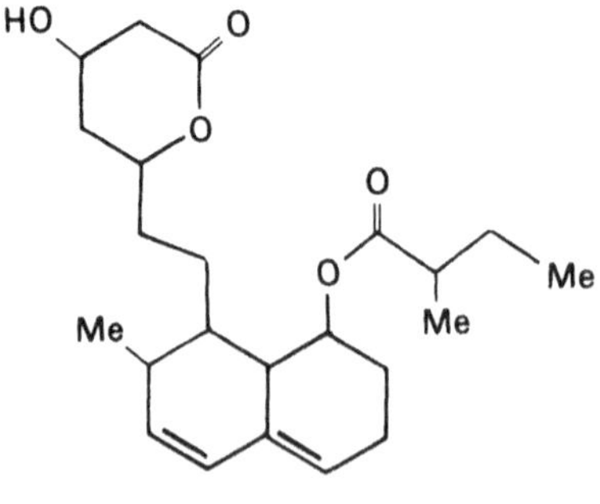

ML-236B (*380*)
Penicillium sp. (Japanese patent, 1985)

3. *ACREMONIUM* METABOLITES

Correct listing of *Acremonium* metabolites is less satisfactory. Very few fungi with secure taxonomic status within the genus have been reported to produce secondary metabolites. *Cephalosporium acremonium* is unique in this respect, since it produces penicillins and cephalosporins though it has more recently been properly included within *Acremonium*. However, other *Cephalosporium* spp., the status within *Acremonium* of which has not been verified and ought not to be assumed to be eligible for inclusion automatically, have been included here as putative members of *Acremonium* and therefore may make the metabolite listing optimistically comprehensive. In notable contrast to the situation regarding the *Penicillia*, until very recently the literature contained no reference to amino-acid-derived metabolites except for the penicillins and cephalosporins. Since this chapter was prepared, Rowan *et al.* (1986) have characterized peramine (49), an insect antifeedant, from *A. loliae*, which is probably derived from proline and arginine and is essentially a modified diketopiperazine. The only other compound that contains nitrogen is cerulenin (*4*), the carbon skeleton of which seems to be hexaketide, and therefore the amino group is presumably added by transamination, possibly from glutamate or glutamine.

3.1. Polyketides

3.1.1. Pentaketides

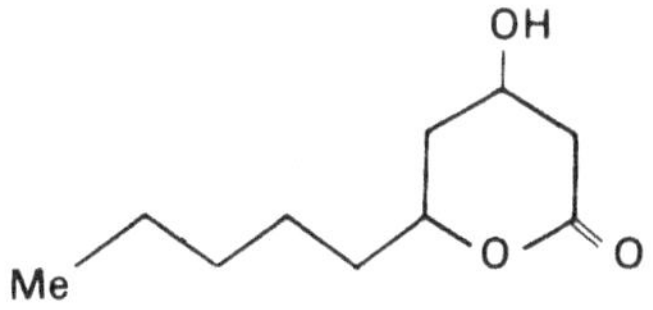

3,5-Dihydroxydecanoic acid δ-lactone (*1*)
A. recifei (Vesonder *et al.*, 1972)

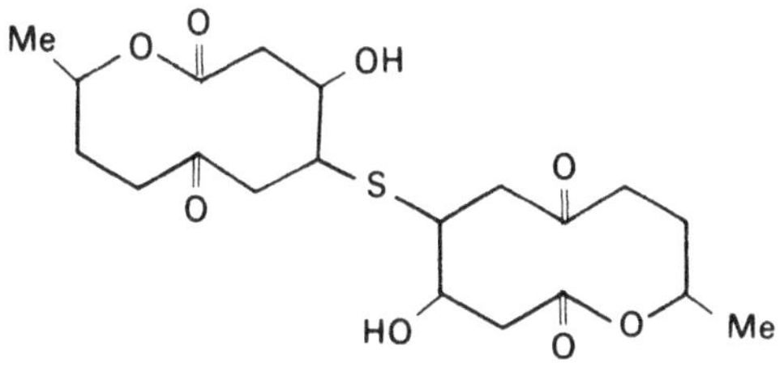

Thiobiscephalosporide A (*2*)
Cephalosporium aphidicola (Mabelis *et al.*, 1981)

3.1.2. Hexaketides

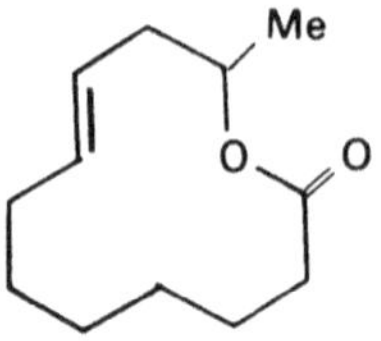

Recifeiolide (*3*)
A. recifei (Vesonder *et al.*, 1971)

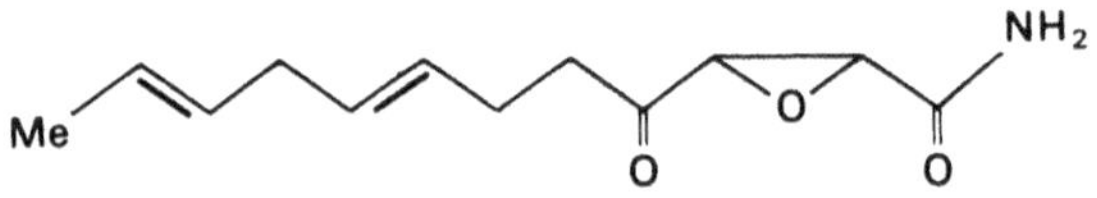

Cerulenin (*4*)
Cephalosporium caerulens (Omura *et al.*, 1969; Awaya *et al.*, 1975)

3.1.3. Heptaketide

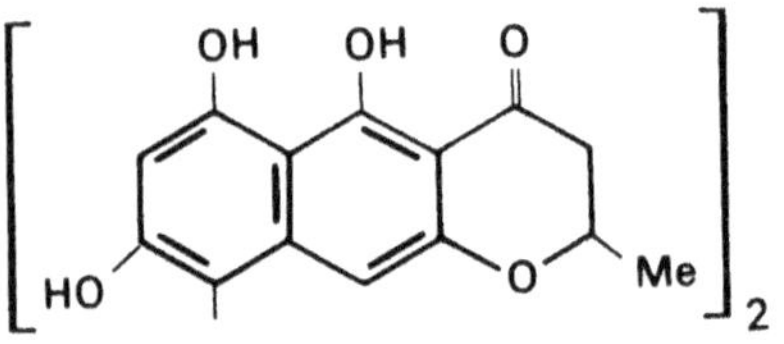

Cephalochromin (*5*)
Acremonium sp. (Tertzakian *et al.*, 1964)

3.2. Isoprenoids

3.2.1. Sesquiterpenes

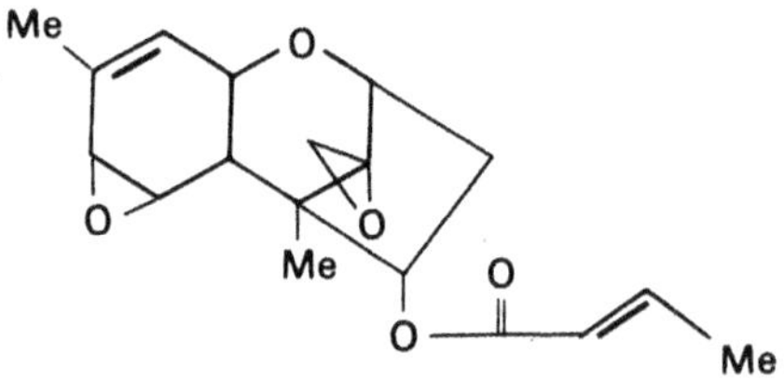

Crotocin (*6*)
A. crotocinigenum (Gyimesi and Melera, 1967)

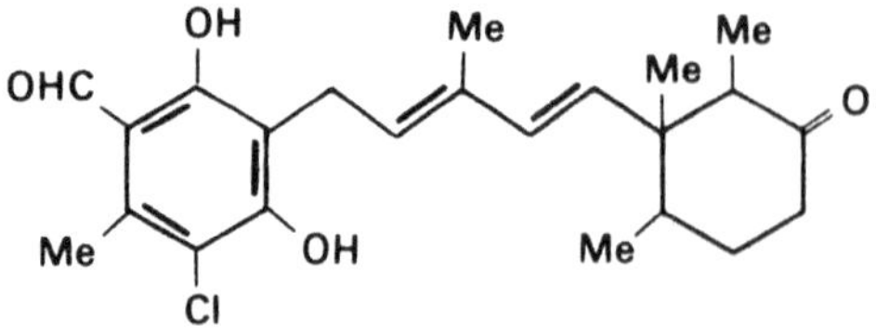

Ascochlorin (*7*)
A. luzulae (Cagnoli-Bellavita *et al.*, 1975)

3.2.2. Diterpenes

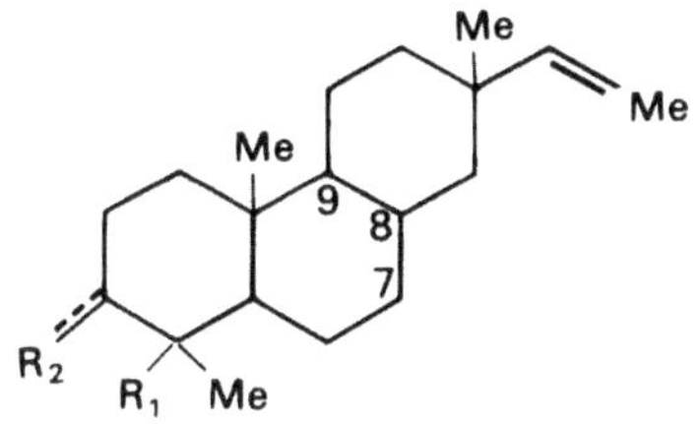

R_1	R_2	
Δ7, CH_2OH	H	Isopimaradienol (*8*) *A. luzulae* (Cagnoli *et al.*, 1982)
Δ7, H	O	19-Norisopimara-7,15-dien-3-one (*9*) *A. luzulae* (Cagnoli *et al.*, 1980)
Δ8, CH_3	H	Pimara-8(9),15-diene (*10*) *Cephalosporium aphidicola* (Adams and Bu'lock, 1975)

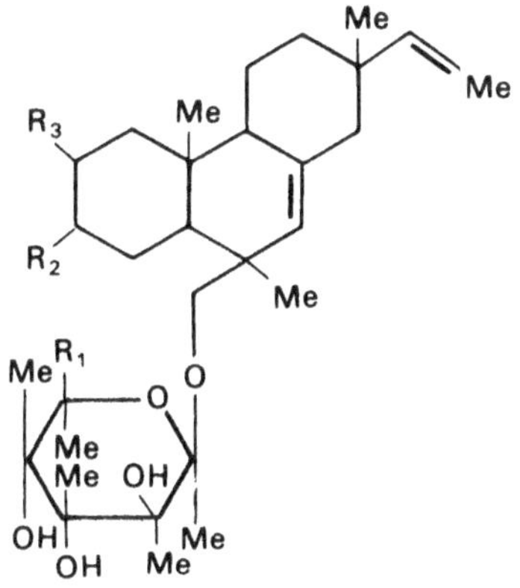

R_{1-3}: Variously H, OH, O, COOH, CH_2OH, CHO, or:

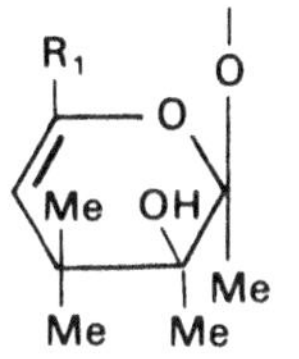

Virescenosides A–L (*11–22*)
A. luzulae (Cagnoli-Bellavita *et al.*, 1977)

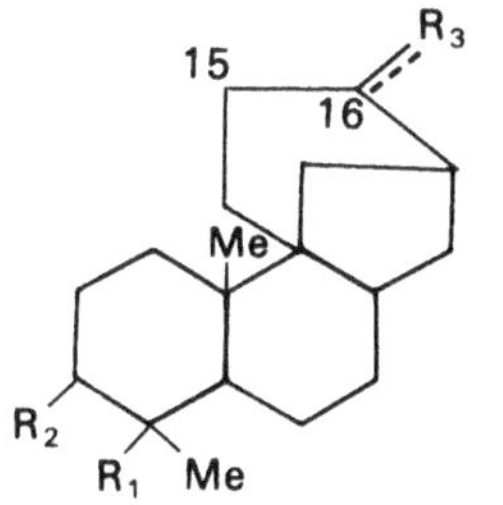

R_1	R_2	R_3	
OH, CH_2OH	OH	CH_2OH	Aphidicolin (*23*)
CH_2	H	Me	Aphidicol-16-ene (*24*)
Me	H	Me	Aphidicol-15-ene (*25*) *Cephalosporium aphidicola* (Dalziel *et al.*, 1973; Ackland *et al.*, 1982)

3.2.3. Sesterterpene

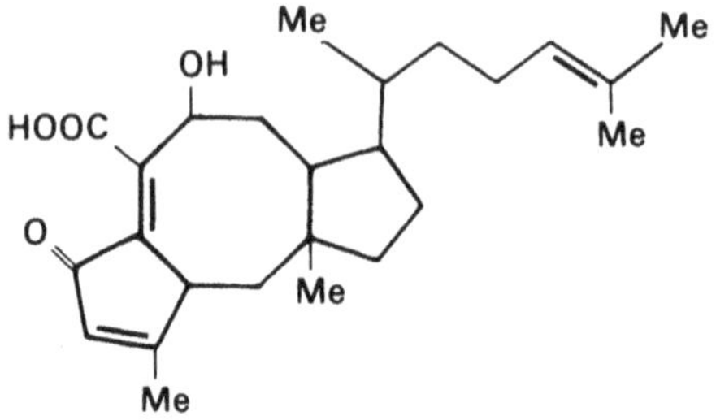

Ophiobolin D (*26*)
Cephalosporium caerulens (Nozoe *et al.*, 1965)

3.2.4. Triterpenes

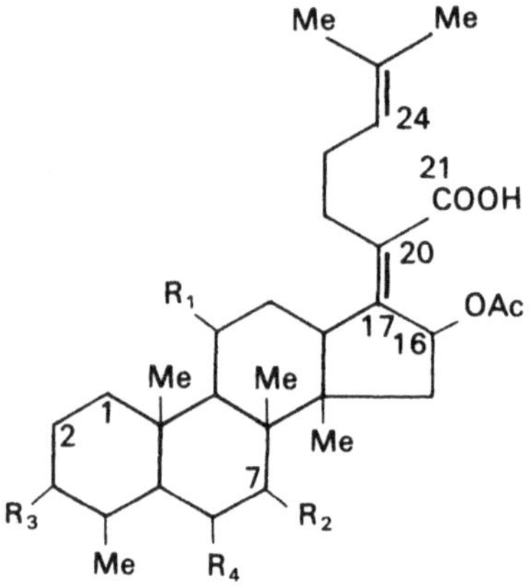

R_1	R_2	R_3	R_4	
OH	H	OH	H	Fusidic acid (*27*) *Acremonium* sp. (Belgian patent, 1962)
H	OH	OH	OAc	Cephalosporin P_1 (*28*) *Acremonium* sp. (Burton and Abraham, 1951)
H	OH	OH	OH	Deacetylcephalosporin P_1 (*29*) *A. strictum* (Chou *et al.*, 1969)
Δ1 H	OAc	O	O	Helvolic acid* (*30*) (Okuda *et al.*, 1964)
Δ1 H	H	O	O	7-Deacetoxyhelvolic acid* (*31*) (Okuda *et al.*, 1966)
Δ1 H	OH	O	O	Helvolinic acid* (*32*) (Okuda *et al.*, 1964)
Δ1 H	H	O	H	3-Oxo-11β-acetoxy fusida-1,17(20)[16,21-cis], 24-trien-21-oic acid* (*33*) (Okuda *et al.*, 1968b)

* *Cephalosporium caerulens*

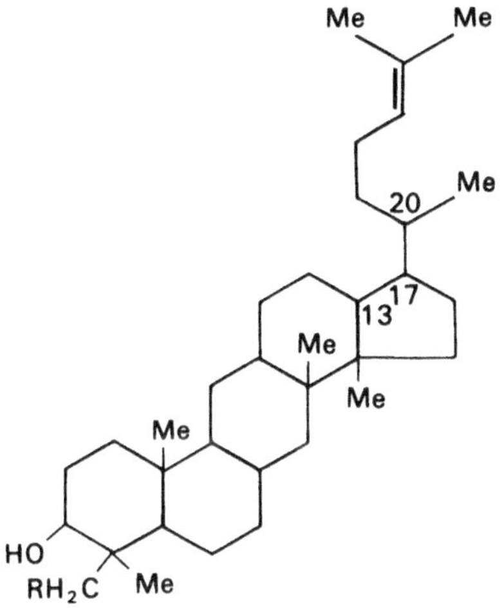

Δ13(17), R = H: Fusisterol (*34*)
Cephalosporium caerulens (Hattori *et al.*, 1969)

Δ17(20), R = H or OH: 3β-Hydroxy-4β-methyl/hydroxymethyl fusida-17(20)[16,21-cis]-24-diene (*35*, *36*)
Cephalosporium caerulens
(Hattori *et al.*, 1969; Okuda *et al.*, 1968a)

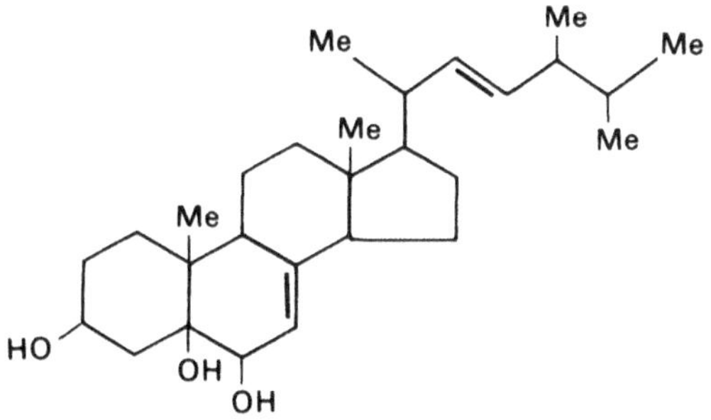

Cerevisterol (*37*)

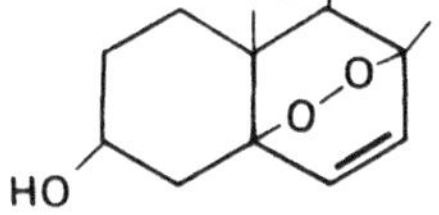

Ergosterol peroxide (*38*)
A. luzulae (Ceccherelli *et al.*, 1975)

3.3. Miscellaneous

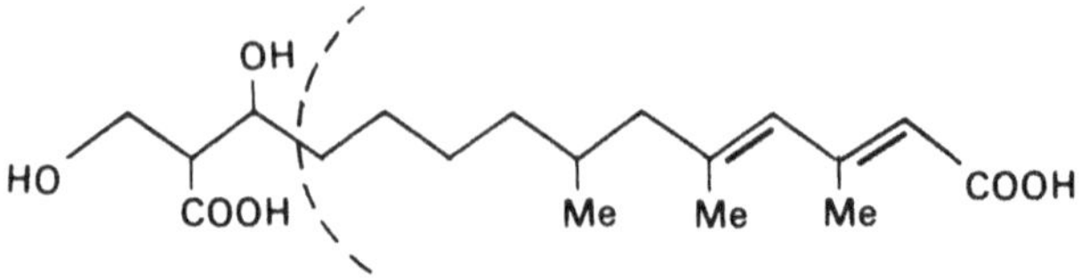

Antibiotic 1233 A (*39*)
Acremonium sp.
(Aldridge *et al.*, 1971)

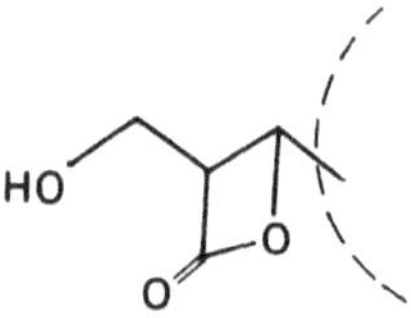

Antibiotic 1233 B (*40*)
Acremonium sp. (Aldridge *et al.*, 1971)

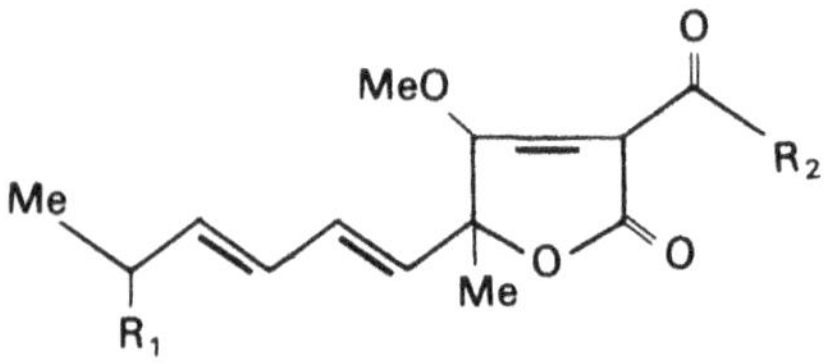

R_1	R_2		
H	CH=CH·Me	Gregatin A (*41*)	
H	Me	Gregatin B (*42*)	*Cephalosporium gregatum*
H	$CH_2 \cdot CH_2(OH) \cdot Me$	Gregatins C and D (*43, 44*) (isomers)	(K. Kobayashi and Ui, 1975)
OH	Me	Gregatin E (*45*)	
H	$CH=CH \cdot CH_2 \cdot CH_2 \cdot Me$	Graminin A (*46*) *Cephalosporium gramineum* (K. Kobayashi and Ui, 1977)	

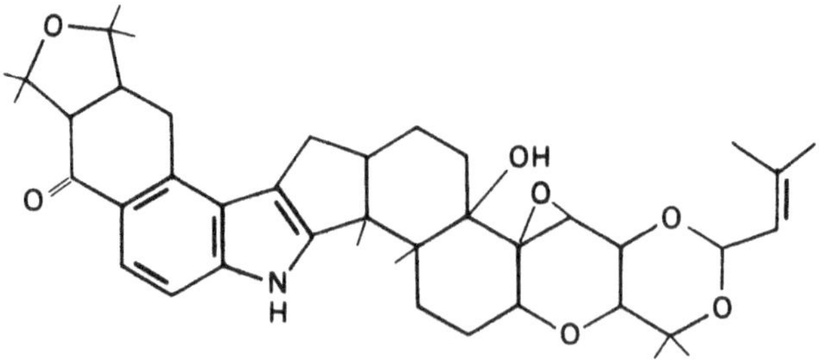

Lolitrem B (*47*)
Probably biosynthesized by *A. loliae.* (See text for reasoning, based partly on the recent finding in the author's laboratory of paxilline [*Penicillium* metabolite *318*], below, as a metabolite also of *A. loliae.*)

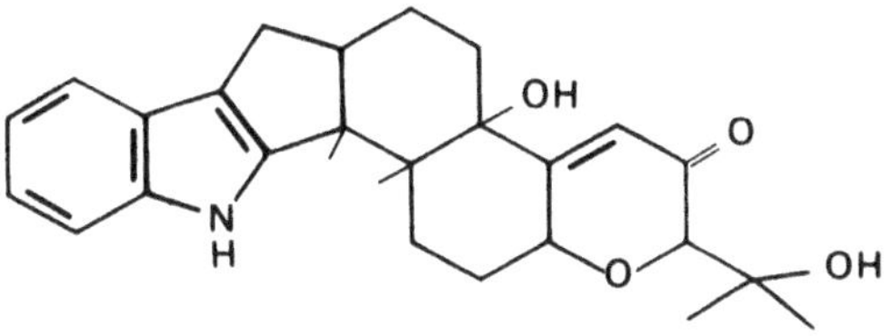

Paxilline (*48*)
A. loliae (Weedon and Mantle, 1987)

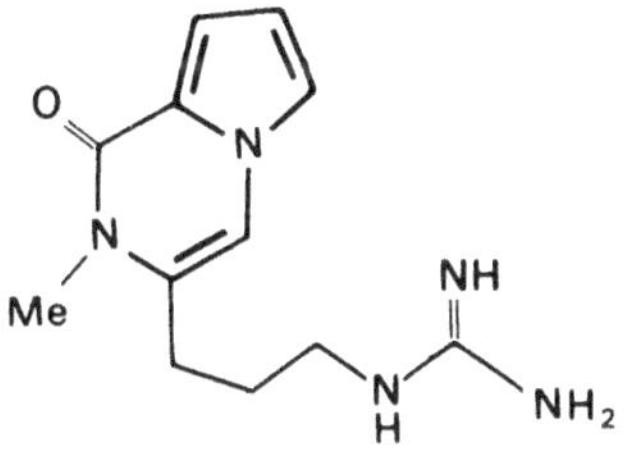

Peramine (*49*)
A. loliae (Rowan *et al.*, 1986)

4. *PENICILLIUM* AND *ACREMONIUM* INTERRELATIONSHIPS AND SOME IMPLICATIONS

Tremorgenic mycotoxins offer a topical link between the genera *Penicillium* and *Acremonium.* These genera have been treated together in this volume not because they are overtly close, judged by current taxonomic criteria, but because the penicillins are almost uniquely produced by representatives of these two genera and the structurally related cephalosporins alone give prominent biosynthetic status to *Acremonium.* Apart from the common thread of β-lactam antibiotic formation, there has seemed little to connect the genera. Even the proportional distribution of known *Acremonium* products among the main groups of metabolites is totally out of phase with the pattern seen in the *Penicillia*; terpenoids are more abundant than polyketides, and apart from the β-lactam antibiotics and the novel metabolite peramine (*49*), amino-acid-derived alkaloids have not been reported. However, as an exercise in biotechnological lateral thinking, and in an attempt to perceive common biosynthetic potential, consider the current attempts to rationalize the etiology of the curious neurological disorder that commonly occurs in sheep and cattle in New Zealand and is known as ryegrass staggers. The cause of this troublesome syndrome evaded rational explanation until a decade ago, when tremorgenic mycotoxins, notably penitrem A (*317*) produced in axenic culture by common soil-borne *Penicillia*, were shown for the first time to reproduce the symptoms of tremors and incoordination typical of ryegrass staggers when given in small doses

(Di Menna *et al.*, 1976). The hypothesis of cause and effect, however, was weak with respect to the acquisition of sufficient toxin by grazing animals. In 1981, novel neurotoxins, the lolitrems, were isolated from ryegrass infected by an endophytic fungus (Gallagher *et al.*, 1981), and independently, endophyte-infected ryegrass was shown to cause ryegrass staggers (Fletcher and Harvey, 1981). The question arises as to whether lolitrem neurotoxins are elaborated by the endophyte (and are thereby tremorgenic mycotoxins), by the host plant, or through biosynthetic collaboration of both. Consideration of the structure of lolitrems (Gallagher *et al.*, 1984) shows a large proportion of the indole–isoprenoid molecule in common with some tremorgenic mycotoxins produced by *Penicillia.* Thus, for example, the whole of paxilline (*316*) is in lolitrems, and lolitrem B and penitrem E (*321*) essentially differ only by the arrangement of "peripheral" isoprenes. Further, the main indole–isoprenoid carbon skeleton of lolitrems is known only as a product of fungal biosynthesis. Thus, although current attempts to explore the biosynthetic potential of the endophyte in axenic culture have not shown immediate ability to elaborate lolitrems, there are compelling grounds to expect that the endophyte is involved *in vivo* in the biosynthesis. Such conviction is strengthened by the description of the endophyte as a new species of *Acremonium* (*A. loliae*) (Latch *et al.*, 1984). While it is evident in the listing of *Acremonium* biosyntheses in this chapter that this is not grounds for immediate optimism concerning biosynthetic potential, several new fungal metabolites, the electron-impact mass spectra of which are consistent with tryptophanol, tyrosol, and a tyrosol dimer, have recently been isolated from axenic culture of *A. loliae* (C. M. Weedon and P. G. Mantle, unpublished observations). A diketopiperazine based on cyclic Phe–Pro is also produced, together with several sterols. Further, *A. loliae* has been found very recently (Weedon and Mantle, 1987) to elaborate paxilline (*48*) when grown in particular conditions in submerged liquid culture. The electron impact mass spectrum and the ^{1}H NMR spectrum of the *A. loliae* product is identical to that of authentic paxilline (*316*) produced by *P. paxilli.* Thus it is now clear that *A. loliae* can biosynthesize an indole–isoprenoid which equates to the principal part of the lolitrem B molecule (*47*). This finding is a major step in assigning the origin of lolitrem biosynthesis in the host/parasite association between perennial ryegrass (*Lolium perenne*) and the endophyte (*A. loliae*). It also identifies another amino-acid-derived secondary metabolic link between *Acremonium* and *Penicillium* and points to a prospect that the biosynthetic potential in *Acremonium* may be richer than is yet evident and that, additional to penicillins, peramine, and paxilline, other biosyntheses in common with *Penicillia* await discovery.

A similar rational basis for expecting similar biosynthesis in groups of fungi that have not been thought to have much phylogenetic connection can be seen between *Penicillium* and *Claviceps. Claviceps paspali* produces the indole–isoprenoid paspalinines (Cole *et al.*, 1977), the main carbon skeleton of which is identical to that of some *Penicillium* metabolites (penitrems and paxilline). There is also intergeneric commonality with respect to ergoline alkaloids, agroclavine being a *Penicillium* metabolite while it is also a normal biosynthetic intermediate in the formation of lysergic acid derivatives by *C. paspali.*

While these transgeneric biosynthetic abilities may allow the expectation of finding interesting biosynthetic potential in unexpected fungi, the situation hardly encourages the prospects for complementary use of metabolite biosynthesis as a phylogenetic criterion in the evolution of taxonomy toward linking like with like on the grounds of what organisms can do, rather than solely on how they appear.

The 1940s to the 1970s were a fruitful period in the search for fungal natural products, and there were beginning to be sufficient examples across the diverse spectrum of structure and function for research to be redirected to the more difficult area of identifying intermediates in biosynthetic pathways and the enzymes that catalyze the steps. The demand for biochemical attention leading ultimately to sequencing of relevant enzymes and identification of the conformation of the functional sites has generally been neglected in view of exciting developments elsewhere in molecular biology that have diverted biochemists almost entirely from the more classic objective of understanding secondary-metabolite biosynthesis and the concomitant potential for the transformation of unusual and unnatural substrates. A rediscovery of the challenges and potential of fungal biosynthesis, exemplified richly by *Penicillium* spp., may prove to be a rewarding approach in the fields of agrochemicals and pharmaceuticals if and when the venture capital for such explorations is made available.

The value of using the natural interdigitation of microbiology, microbial biochemistry, and natural-products chemistry with the animal and plant sciences as a teaching medium in the biological sciences should also be stressed. The rich biosynthetic potential of the *Penicillia*, and the relative ease with which they can be grown in axenic culture, make these microorganisms particularly suitable teaching subjects. Such breadth of integrated understanding in the biological sciences is a vital component for tomorrow's biotechnologist.

REFERENCES

Abe, M., Yamatodani, S., Yamano, T., Kozu, Y., and Yamada, S., 1967, Production of alkaloids and related substances by fungi. I. Examination of filamentous fungi for their ability of producing ergot alkaloids, *J. Agric. Chem. Soc. Jpn.* **41:**68–71.

Abe, M., Ohmomo, S., Ohashi, T., and Tabuchi, T., 1969, Isolation of chanoclavine I and two new interconvertible alkaloids, rugulovasine A and B, from cultures of *Penicillium concavo-rugulosum*, *Agric. Biol. Chem.* **33:**469–471.

Abraham, E. P., and Florey, H. W., 1949, Substances produced by fungi imperfecti and ascomycetes, in: *Antibiotics*, Vol. I (H. W. Florey, E. Chain, N. G. Heatley, M. A. Jennings, A. G. Sanders, E. P. Abraham, and M. E. Florey, eds.), Oxford University Press, London, p. 274.

Achenbach, H., Strittmatter, H., and Kohl, W., 1972, Die Strukturen der Xanthocilline Y1 and Y2, *Chem. Ber.* **105:**3061–3066.

Acker, T. E., Brenneisen, P. E., and Tanenbaum, S. W., 1966, Isolation, structure and radiochemical synthesis of 3,6-dimethyl-4-hydroxy-2-pyrone, *J. Am. Chem. Soc.* **88:**834–837.

Ackland, M. J., Hanson, J. R., Ratcliffe, A. H., and Sadler, I. H., 1982, A ^{2}H NMR study of the rearrangement step in aphidicolin biosynthesis, *J. Chem. Soc. Chem. Commun.* 165–166.

Adam, H. K., Campbell, I. M., and McCorkindale, N. J., 1967, Ergosterol peroxide: A fungal artefact, *Nature (London)* **216:**397.

Adams, M. R., and Bu'Lock, J. D., 1975, Biosynthesis of the diterpene antibiotic, aphidicolin, by radioisotope and ^{13}C nuclear magnetic resonance methods, *J. Chem. Soc. Chem. Commun.* 389.

Aldridge, D. C., and Turner, W. B., 1971, Two new mould metabolites related to avenaciolide, *J. Chem. Soc. C* 2431–2432.

Aldridge, D. C., Giles, D., and Turner, W. B., 1971, Antibiotic 1233A: A fungal β-lactone, *J. Chem. Soc. C* 3888–3891.

Ali, M. S., Shannon, J. S., and Taylor, A., 1968, Isolation and structures of 1,2,3,4-tetrahydro-1,4-dioxopyrazino [1,2-a]-indoles from cultures of *Penicillium terlikowskii*, *J. Chem. Soc. C* 2044.

Ando, K., Kato, A., and Suzuki, S., 1970, Isolation of 2,4-dichlorophenol from a soil fungus and its biological significance, *Biochem. Biophys. Res. Commun.* **39:**1104–1107.

Ando, K., Matsuura, I., Nawata, Y., Endo, H., Sasaki, H., Okytomi, T., Saeki, T., and Tamura, G., 1978, Funiculosin, a new antibiotic. II. Structure elucidation and antifungal activity, *J. Antibiot. (Tokyo)* **31:**533–538.

Angeletti, A., Tappi, G., and Biglino, G., 1952, Composition of the mycelium of *Penicillium notatum*, *Ann. Chim. (Rome)* **42:**502–506.

Anslow, W. K., Breen, J., and Raistrick, H., 1940, Studies in the biochemistry of microorganisms. 64. Emodic acid and hydroxyemodin, metabolic products of a strain of *Penicillium cyclopium* Westling, *Biochem. J.* **34:**159–168.

Arndt, R. R., Holzapfel, C. W., Ferriera, N. P., and Marsh, J. J., 1974, The structure and biogenesis of desoxyverrucarin E, a metabolite of *Eupenicillium hirayamae*, *Phytochemistry* **13:**1865–1870.

Atherton, J., Bycroft, B. W., Roberts, J. C., Roffey, P., and Wilcox, M. E., 1968, Studies in mycological chemistry. 23. The structure of flavomannin, a metabolite of *Penicillium wortmanni*, *J. Chem. Soc. C* 2560–2564.

Aulin-Erdtman, G., 1951, Studies in the tropolone series. IV. Stipitatic, puberulic and puberulonic acids, *Acta Chem. Scand.* **5:**301–315.

Austin, D. J., and Meyers, M. B., 1964, 3-*O*-methylviridicatin, a new metabolite from *Penicillium puberulum*, *J. Chem. Soc.* 1197–1198.

Awaya, J., Kesado, T., Omura, S., and Lukacs, G., 1975, Preparation of ^{13}C- and ^{3}H-labeled cerulenin and biosynthesis with ^{13}C-NMR, *J. Antibiot. (Tokyo)* **28:**824–827.

Bar, H., Zarnack, J., and Pfeifer, S., 1971a, Phenoxazinone in Kulturlösungen von *Penicillium notatum* Westl., *Pharmazie* **26:**314.

Bar, H., Zarnack, J., and Pfeifer, S., 1971b, *N*-Formyl-2-aminophenol—ein neuer Naturstoff, *Pharmazie* **26:**108.

Barrow, K. D., Colley, P. W., and Tribe, D. E., 1979, Biosynthesis of the neurotoxin alkaloid roquefortine, *J. Chem. Soc. Chem. Commun.* 225–226.

Barton, D. H. R., and Sutherland, J. K., 1965, The Nonadrides. I. Introduction and general survey, *J. Chem. Soc.* 1769–1798.

Bassett, E. W., and Tanenbaum, S. W., 1958, The metabolic products of *Penicillium patulum* and their probable interrelation, *Experientia* **14:**38–40.

Bates, M. L., Reid, W. W., and White, J. D., 1976, Duality of pathways in the oxidation of ergosterol to its peroxide *in vivo*, *J. Chem. Soc. Chem. Commun.* 44–45.

Belgian patent, 1962, Fusidic acid, *Chem. Abstr.* **59:**2133.

Bentley, R., and Keil, J. G., 1962, Tetronic acid biosynthesis in moulds. II. Formation of penicillic acid in *Penicillium cyclopium*, *J. Biol. Chem.* **237:**867–873.

Bentley, R., and Thiessen, C. P., 1957, Biosynthesis of itaconic acid in *Aspergillus terreus*, *J. Biol. Chem.* **226:**673–720.

Bentley, R., and Zwitkowits, P. M., 1967, Biosynthesis of tropolones in *Penicillium stipitatum*. VII. The formation of polyketide lactones and other nontropolone compounds as a result of ethionine inhibition, *J. Am. Chem. Soc.* **89:**676–685.

Bentley, R., Bhate, D. S., and Keil, J. G., 1962, Tetronic acid biosynthesis in moulds. I. Formation of carlosic and carolic acids in *Penicillium charlesii*, *J. Biol. Chem.* **237:**859–866.

Betina, V., Nemec, P., Dobias, J., and Barath, Z., 1962, Cyanein, a new antibiotic from *Penicillium cyaneum*, *Folia Microbiol.* **7:**353–357.

Betina, V., Fuska, J., Kjaer, A., Kutkova, M., Nemec, P., and Shapiro, R. H., 1966, Production of cyanein by *Penicillium simplicissimum*, *J. Antibiot. (Tokyo)* **19:**115–117.

Better, J., and Gatenbeck, S., 1976, 1,4-Dihydroxy-2-methoxy-6-methyl-benzene, a metabolite of *Penicillium baarnense*, *Acta Chem. Scand.* **30B:**368.

Biollaz, M., Buchi, G., and Milne, G., 1968, The biosynthesis of the aflatoxins, *J. Am. Chem. Soc.* **92:**1035–1043.

Birch, A. J., and Kocor, M., 1960, Studies in relation to biosynthesis. XXII. Palitantin and cyclopaldic acid, *J. Chem. Soc.* 866–871.

Birch, A. J., and Pride, E., 1962, Studies in relation to biosynthesis. XXVI. 7-Hydroxy-4,6-dimethylphthalide, *J. Chem. Soc.* 370–371.

Birch, A. J., and Russell, R. A., 1972, Studies in relation to biosynthesis. 44. Structural elucidations of brevianamides-B, -C, -D, *Tetrahedron* **28:**2999–3008.

Birch, A. J., and Stapleford, K. S. J., 1967, The structure of nalgiolaxin, *J. Chem. Soc. C* 2570–2571.

Birch, A. J., and Wright, J. J., 1969, Brevianamides: A new class of fungal alkaloid, *Chem. Commun.* 644–645.

Birch, A. J., Kocor, M., Sheppard, N., and Winter, J., 1962, Studies in relation to biosynthesis. 29. The terpenoid chain of mycelianamide, *J. Chem. Soc.* 1502–1505.

Birch, A. J., Qureshi, A. A., and Rickards, R. W., 1968, Metabolites of *Aspergillus indicus*: The structure and some aspects of the biosynthesis of dihydrocanadensolide, *Aust. J. Chem.* **21:**2775–2784.

Birch, A. J., Gager, F., Mo, L., Pelter, A., and Wright, J. J., 1969, Studies in relation to biosynthesis. XLI. Canescin, *Aust. J. Chem.* **22:**2429–2436.

Bird, B. A., and Campbell, I. M., 1982a, Occurrence and biosynthesis of asperphenamate in solid cultures of *Penicillium brevicompactum*, *Phytochemistry* **21:**2405–2406.

Bird, B. A., and Campbell, I. M., 1982b, Disposition of mycophenolic acid, brevianamide A, asperphenamate and ergosterol in solid cultures of *Penicillium brevicompactum*, *Appl. Environ. Microbiol.* **43:**345–348.

Birkinshaw, J. H., and Gowlland, A., 1962, Studies in the biochemistry of microorganisms. 110.

Production and biosynthesis of orsellinic acid by *Penicillium madriti* G. Smith, *Biochem. J.* **84**:342–347.

Birkinshaw, J. H., and Mohammed, Y. S., 1962, Studies in the biochemistry of micro-organisms. 111. The production of l-phenylalanine anhydride (cis-l-3,6-dibenzyl-2,5-dioxopiperazine) by *Penicillium nigricans* (Bainer) Thom, *Biochem. J.* **85**:523–527.

Birkinshaw, J. H., and Raistrick, H., 1934, The metabolic products of *Penicillium minio-luteum* Dierckx, minioluteic acid, *Biochem. J.* **28**:828–836.

Birkinshaw, J. H., and Raistrick, H., 1936, Isolation, properties and constitution of terrestric acid (ethyl carolic acid), a metabolic product of *Penicillium terrestre* Jensen, *Biochem. J.* **30**:2194–2200.

Birkinshaw, J. H., and Samant, M. S., 1960, Metabolites of *Penicillium viridicatum* Westling: Viridicatic acid (ethylcarlosic acid), *Biochem. J.* **74**:369–373.

Birkinshaw, J. H., Luckner, M., Mohammed, Y. S., and Stickings, C. E., 1963a, Studies in the biochemistry of micro-organisms. 114. Viridicatol and cyclopenol, metabolites of *Penicillium viridicatum* Westling and *Penicillium cyclopium* Westling, *Biochem. J.* **89**:196–202.

Birkinshaw, J. H., Kalyanpur, M. G., and Stickings, C. E., 1963b, Studies in the biochemistry of micro-organisms. 113. Pencolide, a nitrogen-containing metabolite of *Penicillium multicolor*, *Biochem. J.* **86**:237–243.

Bloomer, J. L., Moppett, C. E., and Sutherland, J. K., 1968, The Nonadrides. V. Biosynthesis of glauconic acid, *J. Chem. Soc. C* 588–591.

Bodo, M., Massias, M., Molho, L., Molho, D., and Combrisson, S., 1976, Application of carbon-13 NMR to the determination of biosynthesis of a fungal metabolite, spiculisporic acid, by *Penicillium spiculisporum* in shake culture, *Bull. Mus. Natl. Hist. Nat. Sci. Phys.-Chim.* **11**:53–62.

Boeckman, R. K., Fayos, J., and Clardy, J., 1974, A revised structure of vermiculine: A novel macrolide dilactone antibiotic from *Penicillium vermiculatum*, *J. Am. Chem. Soc.* **96**:5954–5956.

Bond, R. F., Boeyens, J. C. A., Holzapfel, C. W., and Steyn, P. S., 1979, Cyclopiamines A and B, novel oxindole metabolites of *Penicillium cyclopium* Westling, *J. Chem. Soc. Perkin Trans. 1* 1751–1761.

Bose, A. K., Das, K. G., Funke, P. T., Kugajevsky, I., Shukla, O. P., Khanchandani, K. S., and Suhadolnik, R. J., 1968, Biosynthetic studies on gliotoxin using stable isotopes and mass spectral methods, *J. Am. Chem. Soc.* **90**:1038–1041.

Bracken, A., and Raistrick, H., 1947, Dehydrocarolic acid, a metabolic product of *Penicillium cinerascens* Biourge, *Biochem. J.* **41**:569–575.

Bracken, A., Pocker, A., and Raistrick, H., 1954, Studies in the biochemistry of micro-organisms. 93. Cyclopenin, a nitrogen-containing metabolic product of *Penicillium cyclopium* Westling, *Biochem. J.* **57**:587–595.

Breen, J., Dacre, J. C., Raistrick, H., and Smith, G., 1955, Studies in the biochemistry of micro-organisms. 95. Rugulosin, a crystalline colouring matter of *Penicillium rugulosum* Thom, *Biochem. J.* **60**:618–626.

Broadbent, D., Mabelis, R. P., and Spencer, H., 1975, 3,6,8-Trihydroxy-1-methylxanthone—an antibacterial metabolite from *Penicillium patulum*, *Phytochemistry* **14**:2082–2083.

Brown, A. G., Smale, T. C., King, T. J., Hasenkamp, R., and Thompson, R. H., 1976, Crystal and molecular structure of compactin, a new antifungal metabolite from *Penicillium brevicompactum*, *J. Chem. Soc. Perkin Trans. 1* 1165–1170.

Brundidge, S. P., Gaeta, F. C. A., Hook, D. J., Sapino, C., Elander, R. P., and Morin, R. B., 1980, Association of 6-oxopiperidine-2-carboxylic acid with penicillin V production in *Penicillium chrysogenum* fermentations, *J. Antibiot.* **33**:1348–1351.

Bryant, R. W., and Light, R. J., 1974, Stipitatonic acid biosynthesis: Incorporation of [formyl-^{14}C]-3-methylorcylaldehyde and [^{14}C]stipitaldehydic acid, a new tropolone metabolite, *Biochemistry* **13**:1516–1522.

Buchi, G., White, J. D., and Wogan, G. N., 1965, The structures of mitorubrin and mitorubrinol, *J. Am. Chem. Soc.* **87:**3484–3489.

Bu'Lock, J. D., 1980, Mycotoxins as secondary metabolites, in: *The Biosynthesis of Mycotoxins* (P. S. Steyn, ed.), Academic Press, New York, pp. 1–16.

Bu'Lock, J. D., and Clay, P. T., 1969, Fatty acid cyclisation in the biosynthesis of brefeldin A: A new route to some fungal metabolites, *Chem. Commun.* 237–238.

Bu'Lock, J. D., and Ryan, A. J., 1958, The biogenesis of patulin, *Proc. Chem. Soc. (London)* 222–223.

Bu'Lock, J. D., and Smith, J. R., 1968, Modified anthraquinones from *Penicillium islandicum*, *J. Chem. Soc. C* 1941–1943.

Burton, H. S., and Abraham, E. P., 1951, Isolation of antibiotics from a species of *Cephalosporium*: Cephalosporins P_1, P_2, P_3, P_4 and P_5, *Biochem. J.* **50:**168–174.

Cagnoli, N., Ceccherelli, P., Curini, M., Spagnoli, N., and Ribaldi, M., 1980, 19-Norisopimara-7,15-dien-3-one: A new norditerpenoid from *Acremonium luzulae* (Fuckel) Gams, *J. Chem. Res. Synop.* **8:**276.

Cagnoli, N., Ceccherelli, P., Curini, M., Madruzza, G. F., and Ribaldi, M., 1982, Isopimara-7,15-dien-19-ol: A new diterpenoid from *Acremonium luzulae* (Fuckel) Gams, *J. Chem. Res. Synop.* **9:**254.

Cagnoli-Bellavita, N., Ceccherelli, P., Fringuelli, R., and Ribaldi, M., 1975, Ascochlorin: A terpenoid metabolite from *Acremonium luzulae*, *Phytochemistry* **14:**807.

Cagnoli-Bellavita, N., Ceccherelli, P., Ribaldi, M., Polonsky, J., Baskevitch-Varon, Z., and Varenne, J., 1977, Structures of virescenosides D and H, new metabolites of *Acremonium luzulae* (Fuckel) Gams, *J. Chem. Soc. Perkin Trans. 1* 351–354.

Campbell, I. M., Calzadilla, C. H., and McCorkindale, N. J., 1966, Some new metabolites related to mycophenolic acid, *Tetrahedron Lett.* **42:**5107–5111.

Canonica, L., Kroszczynski, W., Ranzi, B. M., Rindone, B., and Scolastico, C., 1971, The biosynthesis of mycophenolic acid, *J. Chem. Soc. Chem. Commun.* 257.

Casey, M. L., Paulick, R. C., and Whitlock, H. W., 1978, Carbon-13 nuclear magnetic resonance study of the biosynthesis of daunomycin and islandicin, *J. Org. Chem.* **43:**1627–1634.

Ceccherelli, P., Fringuelli, R., Madruzza, G. F., and Ribaldi, M., 1975, Cerevisterol and ergosterol peroxide from *Acremonium luzulae*, *Phytochemistry* **14:**1434.

Chalmers, A. A., De Jesus, A. E., Gorst-Allman, C. P., and Steyn, P. S., 1981, Biosynthesis of PR toxin by *Penicillium roqueforti*, *J. Chem. Soc. Perkin Trans. 1* 2899–2903.

Chaplen, P., and Thomas, R., 1960, Studies in the biosynthesis of fungal metabolites: The biosynthesis of palitantin, *Biochem. J.* **77:**91–96.

Chen, Y. S., and Haskins, R. H., 1963, Studies on pigments of *Penicillium funiculosum*. I. Production of cholesterol, *Can. J. Chem.* **41:**1647–1650.

Chexal, K. K., Snipes, C., and Tamm, C., 1980, Biosynthesis of the antibiotic verrucarin E: Use of [1-^{13}C]-, [2-^{13}C]-, [1,2-^{13}C]- and [2-^{13}C, 2^2H_3]-acetates, *Helv. Chim. Acta* **63:**761–768.

Chou, T. S., Eisenbraun, E. J., and Rapala, R. T., 1969, The chemistry of steroid acids from *Cephalosporium acremonium*, *Tetrahedron* **25:**3341–3357.

Ciegler, A., and Pitt, J. I., 1970, A survey of the genus *Penicillium* for tremorgenic toxin production, *Mycopathol. Mycol. Appl.* **42:**119–124.

Ciegler, A., Fennell, D. I., Mintzlaff, H.-J., and Leistner, L., 1972, Ochratoxin synthesis by *Penicillium* species, *Naturwissenschaften* **59:**365–366.

Cockrum, P. A., Culvenor, C. C. J., Edgar, J. A., and Payne, A. L., 1979, Chemically different tremorgenic mycotoxins in isolates of *Penicillium paxilli* from Australia and North America, *J. Nat. Prod.* **42:**534–536.

Cole, R. J., Kirksey, J. W., Moore, J. H., Blankenship, B. R., Diener, U. L., and Davis, N. D., 1972, Tremorgenic toxin from *Penicillium verruculosum*, *Appl. Microbiol.* **24:**248–256.

Cole, R. J., Kirksey, J. W., and Wells, J. M., 1974, A new tremorgenic metabolite from *Penicillium paxilli*, *Can. J. Microbiol.* **20:**1159–1162.

Cole, R. J., Kirksey, J. W., Cutler, H. G., Wilson, D. M., and Morgan-Jones, G., 1976a, Two toxic indole alkaloids from *Penicillium islandicum*, *Can. J. Microbiol.* **22:**741–744.

Cole, R. J., Kirksey, J. W., Clardy, J., Eickman, M., Weinreb, S. M., Singh, P., and Kim, D., 1976b, Structure of rugulovasine-A and -B and 8-chlororugulovasine-A and -B, *Tetrahedron Lett.* **43:**3849–3852.

Cole, R. J., Dorner, J. W., Lansden, J. A., Cox, R. H., Pape, C., Cunfer, B., Nicholson, S. S., and Bedell, D. M., 1977, Paspalum staggers: Isolation and identification of tremorgenic metabolites from sclerotia of *C. paspali*, *J. Agric. Food Chem.* **25:**1197–1201.

Cole, R. J., Dorner, J. W., Cox, R. H., Hill, R. A., Cluter, H. G., and Wells, J. M., 1981, Isolation of citreoviridin from *Penicillium charlesii* cultures and molded pecan fragments, *Appl. Environ. Microbiol.* **42:**677–681.

Cole, R. J., Dorner, J. W., Cox, R. H., and Raymond, L. W., 1983, Two classes of alkaloid mycotoxins produced by *Penicillium crustosum* Thom isolated from contaminated beer, *J. Agric. Food Chem.* **31:**655–657.

Collins, R. P., 1976, Terpenes and odoriferous materials from microorganisms, *Lloydia* **39:**20–24.

Colombo, L., Gennari, C., Potenza, D., Scolastico, C., Aragozzini, F., and Gualandris, R., 1982, 6-Farnesyl-5,7-dihydroxy-4-methylphthalide oxidation mechanism in mycophenolic acid biosynthesis, *J. Chem. Soc. Perkin Trans. 1* 365–373.

Corbett, R. E., Johnson, A. W., and Todd, A. R., 1950, Puberulic and puberulonic acids. II. Structure, *J. Chem. Soc.* 6–9.

Cox, R. H., Hernandez, O., Dorner, J. W., Cole, R. J., and Fennell, D. I., 1979, A new isochroman mycotoxin isolated from *Penicillium steckii*, *J. Agric. Food Chem.* **27:**999–1001.

Cram, D. J., 1948, Mold metabolites. II. The structure of sorbicillin, a pigment produced by the mold *Penicillium notatum*, *J. Am. Chem. Soc.* **70:**4240–4243.

Cram, D. J., and Tishler, M., 1948, Mold metabolites. I. Isolation of several compounds from clinical penicillin, *J. Am. Chem. Soc.* **70:**4238–4239.

Cunningham, K. G., and Freeman, G. G., 1953, The isolation and some chemical properties of viridicatin, a metabolic product of *Penicillium viridicatum* Westling, *Biochem. J.* **53:**328–332.

Curtis, R. F., Hassall, C. H., and Nazar, M., 1968, The biosynthesis of phenols. XV. Some metabolites of *Penicillium citrinum* related to citrinin, *J. Chem. Soc. C* 85–93.

Czech patent, 1982, Antibiotic citrinin by fermentation with *Penicillium janthinellum*, *Chem. Abstr.* **97:**22,078.

Dalziel, W., Hesp, B., Stevenson, K. M., and Jarvis, J. A. J., 1973, The structure and absolute configuration of the antibiotic aphidicolin: A tetracyclic diterpenoid containing a new ring system, *J. Chem. Soc. Perkin Trans. 1* 2841–2851.

Damoglou, A. P., Downey, G. A., and Shannon, W., 1984, The production of ochratoxin A and citrinin in barley, *J. Sci. Food Agric.* **35:**395–400.

Day, J. B., and Mantle, P. G., 1982, Biosynthesis of radiolabeled verruculogen by *Penicillium simplicissimum*, *Appl. Environ. Microbiol.* **43:**514–516.

Dean, F. M., Eade, R. A., Moubasher, R., and Robertson, A., 1957, The chemistry of fungi. 27. The structure of fulvic acid from *Carpenteles brefeldianum*, *J. Chem. Soc.* 3497–3510.

De Jesus, A. E., Steyn, P. S., Vleggaar, R., Kirby, G. W., Varley, M. J., and Ferreira, N. P., 1981, Biosynthesis of α-cyclopiazonic acid: Steric course of proton removal during the cyclisation of β-cyclopiazonic acid in *Penicillium griseofulvum*, *J. Chem. Soc. Perkin Trans. 1* 3292–3294.

De Jesus, A. E., Hull, W. E., Steyn, P. S., Van Heerden, F. R., and Vleggaar, R., 1982, Biosynthesis of viridicatumtoxin, a mycotoxin from *Penicillium expansum*, *J. Chem. Soc. Chem. Commun.* 902–904.

De Jesus, A. E., Steyn, P. S., Van Heerden, F. R., Vleggaar, R., Wessels, P. L., and Hull, W. E., 1983a, Tremorgenic mycotoxins from *Penicillium crustosum*: Isolation of penitrems A–F and the structure elucidation and absolute configuration of penitrem A, *J. Chem. Soc. Perkin Trans. 1* 1847–1856.

De Jesus, A. E., Steyn, P. S., Van Heerden, F. R., Vleggaar, R., Wessels, P. L., and Hull, W. E., 1983b, Tremorgenic mycotoxins from *Penicillium crustosum*: Structure elucidation and absolute configuration of penitrems B–F, *J. Chem. Soc. Perkin Trans. 1* 1857–1861.

De Jesus, A. E., Gorst-Allman, C. P., Steyn, P. S., Van Heerden, F. R., Vleggaar, R., Wessels, P. L., and Hull, W. E., 1983c, Tremorgenic mycotoxins from *Penicillium crustosum*: Biosynthesis of penitrem A, *J. Chem. Soc. Perkin Trans. 1* 1863–1868.

De Jesus, A. E., Steyn, P. S., and Van Heerden, F. R., 1983d, Structure of 7-de-*O*-methylsemivioxanthin, a metabolite of *Penicillium janthinellum*, *S. Afr. J. Chem.* **36:**82–83.

De Jesus, A. E., Steyn, P. S., Van Heerden, F. R., and Vleggaar, R., 1984, Structure elucidation of the janthitrems, novel tremorgenic mycotoxins from *Penicillium janthinellum*, *J. Chem. Soc. Perkin Trans. 1* 697–701.

Dewar, M. J. S., 1945, Structure of stipitatic acid, *Nature (London)* **155:**50–51.

Di Menna, M. E., Mantle, P. G., and Mortimer, P. H., 1976, Experimental production of a staggers syndrome in ruminants by a tremorgenic *Penicillium* from soil, *N. Z. Vet. J.* **24:**45–46.

Dimroth, P., Walter, H., and Lynen, F., 1970, Biosynthese von 6-Methylsalicylsäure, *Eur. J. Biochem.* **13:**98–110.

Divekar, P. V., Brenneisen, P. E., and Tanenbaum, S. W., 1961, Stipitatic acid ethyl ester: A naturally occurring tropolone derivative, *Biochim. Biophys. Acta* **50:**588–589.

Dix, D. T., Martin, J., and Moppett, C. E., 1972, Molecular structure of the metabolite lanosulin, *J. Chem. Soc. Chem. Commun.* 1168.

Doerfler, D. L., Bird, B. A., and Campbell, I. M., 1981, *N*-Benzoylphenylalanine and *N*-benzoylphenylalaninol, and their biosynthesis in *Penicillium brevicompactum*, *Phytochemistry* **20:**2303–2304.

Dorner, J. W., Cole, R. J., Hill, R., Wicklow, D., and Cox, R. H., 1980, *Penicillium rubrum* and *Penicillium biforme*, new sources of rugulovasins A and B, *Appl. Environ. Microbiol.* **40:**685–687.

Endo, A., Kuroda, M., and Tsujita, Y., 1976, ML-236 A, B, C, new inhibitors of cholesterogenesis produced by *Penicillium citrinum*, *J. Antibiot.* **29:**1346–1348.

Fayos, J., Lokensgard, D., Clardy, J., Cole, R. J., and Kirksey, J. W., 1974, Structure of verruculogen, a tremor producing peroxide from *Penicillium verruculosum*, *J. Am. Chem. Soc.* **96:**6785–6787.

Fletcher, L. R., and Harvey, I. C., 1981, An association of a *Lolium* endophyte with ryegrass staggers, *N. Z. Vet. J.* **29:**185–186.

Forrester, P. I., and Gaucher, G. M., 1972a, Conversion of 6-methylsalicylic acid into patulin by *Penicillium urticae*, *Biochemistry* **11:**1102–1107.

Forrester, P. I., and Gaucher, G. M., 1972b, *m*-Hydroxy benzyl alcohol dehydrogenase from *Penicillium urticae*, *Biochemistry* **11:**1108–1114.

Framm, J., Nover, L., El Azzouny, A., Richter, H., Winter, K., Werner, S., and Luckner, M., 1973, Cyclopeptin und Dehydrocyclopeptin: Zwischenprodukte der Biosynthese von Alkaloiden der Cyclopenin–Viridicatin-Gruppe bei *Penicillium cyclopium* Westling, *Eur. J. Biochem.* **37:**78–85.

Franck, B., 1980, The biosynthesis of the ergochromes, in: *The Biosynthesis of Mycotoxins: A Study in Secondary Metabolism* (P. S. Steyn, ed.), Academic Press, New York, pp. 157–191.

Fujimoto, Y., Kamiya, M., Tsunoda, H., Ohtsubo, K., and Tatsuno, T., 1980, Recherche toxicologique des substances métaboliques de *Penicillium carneo-lutescens*, *Chem. Pharm. Bull.* **28:**1062–1066.

Fujimoto, Y., Tsunoda, H., Uzawa, J., and Tatsuno, T., 1982, The structure of islandic acid, a new metabolite from *Penicillium islandicum* Sopp., *J. Chem. Soc. Chem. Commun.* 83–84.

Fujimoto, Y., Takahashi, T., Yokoyama, E., Uzawa, J., Tsunoda, H., and Tatsuno, T., 1983,

The isolation and structural elucidation of the antitumor metabolites produced by *Penicillium diversum* var. *aureum*, *Tennen Yuki Kagobutsu Toronkai Koen Yoshishu* **26:**166–172.

Furusaki, A., Matsumoto, T., Ogura, H., Takayanagi, H., Hirano, A., and Omura, S., 1980, X-ray crystal structure of herquline, a new biologically active piperazine from *Penicillium herquei* Fg-372, *J. Chem. Soc. Chem. Commun.* 698.

Fuska, J., and Proksa, B., 1983, Chromatographic determination of vermiculine, *Pharmazie* **38:**634–635.

Fuska, J., Nemec, P., and Fuskova, A., 1979, Vermicillin, a new metabolite from *Penicillium vermiculatum* inhibiting tumor cells *in vitro*, *J. Antibiot.* **32:**667–669.

Gallagher, R. T., and Latch, G. C. M., 1977, Production of the tremorgenic mycotoxins verruculogen and fumitremorgin B by *Penicillium piscarium* Westling, *Appl. Environ. Microbiol.* **33:**730–731.

Gallagher, R. T., White, E. P., and Mortimer, P. H., 1981, Ryegrass staggers: Isolation of potent neurotoxins lolitrem A and lolitrem B from staggers-producing pastures, *N. Z. Vet. J.* **29:**189–190.

Gallagher, R. T., Hawkes, A. D., Steyn, P. S., and Vleggaar, R., 1984, Tremorgenic neurotoxins from perennial ryegrass causing ryegrass staggers disorder of livestock: Structure elucidation of lolitrem B, *J. Chem. Soc. Chem. Commun.* 614–616.

Ganguli, M., Burka, L. T., and Harris, T. M., 1984, Structural studies of the mycotoxin verrucosidin, *J. Org. Chem.* **49:**3762–3766.

Gatenbeck, S., 1957, 3-Hydroxyphthalic acid, a metabolite in *Penicillium islandicum* Sopp, *Acta Chem. Scand.* **11:**555–557.

Gatenbeck, S., 1959a, The occurrence of endocrocin in *Penicillium islandicum*, *Acta Chem. Scand.* **13:**386–387.

Gatenbeck, S., 1959b, Studies of mono-*C*-methylquinalizarins in relation to a phenolic metabolite of *Penicillium islandicum*, *Acta Chem. Scand.* **13:**705–710.

Gatenbeck, S., 1962, The mechanism of the biological formation of anthraquinones, *Acta Chem. Scand.* **16:**1053–1054.

Gatenbeck, S., and Lonnroth, I., 1962, The biosynthesis of gentisic acid, *Acta Chem. Scand.* **16:**2298–2299.

Gatenbeck, S., and Mahlen, A., 1964, The enzymic synthesis of spiculisporic acid, in: Proc. Congress Antibiotics, Prague, Butterworth, London, p. 540.

Gatenbeck, S., and Mahlen, A., 1968, A metabolic variation in *Penicillium spiculisporum* Lehman. I. Production of (+) and (−)-decylcitric acids, *Acta Chem. Scand.* **22:**2613–2616.

Gatenbeck, S., and Mosbach, K., 1963, The mechanism of biosynthesis of citromycetin, *Biochem. Biophys. Res. Commun.* **11:**166–169.

German patent, 1984, Antibiotic compound, *Chem. Abstr.* **100:**137, 374.

Ghosh, A. C., and Ramgopal, M., 1980, Cyclic peptides from *Penicillium islandicum*: A review and a reevaluation of the structure of islanditoxin, *J. Heterocycl. Chem.* **18:**1809–1812.

Ghosh, A. C., Manmade, A., Kobbe, B., Townsend, J. M., and Demain, A. L., 1978, Production of luteoskyrin and isolation of a new metabolite, pibasterol, from *Penicillium islandicum* Sopp., *Appl. Environ. Microbiol.* **35:**563–566.

Godin, P., 1955, Separation of chromatography on column of cellulose of phenolic substances produced by *Penicillium brevi-compactum* and the perfunctory chemical analysis of one of them, *Antonie van Leeuwenhoek J. Microbiol. Serol.* **21:**362–366.

Gorst-Allman, C. P., and Steyn, P. S., 1983, Biosynthesis of 5,6-dihydro-4-methoxy-2H-pyran-2-one in *Penicillium italicum*, *S. Afr. J. Chem.* **36:**83–84.

Gorst-Allman, C. P., Steyn, P. S., and Vleggaar, R., 1982, The biosynthesis of roquefortine: An investigation of acetate and mevalonate incorporation using high field NMR spectroscopy, *J. Chem. Soc. Chem. Commun.* 652–653.

Gray, R. W., and Whalley, W. B., 1971, The chemistry of fungi. 63. Rubrorotiorin, a metabolite of *Penicillium hirayamae*, *J. Chem. Soc. C* 3575–3577.

Gray, R. A., Gauger, G. W., Dulaney, E. L., Kaczka, E. A., and Woodruff, H. B., 1964, Hadacidin, a new plant-growth inhibitor produced by fermentation, *Plant Physiol.* **39:**204–207.

Gudgeon, J. A., Holker, J. S. E., Simpson, T. J., and Young, K., 1979, The structures and biosynthesis of multicolanic, multicolic and multicolosic acids, novel tetronic acid metabolites of *Penicillium multicolor*, *Bioorg. Chem.* **8:**311–322.

Gyimesi, J., and Melera, A., 1967, On the structure of crotocin, an antifungal antibiotic, *Tetrahedron Lett.* **17:**1665–1673.

Haefliger, W., and Hauser, D., 1973, Isolierung und Strukturaufklärung von 11-Desacetoxy-wortmannin, *Helv. Chim. Acta* **56:**2901–2904.

Hagedorn, I., Eholzer, U., and Luttringhaus, A., 1960, Beiträge zur Konstitutionsermittlung des Antibiotikums Xanthocillin, *Chem. Ber.* **93:**1584–1590.

Halsall, T. G., and Sayer, G. C., 1959, The chemistry of the triterpenes and related compounds. 35. Some non-acidic constituents, *J. Chem. Soc.* 2031–2036.

Harris, C. M., Robertson, J. S., and Harris, T. M., 1976, Biosynthesis of griseofulvin, *J. Am. Chem. Soc.* **98:**5380–5386.

Hattori, T., Igarashi, H., Iwasaki, S., and Okuda, S., 1969, Isolation of 3β-hydroxy-4β-methylfusida-17(20)[16,21-cis],24-diene(3β-hydroxy-protosta-17(20)[16,21-cis],24 diene) and a related triterpene alcohol, *Tetrahedron Lett.* **13:**1023–1026.

Hermansen, K., Frisvad, J. C., Emborg, C., and Hansen, J., 1984, Cyclopiazonic acid production by submerged cultures of *Penicillium* and *Aspergillus* strains, *FEMS Microbiol. Lett.* **21:**253–261.

Hikino, H., Nabetani, S., and Takemoto, T., 1973, Structure and biosynthesis of chrysogine, a metabolite of *Penicillium chrysogenum*, *Yakugaku Zasshi* **93:**619–623.

Hind, H. G., 1940a, The colouring matters of *Penicillium carmino-violaceum* Biourge, with a note on the production of ergosterol by this mould, *Biochem. J.* **34:**67–72.

Hind, H. G., 1940b, The constitution of carviolin: A colouring matter of *Penicillium carmino-violaceum* Biourge, *Biochem. J.* **34:**577–579.

Hodges, F. A., Zust, J. R., Smith, H. R., Nelson, A. A., Armbrecht, B. H., and Campbell, A. D., 1964, Mycotoxins: Aflatoxin isolated from *Penicillium puberulum*, *Science* **145:**1439.

Hofle, G., and Roser, K., 1978, Structure of xanthomegnin and related pigments: Reinvestigation by ^{13}C nuclear magnetic resonance spectroscopy, *J. Chem. Soc. Chem. Commun.* 611–612.

Holker, J. S. E., Ross, W. J., Staunton, J., and Whalley, W. B., 1962, The chemistry of fungi. 40. Further evidence for the structure of sclerotiorin, *J. Chem. Soc.* 4150–4154.

Holker, J. S. E., Staunton, J., and Whalley, W. B., 1964, The biosynthesis of fungal metabolites. 1. Two different pathways to β-ketide chains in rotiorin, *J. Chem. Soc.* 16–22.

Holzapfel, C. W., 1968, The isolation and structure of cyclopiazonic acid, a toxic metabolite of *Penicillium cyclopium* Westling, *Tetrahedron* **24:**2101–2119.

Holzapfel, C. W., and Marsh, J. J., 1977, Isolation and structure of viridamine, a new nitrogenous metabolite of *Penicillium viridicatum* Westling, *S. Afr. J. Chem.* **30:**197–204.

Holzapfel, C. W., Hutchison, R. D., and Wilkins, D. C., 1970, The isolation and structure of two new indole derivatives from *Penicillium cyclopium* Westling, *Tetrahedron* **26:**5239–5246.

Hossain, M. B., Eng-Wilmot, D. L., Loghry, R. A., and Van der Helm, D., 1980, Circular dichroism, crystal structure and absolute configuration of the siderophore ferric *N*,*N*,*N*-triacetylfusarinine, $FeC_{39}H_{57}N_6O_{15}$, *J. Am. Chem. Soc.* **102:**5766.

Howard, B. H., and Raistrick, H., 1950, Studies in the biochemistry of microorganisms. 81. The colouring matters of *Penicillium islandicum* Sopp. 2. Chrysophanic acid, 4,5-dihydroxy-2-methylanthraquinone, *Biochem. J.* **46:**49–53.

Howard, C. C., Johnstone, R. A. W., King, T. J., and Lessinger, L., 1976, Fungal metabolites. VI. Crystal and molecular structure of secalonic acid A, *J. Chem. Soc. Perkin Trans 1* 1820–1822.

Hutchison, R. D., Steyn, P. S., and Thompson, D. L., 1971, The isolation and structure of 4-hydroxyochratoxin A and 7-carboxy-3,4-dihydro-8-hydroxy-3-methylisocoumarin from *Penicillium viridicatum*, *Tetrahedron Lett.* 4033–4036.

Isogai, A., Washizu, M., Murakoshi, S., and Suzuki, A., 1985, A new shikimate derivative, methyl 5-lactyl shikimate lactone, from *Penicillium* sp., *Agric. Biol. Chem.* **49:**167–169.

Japanese patent, 1984, Oxathinecarboxylic acid and its derivatives, *Chem. Abstr.* **100:**4785.

Japanese patent, 1985, Production of the anti-cholesteremic agent ML-236b, *Chem. Abstr.* **102:**77,263.

Johnson, J. R., Kidwai, A. R., and Warner, J. S., 1953, Gliotoxin. XI. A related antibiotic from *Penicillium terlikowski*: Gliotoxin monoacetate, *J. Am. Chem. Soc.* **75:**2110–2112.

Kaczka, E. A., Gitterman, C. O., Dulaney, E. L., and Folkers, K., 1962, Hadacidin, a new growth-inhibitory substance in human tumor systems, *Biochemistry* **1:**340–343.

Kalle, G. P., and Deo, Y. M., 1983, Effect of calcium on synthesis of dipicolinic acid in *Penicillium citreoviride* and its feedback resistant mutant, *J. Biosci.* **5:**321–330.

Kamal, A., Husain, S. A., Murtaza, N., Noorani, R., Qureshi, I. H., and Qureshi, A. A., 1970a, Studies in the biochemistry of microorganisms. 9. Structure of amudane, amudene and amujane, metabolic products of *Penicillium martensii*, *Pak. J. Sci. Ind. Res.* **13:**240–243.

Kamal, A., Jarboe, C. H., Qureshi, I. H., Husain, S. A., Murtaza, N., Noorani, R., and Qureshi, A. A., 1970b, Studies in the biochemistry of microorganisms. VIII. Isolation and characterisation of *Penicillium martensii* Biourge metabolic products: The structure of amudol, *Pak. J. Sci. Ind. Res.* **13:**236–239.

Kimura, Y., Suzuki, A., and Tamura, S., 1980, ^{13}C-NMR spectra of pestalotin and its analogues, *Agric. Biol. Chem.* **44:**451–452.

King, G. S., Waight, E. S., Mantle, P. G., and Szczyrbak, C. A., 1977, The structure of clavicipitic acid, an azepinoindole derivative from *Claviceps fusiformis*, *J. Chem. Soc. Perkin Trans. 1* 2099–2103.

King, T. J., Roberts, J. C., and Thompson, D. J., 1973, Studies in mycological chemistry. XXX. Isolation and structure of purpuride, a metabolite of *Penicillium purpurogenum* Stoll, *J. Chem. Soc. Perkin Trans. 1* 78–80.

Kobayashi, K., and Ui, T., 1975, Isolation of phytotoxic substances produced by *Cephalosporium gregatum* Allington and Chamberlain, *Tetrahedron Lett.* **47:**4119–4122.

Kobayashi, K., and Ui, T., 1977, Graminin A, a new toxic metabolite from *Cephalosporium gramineum* Nisikado and Ikata, *J. Chem. Soc. Chem. Commun.* 774.

Kobayashi, N., Iitaka, Y., Sankawa, U., Ogihara, Y., and Shibata, S., 1968, The crystal and molecular structure of a bromination product of tetrahydrorugulosin, *Tetrahedron Lett.* **58:**6135–6138.

Kozlovskii, A. G., and Reshetilova, T. A., 1984, Roquefortine biosynthesis in *Penicillium* culture, *Mikrobiologiya* **53:**81–84.

Kozlovskii, A. G., Soloveva, T. F., Reshetilova, T. A., and Skryabin, G. K., 1981a, Biosynthesis of roquefortine and 3,12-dihydroroquefortine by the culture of *Penicillium farinosum*, *Experientia* **37:**472–473.

Kozlovskii, A. G., Soloveva, T. F., Sakharovskii, V. G., and Adanin, V. M., 1981b, Biosynthesis of unusual ergot alkaloids by the mold *Penicillium aurantiovirens*, *Dokl. Akad. Nauk SSSR* **260:**230–233.

Kozlovskii, A. G., Stefanova-Avramova, L., Reshetilova, T. A., Sakharovskii, V. G., and Adanin, V. M., 1981c, Clavine ergoalkaloids—metabolites of *Penicillium gorlenkoanum*, *Prikl. Biokhim. Mikrobiol.* **17:**806–812.

Kozlovskii, A. G., Soloveva, T. F., Sakharovskii, V. G., and Adanin, V. M., 1982, Ergoalkaloids of agroclavine and epoxyagroclavine, metabolites of *Penicillium corylophilum*, *Prikl. Biokhim. Mikrobiol.* **18:**535–541.

Kriegler, A. B., and Thomas, R., 1971, The biosynthetic interrelationships of fungal phenalenones, *Chem. Commun.* 738–739.

Kuhr, I., Fuska, J., Sedmera, P., Podojil, M., Vokoun, J., and Vanek, Z., 1973, Antitumor antibiotic produced by *Penicillium stipitatum*: Its identity with duclauxin, *J. Antibiot.* **26**:535–536.

Kurobane, I., Hutchinson, C. R., and Vining, L. C., 1981, The biosynthesis of fulvic acid, a fungal metabolite of heptaketide origin, *Tetrahedron Lett.* **22**:493–496.

Lafont, P., Debeaupuis, J.-P., Gaillardin, M., and Payen, J., 1979, Production of mycophenolic acid by *Penicillium roqueforti* strains, *Appl. Environ. Microbiol.* **37**:365–368.

Lanigan, G. W., Payne, A. L., and Cockrum, P. A., 1979, Production of tremorgenic toxins by *Penicillium janthinellum* Biourge: A possible aetiological factor in ryegrass staggers, *Aust. J. Exp. Biol. Med. Sci.* **57**:31–37.

Latch, G. C. M., Christensen, M. J., and Samuels, G. J., 1984, Five endophytes of *Lolium* and *Festuca* in New Zealand, *Mycotaxon* **20**:535–550.

Laws, I., 1985, Biosynthesis and metabolism of indolic fungal metabolites, Ph.D. thesis, University of London.

Laws, I., and Mantle, P. G., 1985, Nigrifortine, a diketopiperazine metabolite of *Penicillium nigricans*, *Phytochemistry* **24**:1395–1397.

Leistner, L., and Pitt, J. I., 1977, Miscellaneous *Penicillium* toxins in: *Mycotoxins in Human and Animal Health* (J. V. Rodricks, C. W. Hesseltine, and M. A. Mehlman, eds.), Pathotox, Illinois, pp. 645–649.

Ljungcrantz, I., and Mosbach, K., 1964, Synthesis of four ethyl-dimethyl-benzenetriols in relation to a new phenolic metabolite of *Penicillium baarnense*, *Acta Chem. Scand.* **18**:638–642.

Locci, R., Merlini, L., Hasini, G., and Locci, J. R., 1967, Mitorubrinic acid and related compounds from a strain of *Penicillium funiculosum* Thom, *G. Microbiol.* **15**:93–102.

Lowe, G., Taylor, A., and Vining, L. C., 1966, Sporidesmins. IV. Isolation and structure of dehydrogliotoxin, a metabolite of *Penicillium terlikowskii*, *J. Chem. Soc. C* 1799–1803.

Luckner, M., 1967, Zur Bildung von Chinolinalkaloiden in Pflanzen. 2. Die fermentative Umwandlung der *Penicillium*-Alkaloide Cyclopenin und Cyclopenol in Viridicatin und Viridicatol, *Eur. J. Biochem.* **2**:74–78.

Lybing, S., and Reio, L., 1958, Degradation of ^{14}C-labelled carolic and carlosic acids from *Penicillium charlesii* G. Smith, *Acta Chem. Scand.* **12**:1575–1584.

Mabelis, R. P., Ratcliffe, A. H., Ackland, M. J., Hanson, J. R., and Hitchcock, P. B., 1981, Structure of thiobiscephalosporide-A, a macrolide from *Cephalosporium aphidicola*, *J. Chem. Soc. Chem. Commun.* 1006–1007.

MacMillan, J., 1953, Griseofulvin. 7. Dechlorogriseofulvin, *J. Chem. Soc.* 1967–1702.

MacMillan, J., 1954, Griseofulvin. 9. Isolation of the bromo-analogue from *Penicillium griseofulvum* and *Penicillium nigricans*, *J. Chem. Soc.* 2585–2587.

MacMillan, J., Vanstone, A. E., and Yeboah, S. K., 1972, Fungal products. III. Structure of wortmannin and some hydrolysis products, *J. Chem. Soc. Perkin Trans. 1* 2898–2903.

MacMillan, J. G., Springer, J. P., Clardy, J., Cole, R. J., and Kirksey, J. W., 1976, Structure and synthesis of verruculotoxin, a new mycotoxin from *Penicillium verruculosum* Peyronel, *J. Am. Chem. Soc.* **98**:246–247.

Mahmoodian, A., and Stickings, C. E., 1964, Studies in the biochemistry of micro-organisms. 115. Metabolites of *Penicillium frequentans* Westling: Isolation of sulochrin, asterric acid, (+)-bisdechlorogeodin and two new substituted anthraquinones, questin and questinol, *Biochem. J.* **92**:369–378.

Mantle, P. G., Day, J. B., Haigh, C. R., and Penny, R. H. C., 1978, Tremorgenic mycotoxins and incoordination syndromes, *Vet. Rec.* **103**:403.

Mantle, P. G., Perera, K. P. W. C., Maishman, N. J., and Mundy, G. R., 1983, Biosynthesis of penitrems and roquefortine by *Penicillium crustosum*, *Appl. Environ. Microbiol.* **45**:1486–1490.

Mantle, P. G., Laws, I., Tan, M. J. L., and Tizard, M., 1984, A novel process for the production of penitrem mycotoxins by submerged fermentation of *Penicillium nigricans*, *J. Gen. Microbiol.* **130**:1293–1298.

Mantle, P. G., and Shipston, N. F., 1987, Temporal separation of steps in the biosynthesis of verruculogen, *Biochem. Int.* (in press).

Marumo, S., 1959, Islanditoxin, a toxic metabolite produced by *Penicillium islandicum* Sopp, *Bull. Agric. Chem. Soc. Jpn.* **23**:428–437.

McCapra, F., Scott, A. I., Delmotte, P., Delmotte-Plaquee, J., and Bhacca, N. S., 1964, The constitution of monorden, an antibiotic with tranquilising action, *Tetrahedron Lett.* **15**:869–875.

McCorkindale, N. J., and Baxter, R. L., 1981, Brevigellin, a benzoylated cyclodepsipeptide from *Penicillium brevicompactum*, *Tetrahedron* **37**:1795–1801.

McCorkindale, N. J., Wright, J. L. C., Brain, P. W., Clarke, S. M., and Hutchinson, S. A., 1968, Canadensolide—an antifungal metabolite of *Penicillium canadense*, *Tetrahedron Lett.* **6**:727–730.

McCorkindale, N. J., Roy, T. P., and Hutchinson, S. A., 1972, Isolation and synthesis of 3-chlorogentisyl alcohol—a metabolite of *Penicillium canadense*, *Tetrahedron* **28**:1107–1111.

McCorkindale, N. J., Baxter, R. L., Roy, T. P., Shields, H. S., Stewart, R. M., and Hutchinson, S. A., 1978a, Synthesis and chemistry of *N*-benzoyl-*o*-[*N'*-benzoyl-l-phenylalanyl]-l-phenylalaninol, the major mycelial metabolite of *Penicillium canadense*, *Tetrahedron* **34**:2791–2795.

McCorkindale, N. J., Blackstock, W. P., Johnston, G. A., Roy, T. P., and Troke, J. A., 1978b, The biosynthesis of canadensolide, ethisolide and related metabolites, in: *11th IUPAC International Symposium on Chemistry of Natural Products: Symposium Papers*, Vol. 1 (R. Vlahov, ed.), Bulgarian Academy of Sciences, Bulgaria, pp. 151–154.

McCorkindale, N. J., Calzadilla, C. H., and Baxter, R. L., 1981, Biosynthesis of pebrolide, *Tetrahedron* **37**:1991–1993.

McGahren, W. J., Ellestad, G. A., Morton, G. O., Kunstmann, M. P., and Mullen, P., 1973, A new fungal lactone, LL-P880β, and a new pyrone, LL-P880γ, from a *Penicillium* sp., *J. Org. Chem.* **38**:3542–3544.

McMaster, W. J., Scott, A. I., and Trippett, S., 1960, Metabolic products of *Penicillium patulum*, *J. Chem. Soc.* 4628–4631.

Merlini, L., Nasini, G., and Selva, A., 1970, The structure of funicone, a new metabolite from *Penicillium funiculosum* Thom, *Tetrahedron* **26**:2739–2749.

Michel, K. H., Chaney, M. O., Jones, N. D., Hoehn, M. M., and Nagarajan, R., 1974, Epipolythiopiperazinedione antibiotics from *Penicillium turbatum*, *J. Antibiot.* **27**:57–64.

Michel, K. H., Demarco, P. V., and Nagarajan, R., 1977, The isolation and structure elucidation of macrocyclic lactone antibiotic A26771B, *J. Antibiot.* **30**:571–575.

Minato, S., 1979, Isolation of anthglutin, an inhibitor of γ-glutamyl transpeptidase from *Penicillium oxalicum*, *Arch. Biochem. Biophys.* **192**:235–240.

Miyao, K., 1960, The structure of fungisporin, *Bull. Agric. Chem. Soc. Jpn.* **24**:23–30.

Mizuno, K., Tsujino, M., Takada, M., Hayashi, M., Atsumi, K., Asano, K., and Matsuda, T., 1974, Studies on bredinin. I. Isolation, characterisation and biological properties, *J. Antibiot. (Tokyo)* **27**:775–782.

Mohammed, Y. S., and Luckner, M., 1963, The structure of cyclopenin and cyclopenol, metabolic products from *Penicillium cyclopium* Westling and *Penicillium viridicatum* Westling, *Tetrahedron Lett.* **28**:1953–1958.

Moore, R. E., and Emery, T., 1976, *N*-Acetylfusarinines: Isolation, characterisation and properties, *Biochemistry* **15**:2719–2733.

Moreau, S., Biguet, J., Lablache-Combier, A., Baert, F., Foulon, M., and Delfosse, C., 1980, Structures et stéréochimie des sesquiterpenes de *Penicillium roqueforti*, PR toxine et eremefortines A, B, C, D, E, *Tetrahedron* **36**:2989–2997.

Mosbach, K., 1960, Die Biosynthese der Orsellinsäure und Penicillinsäure, *Acta Chem. Scand.* **14**:457–464.

Moss, M. O., Wood, A. B., and Robinson, F. V., 1969, The structure of rubratoxin A, a toxic metabolite of *Penicillium rubrum*, *Tetrahedron Lett.* **5**:367–370.

Mull, R. P., Townley, R. W., and Scholz, C. R., 1945, Production of gliotoxin and a second active isolate by *Penicillium obscurum* Biourge, *J. Am. Chem. Soc.* **67**:1626–1627.

Murphy, G., and Lynen, F., 1975, Patulin biosynthesis: The metabolism of *m*-hydroxybenzyl alcohol and *m*-hydroxybenzaldehyde, *Eur. J. Biochem.* **58**:467–475.

Murphy, G., Vogel, G., Krippahl, G., and Lynen, F., 1974, Patulin biosynthesis: The role of mixed function oxidases in the hydroxylation of *m*-cresol, *Eur. J. Biochem.* **49**:443–455.

Nagel, D. W., Steyn, P. S., and Scott, D. B., 1972, Production of citreoviridin by *Penicillium pulvillorum*, *Phytochemistry* **11**:627–630.

Nagel, D. W., Pachler, K. G. R., Steyn, P. S., Vleggaar, R., and Wessels, P. L., 1976, The chemistry and ^{13}C NMR assignments of oxaline, a novel alkaloid from *Penicillium oxalicum*, *Tetrahedron* **32**:2625.

Nakajima, S., and Nozawa, K., 1979, Isolation in high yield of citrinin from *Penicillium odoratum* and of mycophenolic acid from *Penicillium brunneo-stoloniferum*, *J. Nat. Prod.* **42**:423–426.

Nakakita, Y., Shima, S., and Sakai, H., 1984a, Isolation of curvulic acid as an antimicrobial substance from *Penicillium janthinellum* C-268, *Agric. Biol. Chem.* **48**:1899–1900.

Nakakita, Y., Yomosa, K., Hirota, A., and Sakai, H., 1984b, Isolation of a novel phenolic compound from *Penicillium janthinellum* Biourge, *Agric. Biol. Chem.* **48**:239–240.

Narasimhachari, N., and Vining, L. C., 1963, Studies on the pigments of *Penicillium herquei*, *Can. J. Chem.* **41**:641–648.

Narasimhachari, N., Gopalkrishnan, K. S., Haskins, R. H., and Vining, L. C., 1963, The production of the antibiotic atrovenetin by a strain of *Penicillium herquei*, *Can. J. Microbiol.* **9**:134–136.

Niwa, M., Ogiso, S., Endo, T., Furukawa, H., and Yamamura, S., 1980, Isolation and structure of citreopyrone, a metabolite of *Penicillium citreo-viride* Biourge, *Tetrahedron Lett.* **21**:4481–4482.

Niwa, M., Endo, T., Ogiso, S., Furukawa, H., and Yamamura, S., 1981, Two new pyrones, metabolites of *Penicillium citreoviride* Biourge, *Chem. Lett.* 1285–1288.

Nozawa, K., and Nakajima, S., 1979, Isolation of radicicol from *Penicillium luteo-aurantium*, and meleagrin, a new metabolite, from *Penicillium meleagrinum*, *J. Nat. Prod.* **42**:374–377.

Nozoe, S., Morisaki, M., Tsuda, K., Iitaka, Y., Takahashi, N., Tamura, S., Ishibashi, K., and Shirasaka, M., 1965, The structure of ophiobolin, a C_{25} terpenoid having a novel skeleton, *J. Am. Chem. Soc.* **87**:4968–4970.

Ogihara, Y., Kobayashi, N., and Shibata, S., 1968, Further studies on the bianthraquinones of *Penicillium islandicum* Sopp, *Tetrahedron Lett.* **15**:1881–1886.

Ohmomo, S., Sato, T., Utagawa, T., and Abe, M., 1975, Isolation of festuclavine and three new indole alkaloids, roquefortine A, B, and C, from the cultures of *Penicillium roqueforti*, *Agric. Biol. Chem.* **39**:1333–1334.

Ohmomo, S., Miyazaki, K., Ohashi, T., and Abe, M., 1977, On the mechanism for the formation of indole alkaloids in *Penicillium concavo-rugulosum*, *Agric. Biol. Chem.* **41**:1707–1710.

Ohmomo, S., Oguma, K., Ohashi, T., and Abe, M., 1978, Isolation of a new indole alkaloid, roquefortine D, from cultures of *Penicillium roqueforti*, *Agric. Biol. Chem.* **42**:2387–2389.

Ohmomo, S., Oshashi, T., and Abe, M., 1980, Isolation of biogenetically correlated four alkaloids from the cultures of *Penicillium corymbiferum*, *Agric. Biol. Chem.* **44**:1929–1930.

Okuda, S., Iwasaki, S., Tsuda, K., Sano, Y., Hata, T., Udagawa, S., Nakayama, Y., and Yamaguchi, H., 1964, The structure of helvolic acid, *Chem. Pharm. Bull. (Tokyo)* **12**:121–124.

Okuda, S., Nak. yama, Y., and Tsuda, K., 1966, Studies on microbial products I: Helvolic acid and related compounds. I. 7-Desacetoxyhelvolic acid and helvolinic acid, *Chem. Pharm. Bull. (Tokyo)* **14**:436–441.

Okuda, S., Sato, Y., Hattori, T., and Igarashi, H., 1968a, Isolation of 3β-hydroxy-4β-hydroxymethylfusida-17(20),[16,21-cis],24-diene, *Tetrahedron Lett.* **47**:4769–4772.

Okuda, S., Sato, Y., Hattori, T., and Wakabayashi, M., 1968b, Isolation and structural elucidation of 3-oxo-16β-acetoxyfusida-1,17(20)[16,21-cis],24-trien-21-oic acid, *Tetrahedron Lett.* **47**:4847–4850.

Okuda, T., Yokose, K., Furumai, T., and Maruyama, H. B., 1984, Penitricin, a new class of antibiotic produced by *Penicillium aculeatum*. II. Isolation and characterisation, *J. Antibiot.* **37**:718–722.

Okuyama, E., Yamazaki, M., Kobayashi, K., and Sakurai, T., 1983, Paraherquonin, a new meroterpenoid from *Penicillium paraherquei*, *Tetrahedron Lett.* **24**:3113–3114.

Olivigni, F. J., and Bullerman, L. B., 1978, Production of penicillic acid and patulin by an atypical *Penicillium roqueforti* isolate, *Appl. Environ. Microbiol.* **35**:435–438.

Omura, S., Nakagawa, A., Sekikawa, K., Otani, M., and Hata, T., 1969, Studies on cerulenin. VI. Some spectroscopic features of cerulenin, *Chem. Pharm. Bull.* **17**:2361–2363.

Ooyama, J., Nakamura, N., and Tanabe, O., 1960, Biosynthesis of dipicolinic acid by a *Penicillium* sp., *Bull. Agric. Chem. Soc. Jpn.* **24**:743–744.

Oxford, A. E., and Raistrick, H., 1933, Studies in the biochemistry of microorganisms XXX: The molecular constitution of the metabolic products of *Penicillium brevi-compactum* and related species. I. The acids $C_{10}H_{10}O_5$, $C_{10}H_{10}O_6$ and $C_{10}H_{10}O_7$, *Biochem. J.* **27**:634–653.

Oxford, A. E., and Raistrick, H., 1948, Studies in the biochemistry of microorganisms. 76. Mycelianamide, $C_{22}H_{28}O_5N_2$, a metabolic product of *Penicillium griseofulvum* Dierckx. I. Preparation, properties and breakdown products, *Biochem. J.* **42**:323–329.

Oyama, H., Sassa, T., and Ikeda, M., 1978, Structures of new plant growth inhibitors, *trans*- and *cis*-resorcylide, *Agric. Biol. Chem.* **42**:2407–2409.

Patterson, D. S. P., Roberts, B. A., Shreeve, B. J., McDonald, S. M., and Hayes, A. W., 1979, Tremorgenic toxins produced by soil fungi, *Appl. Environ. Microbiol.* **37**:172–173.

Patterson, D. S. P., Shreeve, B. J., Roberts, B. A., and MacDonald, S. M., 1981, Verruculogen produced by soil fungi in England and Wales, *Appl. Environ. Microbiol.* **42**:916–917.

Pettersson, G., 1965a, Biosynthesis of spinulosin in *Penicillium spinulosum*, *Acta Chem. Scand.* **19**:1016–1017.

Pettersson, G., 1965b, The biosynthesis of flavipin. II. Incorporation of aromatic precursors, *Acta Chem. Scand.* **19**:1724–1732.

Pfeifer, S., Bar, H., and Zarnack, J., 1972, Uber Stoffwechselprodukte der Xanthocillin bildenden Mutante von *Penicillium notatum* Westl., *Pharmazie* **27**:536–542.

Pitt, J. I., 1979a, *The Genus Penicillium and Its Teleomorphic States Eupenicillium and Talaromyces*, Academic Press, London.

Pitt, J. I., 1979b, *Penicillium crustosum* and *P. simplicissimum*, the correct names for two common species producing tremorgenic mycotoxins, *Mycologia* **71**:1166–1177.

Polonsky, J., Merrien, M.-A., Prange, T., Pascard, C., and Moreau, S., 1980, Isolation and structure (X-ray analysis) of marcfortine A, a new alkaloid from *Penicillium roqueforti*, *J. Chem. Soc. Chem. Commun.* 601–602.

Posternak, T., 1940, Recherches sur la biochemie des champignons inférieurs. 4. Sur le pigment de *Penicillium roseopurpureum*, *Helv. Chim. Acta* **23**:1046–1053.

Prange, T., Billion, M.-A., Vuilhorgne, M., Pascard, C., and Polonsky, J., 1981, Structures of marcfortine B and C (X-ray analysis), alkaloids from *Penicillium roqueforti*, *Tetrahedron Lett.* **22**:1977–1980.

Price, M. J., and Worth, G. K., 1974, The occurrence of ergosta-4,6,8(14),22-tetraen-3-one in several fungi, *Aust. J. Chem.* **27**:2505–2507.

Probst, A., and Tamm, C., 1981, Biosynthesis of cytochalasin: Biosynthetic studies on chaetoglobosin A and 19-*O*-acetylchaetoglobosin A, *Helv. Chim. Acta* **64**:2065–2077.

Quick, A., Thomas, R., and Williams, D. J., 1980, X-ray crystal structure and absolute configuration of the fungal phenalenone herqueinone, *J. Chem. Soc. Chem. Commun.* 1051–1053.

Qureshi, I. H., Begum, T., and Murtaza, N., 1980, Isolation and identification of the metabolic products of *Penicillium funiculosum* Thom. The chemistry of funiculosic acid, *Pak. J. Sci. Ind. Res.* **23**:16–20.

Raistrick, H., and Rice, F. A. H., 1971, 2,3-Dihydro-3,6-dihydroxy-2-methyl-4-pyrone and curvularin from *Penicillium gilmanii*, *J. Chem. Soc. C* 3069–3070.

Raistrick, H., and Ross, D. J., 1952, Studies in the biochemistry of micro-organisms. 87. Dihydrogladiolic acid, a metabolic product of *Penicillium gladioli* Machacek, *Biochem. J.* **50**:635–647.

Raistrick, H., and Stossl, A., 1958, Studies in the biochemistry of micro-organisms. 104. Metabolites of *Penicillium atrovenetum* G. Smith: β-Nitropropionic acid, a major metabolite, *Biochem. J.* **68**:647–653.

Raju, M. S., Wu, G.-S., Gard, A., and Rosazza, J. P., 1982, Microbial transformations of natural antitumor agents. 20. Glucosylation of viridicatum toxin, *J. Nat. Prod.* **45**:321–327.

Rebstock, M. C., 1964, A new metabolite of patulin-producing *Penicillia*, *Arch. Biochem. Biophys.* **104**:156–159.

Rebuffat, S., Davoust, D., Molho, L., and Molho, D., 1980, La citreomontanine, nouvelle α-pyrone polyéthylénique isolée de *Penicillium pedemontanum*, *Phytochemistry* **19**:427–431.

Rebuffat, S., Davoust, D., and Molho, D., 1981, Biosynthesis of citreomontanin in *Penicillium pedemontanum*, *Phytochemistry* **20**:1279–1281.

Renauld, F., Moreau, S., and Lablache-Combier, A., 1984, Biosynthese de la botryodiplodine, mycotoxine de *Penicillium roqueforti*: Incorporations d'acetate [1-^{13}C], [2-^{13}C], [1-2-^{13}C] et d'acide orsellinique [2-^{13}C-carboxyle ^{13}C], [3-4^{13}C], *Tetrahedron* **40**:1823–1834.

Rhodes, A., Boothroyd, B., McGonagle, M. P., and Somerfield, G. A., 1961, Biosynthesis of griseofulvin: The methylated benzophenone intermediates, *Biochem. J.* **81**:28–37.

Rice, F. A. H., 1971, The structure of leucogenol, *J. Chem. Soc. C* 2599–2606.

Robbers, J. E., Straus, J. W., and Tuite, J., 1975, The isolation of brevianamide A from *Penicillium ochraceum*, *Lloydia* **38**:355–366.

Roberts, J. C., and Thompson, D. J., 1971, Studies in mycological chemistry. 27. Reinvestigation of the structure of purpurogenone, a metabolite of *Penicillium purpurogenum* Stoll, *J. Chem. Soc. C* 3488–3495.

Rowan, D. D., Hunt, M. B., and Gaynor, D. L., 1986, Peramine, a novel insect feeding deterrent from ryegrass infected with the endophyte *Acremonium loliae*, *J. Chem. Soc. Chem. Commun.* 935–936.

Sakabe, N., Goto, T., and Hirata, Y., 1977, Structure of citreoviridin, a mycotoxin produced by *Penicillium citreoviride* molded on rice, *Tetrahedron* **33**:3077–3081.

Sankawa, U., Taguchi, H., Ogihara, Y., and Shibata, S., 1966, Biosynthesis of duclauxin, *Tetrahedron Lett.* **25**:2883–2886.

Sankawa, U., Shimada, H., Sato, T., Kinoshita, T., and Yamasaki, K., 1977, Biosynthesis of scytalone, *Tetrahedron Lett.* **5**:483–486.

Sassa, T., Takemura, T., Ikeda, M., and Miura, Y., 1973, Structure of radiclonic acid, a new plant growth-regulator produced by a fungus, *Tetrahedron Lett.* **26**:2333–2334.

Sassa, T., Niwa, G., Unno, H., Ikeda, M., and Miura, Y., 1974, Structure of penicillide, a new metabolite produced by a *Penicillium* sp., *Tetrahedron Lett.* **45**:3941–3942.

Sassa, T., Nakano, K., and Miura, Y., 1975, Isolation and identification of *O*-acetyl and 12-hydroxyradiclonic acids, *Agric. Biol. Chem.* **39**:1899–1900.

Scott, A. I., and Beadling, L., 1974, Biosynthesis of patulin: Dehydrogenase and dioxygenase enzymes of *Penicillium patulum*, *Bioorg. Chem.* **3**:281–301.

Scott, A. I., Phillips, G. T., and Kircheis, U., 1971, Biosynthesis of polyketides: Synthesis of 6-methylsalicyclic acid and triacetic acid lactone in *Penicillium patulum*, *Bioorg. Chem.* **1**:380–399.

Scott, A. I., Zamir, L., Phillips, G. T., and Yalpani, M., 1973, Biosynthesis of patulin, *Bioorg. Chem.* **2**:124–139.

Scott, P. M., 1977, Penicillium mycotoxins, in: *Mycotoxic Fungi, Mycotoxins, Mycotoxicoses*, Vol. I (T. D. Wyllie and L. G. Morehouse, eds.), Marcel Dekker, New York, pp. 283–356.

Scott, P. M., Kennedy, B., and Van Walbeek, W., 1972, Desoxypatulinic acid from a patulin-producing strain of *Penicillium patulum*, *Experientia* **28:**1252.

Scott, P. M., Kennedy, B. P. C., Harwig, J., and Chen, Y. K., 1974, Formation of diketopiperazines by *Penicillium italicum* isolated from oranges, *Appl. Microbiol.* **28:**892.

Scott, P. M., Merrien, M.-A., and Polonsky, J., 1976, Roquefortine and isofumigaclavine A, metabolites of *Penicillium roqueforti*, *Experientia* **32:**140–142.

Sedmera, P., Podojil, M., Cokown, J., Bellina, V., and Memec, P., 1978, 2,2′-Dimethoxy-4a,4a′-dehydrorugulosin (rugulin), *Folia Microbiol. (Prague)* **23:**64–67.

Segal, W., 1959, Stipitatonic acid: A new mould tropolone from *Penicillium stipitatum* Thom, *J. Chem. Soc.* 2847–2851.

Sekiguchi, J., and Gaucher, G. M., 1979a, Isoepoxydon, a new metabolite of the patulin pathway in *Penicillium urticae*, *Biochem. J.* **182:**445–453.

Sekiguchi, J., and Gaucher, G. M., 1979b, Patulin biosynthesis: The metabolism of phyllostine and isoepoxydon by cell-free preparations from *Penicillium urticae*, *Can. J. Microbiol.* **25:**881–887.

Sekiguchi, J., Gaucher, G. M., and Yamada, Y., 1979, Biosynthesis of patulin in *Penicillium urticae*: Identification of isopatulin as a new intermediate, *Tetrahedron Lett.* **1:**41–42.

Sekiguchi, J., Shimamato, T., Yamada, Y., and Gaucher, G. M., 1983, Patulin biosynthesis: Enzymatic and nonenzymatic transformations of the mycotoxin (E)-ascladiol, *Appl. Environ. Microbiol.* **45:**1939–1942.

Sekiguchi, J., Kuroda, H., Yamada, Y., and Okada, H., 1985, Structure of patulolide A, a new metabolite from *Penicillium urticae* mutants, *Tetrahedron Lett.* **26:**2341–2342.

Seo, S., Sankawa, U., Ogihara, Y., Iitaka, Y., and Shibata, S., 1973, Studies on fungal metabolites. 32. A renewed investigation on flavoskyrin and its analogues, *Tetrahedron* **29:**3721–3726.

Seto, H., and Tanabe, M., 1974, Utilization of ^{13}C–^{13}C coupling in structural and biosynthetic studies: Ochrephilone—a new fungal metabolite, *Tetrahedron Lett.* **8:**651–654.

Shibata, S., and Udagawa, S., 1963, Metabolic products of fungi. 19. Isolation of rugulosin from *Penicillium brunneum*, *Chem. Pharm. Bull.* **11:**402–403.

Shibata, S., Sankawa, U., Taguchi, H., and Yamasaki, K., 1966, Biosynthesis of natural products. 3. Biosynthesis of erythroskyrin, a coloring matter of *Penicillium islandicum* Sopp., *Chem. Pharm. Bull.* **14:**474–478.

Shizuri, Y., Niwa, M., Furukawa, H., and Yamamura, S., 1983, Isolation and structure of citreothiolactone, a novel metabolite of *Penicillium citreoviride*, *Tetrahedron Lett.* **24:**1053–1054.

Shizuri, Y., Kosemura, S., Yamamura, S., Furukawa, H., Kawai, K., and Okada, N., 1984a, Biosynthesis of citreothiolactone, citreopyrone and pyrenocine B, *Tetrahedron Lett.* **25:**1583–1584.

Shizuri, Y., Nishiyama, S., Imai, D., Yamamura, S., Furukawa, H., Kawai, K., and Okada, N., 1984b, Isolation and stereostructures of citreoviral, citreodiol and epicitreodiol, *Tetrahedron Lett.* **25:**4771–4774.

Sigg, H. P., 1963, Die Konstitution von Frequentin, *Helv. Chim. Acta* **46:**1061–1065.

Sigg, H. P., 1964, Die Konstitution von Brefeldin A, *Helv. Chim. Acta* **47:**1401–1415.

Simonart, P., and Wiaux, A., 1959, Biochemical study of *Penicillium griseofulvum*. I. Presence of *o*-, *m*-, and *p*-hydroxybenzoic acids, *Bull. Soc. Chim. Biol.* **41:**537–540.

Simpson, T. J., 1979, Carbon-13 nuclear magnetic resonance structural and biosynthetic studies on deoxyherqueinone and herqueichrysin, phenolenone metabolites of *Penicillium herquei*, *J. Chem. Soc. Perkin Trans. 1* 1233–1238.

Simpson, T. J., Lunnon, M. W., and MacMillan, J., 1979, Fungal products. 21. Biosynthesis of the fungal metabolite, wortmannin, from [1,2$^{13}C_2$]-acetate, *J. Chem. Soc. Perkin Trans 1* 931–934.

Simpson, T. J., Stenzel, D. J., Bartlett, A. J., O'Brien, E., and Holker, J. S. E., 1982, Studies on fungal metabolites. 3. ^{13}C NMR spectral and structural studies on austin and new related meroterpenoids from *Aspergillus ustus, Aspergillus variecolor* and *Penicillium diversum, J. Chem. Soc. Perkin Trans. 1* 2687–2692.

Singleton, V. L., Bohonos, N., and Ullstrup, A. J., 1958, Decumbin, a new compound from a species of *Penicillium, Nature (London)* **181:**1072–1073.

Springer, J. P., Clardy, J., Wells, J. M., Cole, R. J., and Kirksey, J. W., 1975, The structure of paxilline, a tremorgenic metabolite of *Penicillium paxilli, Tetrahedron Lett.* **30:**2531–2534.

Stack, M. E., and Mislivec, P. B., 1978, Production of xanthomegnin and viomellein by isolates of *Aspergillus ochraceous, Penicillium cyclopium* and *Penicillium viridicatum, Appl. Environ. Microbiol.* **36:**552–554.

Stack, M. E., Eppley, R. M., Dreifuss, P. A., and Pohland, A. E., 1977, Isolation and identification of xanthomegnin, viomellein, rubrosulphin and viopurpurin as metabolites of *Penicillium viridicatum, Appl. Environ. Microbiol.* **33:**351–355.

Stack, M. E., Mazzola, E. P., and Eppley, R. M., 1979, Structures of xanthoviridicatin D and xanthoviridicatin G, metabolites of *Penicillium viridicatum*: Application of proton and carbon-13 NMR spectroscopy, *Tetrahedron Lett.* **52:**4989–4992.

Steiner, E., Kalamar, J., Charollais, E., and Posternak, T., 1974, Recherches sur la biochemie des champignons inférieurs. IX. Synthèse de précurseurs marqués et biosynthèse de la phoenicine et de l'oosporeine, *Helv. Chim. Acta* **57:**2377–2387.

Stevens, R. L., and Emery, T. F., 1966, The biosynthesis of hadacidin, *Biochemistry* **5:**74–81.

Steyn, P. S., and Vleggaar, R., 1983, Roquefortine, an intermediate in the biosynthesis of oxaline in cultures of *Penicillium oxalicum, J. Chem. Soc. Chem. Commun.* 560–561.

Steyn, P. S., Vleggaar, R., Wessels, P. L., and Woudenberg, M., 1982, Biosynthesis of citreoviridin: A carbon-13 NMR study, *J. Chem. Soc. Perkin Trans. 1* 2175–2178.

Studer, R. O., 1969, Synthesis and structure of fungisporin, *Experientia* **25:**899.

Sumiki, Y., and Miyao, K., 1952, Studies on fungisporin. I, *J. Agric. Chem. Soc. Jpn.* **26:**27–31.

Sunagawa, M., Ohta, T., and Nozoe, S., 1979, Isolation and structure of brefeldin C, *Heterocycles* **13:**267–270.

Suter, P. J., and Turner, W. B., 1967, 2-Pyruvoylaminobenzamide, a metabolite of *Penicillium chrysogenum, J. Chem. Soc. C* 2240–2242.

Tabuchi, T., Nakamura, I., and Kobayashi, T., 1977, Accumulation of the open-ring acid of spiculisporic acid by *Penicillium spiculisporum, J. Ferment. Technol.* **55:**37–49.

Takeda, N., Seo, S., Ogihara, Y., Sankawa, U., Iitaka, Y., Kitagawa, I., and Shibata, S., 1973, Studies on fungal metabolites. 31. Anthraquinoid colouring matters of *Penicillium islandicum* Sopp and some other fungi, luteoskyrin, rubroskyrin, rugulosin and their related compounds, *Tetrahedron* **29:**3703–3719.

Tanenbaum, S. W., and Bassett, E. W., 1959, The biosynthesis of patulin. III. Evidence for a molecular rearrangement of the aromatic ring, *J. Biol. Chem.* **234:**1861–1866.

Tanenbaum, S. W., and Nakajima, S., 1969, The biosynthesis of pulvilloric acid, *Biochemistry* **8:**4622–4631.

Tatsuno, T., Kobayashi, N., Okubo, K., and Tsunoda, H., 1975, Recherches toxicologiques sur les substances toxiques de *Penicillium tardum.* 1. Isolement et identification des substances cytotoxiques, *Chem. Pharm. Bull.* **23:**351–354.

Tertzakian, G., Haskins, R. H., Slater, G. P., and Nesbitt, L. R., 1964, The structure of cephalochromin, *Proc. Chem. Soc.* 195–196.

Thomas, R., 1961, Studies in the biosynthesis of fungal metabolites. 3. The biosynthesis of fungal perinaphthenones, *Biochem. J.* **78:**807–813.

Tony Lam, Y. K., Gullo, U. P., Goegelman, R. T., Jorn, D., Huang, L., De Riso, C., Monaghan, R. L., and Putter, I., 1981, Dihydrocompactin, a new potent inhibitor of 3-hydroxy-3-methylglutaryl coenzyme A reductase from *Penicillium citrinum*, *J. Antibiot.* **34:**614–616.

Turner, W. B., 1971, *Fungal Metabolites*, Academic Press, London.

Turner, W. B., 1978, The isolation and structures of the fungal metabolites lapidosin and diversonol, *J. Chem. Soc. Perkin Trans. 1* 1621.

Turner, W. B., and Aldridge, D. C., 1983, *Fungal Metabolites II*, Academic Press, New York.

Udagawa, S., 1963, Sclerotiorin, a major metabolite of *Penicillium hirayamae*, *Chem. Pharm. Bull.* **11:**366–367.

Uramoto, M., Tanabe, M., Hirotsu, K., and Clardy, J., 1982, A new tremorgenic metabolite related to verruculogen from *Penicillium verruculosum*, *Heterocycles* **17:**349–354.

Van der Merwe, K. J., Steyn, P. S., and Fourie, L., 1965, The constitutions of ochratoxins A, B, and C, metabolites of *Aspergillus ochraceous* Wilh., *J. Chem. Soc.* 7083–7088.

Van Eijk, G. W., 1973, Anthraquinone in the fungus *Talaromyces stipitatus*, *Experientia* **29:**522–523.

Van Walbeek, W., Scott, P. M., Harwig, J., and Lawrence, J. W., 1969, *Penicillium viridicatum* Westling: A new source of ochratoxin A, *Can. J. Microbiol.* **15:**1281–1285.

Verachtert, H., and Hanssens, L., 1975, Isolation and identification of phenolic substances from the fungus *Penicillium spinulosum*, *Ann. Microbiol. (Inst. Pasteur)* **126A:**143–149.

Vesonder, R. F., 1979, Xanthocillin, a metabolite of *Eupenicillium egyptiacum* NRRL 1022, *J. Nat. Prod.* **42:**232–233.

Vesonder, R. F., Stodola, F. H., Wickerham, L. J., Ellis, J. J., and Rohwedder, W. K., 1971, 11-Hydroxy-*trans*-9-dodecanoic acid lactone, a 12-membered-ring compound from a fungus, *Can. J. Chem.* **49:**2029–2032.

Vesonder, R. F., Stodola, F. H., and Rohwedder, W. K., 1972, Formation of the δ-lactone of 3,5-dihydroxydecanoic acid by the fungus *Cephalosporium recifei*, *Can. J. Biochem.* **50:**363–365.

Vesonder, R. F., Ciegler, A., Fennell, D., Tjarks, L. W., and Jensen, A. H., 1976, Curvularin from *Penicillium baradicum* Baghdadi NRRL 3754, and biological effects, *J. Environ. Sci. Health* **B11:**289–297.

Vesonder, R. F., Tjarks, L., Rohwedder, W. K., and Kieswetter, D. O., 1980, Indole metabolites of *Penicillium cyclopium* NRRL 6093, *Experientia* **36:**1308.

Vining, L. C., McInnes, A. G., Smith, D. G., Wright, J. L. C., and Taber, W. A., 1982, Dimeric clavine alkaloids produced by *Penicillium citreo-viride*, *FEMS Symp.* **13:**243–251.

Vleggaar, R., and Wessels, P. L., 1980, Stereochemistry of the dehydrogenation of (2*S*)-histidine in the biosynthesis of roquefortine and oxaline, *J. Chem. Soc. Chem. Commun.* 160.

Wagener, R. E., Davis, N. D., and Diener, U. L., 1980, Penitrem A and roquefortine production by *Penicillium commune*, *Appl. Environ. Microbiol.* **39:**882–887.

Weedon, C. M., and Mantle, P. G., 1987, Paxilline biosynthesis by *Acremonium loliae*; a step towards defining the origin of lolitrem neurotoxins. *Phytochemistry* **26:**969–971.

Weete, J. D., and Laseter, J. L., 1974, Distribution of sterols in fungi. I. Fungal spores, *Lipids* **9:**575–581.

Wei, R.-D., Schnoes, H. K., Hart, P. A., and Strong, F. M., 1975, The structure of PR toxin, a mycotoxin from *Penicillium roqueforti*, *Tetrahedron* **31:**109–114.

White, J. D., and Taylor, S. I., 1970, Biosynthesis of ergosta-4,6,8(14),22-tetraen-3-one: *In vivo* incorporation of a 1,4-dioxide, *J. Am. Chem. Soc.* **92:**5811–5813.

White, J. D., Perkins, D. W., and Taylor, S. I., 1973, Biosynthesis of ergosta-4,6,8(14),22-tetraen-3-one: A novel oxygenative pathway, *Bio-org. Chem.* **2:**163–175.

Wijkman, N., 1931, Uber einige neue, durch Schimmelpilze gebildete Substanzen, *Ann. Chim.* **485:**61–73.

Willingale, J., Atwell, S. M., and Mantle, P. G., 1983a, Unusual ergot alkaloid biosynthesis in sclerotia of a *Claviceps purpurea* mutant, *J. Gen. Microbiol.* **129:**2109–2115.

Willingale, J., Perera, K. P. W. C., and Mantle, P. G., 1983b, An intermediary role for the tremorgenic mycotoxin TR-2 in the biosynthesis of verruculogen, *Biochem. J.* **214:**991–993.

Wilson, B. J., Wilson, C. H., and Hayes, A. W., 1968, Tremorgenic toxin from *Penicillium cyclopium* grown on food materials, *Nature (London)* **220:**77–78.

Wilson, B. J., Yang, D. T. C., and Harris, T. M., 1973, Production, isolation and preliminary toxicity studies of brevianamide A from cultures of *Penicillium viridicatum*, *Appl. Microbiol.* **26:**633–635.

Yamamoto, I., Mizuta, E., Henmi, T., Yamono, T., and Yamatodani, S., 1973, Epoformin, a new antibiotic produced by *Penicillium claviforme*, *Takeda Kenkynsho Ho* **32:**532–538.

Yamatodani, S., Asahi, Y., Matsukura, A., Ohmomo, S., and Abe, M., 1970, Structure of rugulovasine A, B and their derivatives, *Agric. Biol. Chem.* **34:**485–487.

Yamazaki, M., Fujimoto, H., and Miyaki, K., 1972, Metabolites of some strains of *Penicillium* isolated from foods, *Yakugaku Zasshi* **91:**101–104.

Yamazaki, M., Okuyama, E., Kobayashi, M., and Inoue, H., 1981, The structure of paraherquamide, a toxic metabolite from *Penicillium paraherquei*, *Tetrahedron Lett.* **22:**135–136.

Yeulet, S. E., Mantle, P. G., Bilton, J. N., Rzepa, H. S., and Sheppard, R. N., 1986, Auranthine, a new benzodiazepinone metabolite of *Penicillium aurantiogriseum*, *J. Chem. Soc. Perkin Trans. 1* 1891–1894.

Yoshihira, K., Takahashi, C., Sekita, S., and Natori, S., 1972, Tetrahydroauroglaucin from *Penicillium charlesii*, *Chem. Pharm. Bull.* **20:**2727–2728.

Yoshioka, H., Nakatsu, K., Sato, M., and Tatsuno, T., 1973, The molecular structure of cyclochlorotine, a toxic chlorine-containing cyclic pentapeptide, *Chem. Lett.* 1319–1322.

Zeeck, A., Rub, P., Laatsch, H., Loeffler, W., Wehrle, H., Zahner, H., and Holst, H., 1979, Isolierung des Antibioticums Semi-vioxanthin aus *Penicillium citreoviride* und Synthese des Xanthomegnins, *Chem. Ber.* **112:**957–978.

Extracellular Enzymes of *Penicillium* 7

PAUL F. HAMLYN, DAVID S. WALES, and
BRIAN F. SAGAR

1. INTRODUCTION

The wide range of extracellular enzymes produced by species of *Penicillium* play an important role in the microbiological breakdown of organic materials. Notable examples of hydrolases of *Penicillium* include various cellulolytic enzymes and other polysaccharases, such as α- and β-glucanases, hemicellulases, and pectic enzymes, together with a variety of lipases and proteolytic enzymes that are recognized as being responsible for the development of the characteristic flavors in ripened cheeses. Despite this plethora of extracellular enzymes, surprisingly few species of *Penicillium* have so far been adopted for industrial enzyme production. Indeed, Godfrey and Reichelt (1983) point out that most industrial microbial enzymes are produced from no more than 11 fungi, 8 bacteria, and 4 yeasts. This restriction is largely explained by the fact that only a few microorganisms are approved for use in the food industry, and enzyme manufacturers are usually keen to include food processing as a potential market for their products. Nevertheless, the use of microbial enzymes in industrial processes has expanded considerably in recent years, and the *Penicillia* would seem to be particularly well placed to have an increasing part to play in the future manufacture of these enzymes.

2. CELLULOSE-DEGRADING ENZYMES

Many microorganisms produce extracellular enzymes that catalyze the hydrolysis of water-soluble cellulose derivatives, such as carboxymethyl- or hydroxyethylcellulose of relatively low degrees of substitution, and of highly

---

PAUL F. HAMLYN, DAVID S. WALES, and BRIAN F. SAGAR • Biotechnology Group, Shirley Institute, Didsbury, Manchester M20 8RX, England.

swollen forms of cellulose, e.g., phosphoric-acid-swollen cellulose, but relatively few have the ability to hydrolyze highly ordered crystalline cellulose, as typified by cotton (Wood *et al.*, 1980). Included among the cellulolytic microfungi that are able to hydrolyze highly ordered cellulose are members of the genus *Penicillium* (Selby, 1968; Wood and McCrae, 1977; Somani and Wangikar, 1979).

2.1. The Cellulase Complex

As long ago as 1950, Reese *et al.* (1950) proposed that the cellulase complex contains three major enzyme components. The first of these components, which is produced only by organisms that are able to degrade crystalline cellulose, was termed C_1, so named because it was believed to act first on the highly ordered arrays of cellulose molecules in the elementary fibrils by deaggregating the cellulose chains, making them available for subsequent hydrolysis by the β-1,4-glucanases. This latter group of hydrolytic enzymes forms the second major component of the cellulase complex, generally referred to as C_x, the x reflecting the fact that normally there are several of these enzymes present in culture filtrates of cellulolytic microorganisms. The C_x component contains both endo- and exo-β-1,4-glucanases that are responsible for hydrolysis of soluble derivatives of cellulose or swollen and partially degraded cellulose, but are without effect on crystalline cellulose in the absence of the C_1 component. The products of the action of the C_x enzymes are short-chain cellooligosaccharides, cellobiose, and glucose. The third component of the cellulase complex, the *β-glucosidases* (β-D-glucoside glucohydrolases), convert the cellooligosaccharides and cellobiose to glucose. The proposed actions of the various components are summarized in Fig. 1.

The cellulase complex of *P. funiculosum* was shown by Selby (1968) to conform to Reese's concept. Selby also demonstrated cross-synergism between the C_1 and C_x components of *P. funiculosum* and *Trichoderma viride*; the C_1 of *P. funiculosum*, when recombined with the C_x of *T. viride*, was almost as effective at solubilizing crystalline cellulose as the recombined C_1 and C_x from *P. funiculosum* only.

While the C_x components from *P. funiculosum* exhibit the classic exo- and endo-β-1,4-glucanase activities (Wood and McCrae, 1977), the action of the C_1 component remains unclear. Following the development of improved fractionation methods and the discovery that the C_1 fraction appears to possess cellobiohydrolase activity, several groups of workers reversed the original concept of Reese *et al.* (1950) by suggesting that it is the endo-β-1,4-glucanase that attacks first, in a random manner, to produce chain ends for subsequent attack by the exoenzyme, β-1,4-glucan cellobiohydrolase. However, Reese (1976) critically questioned the implication that C_1 and this

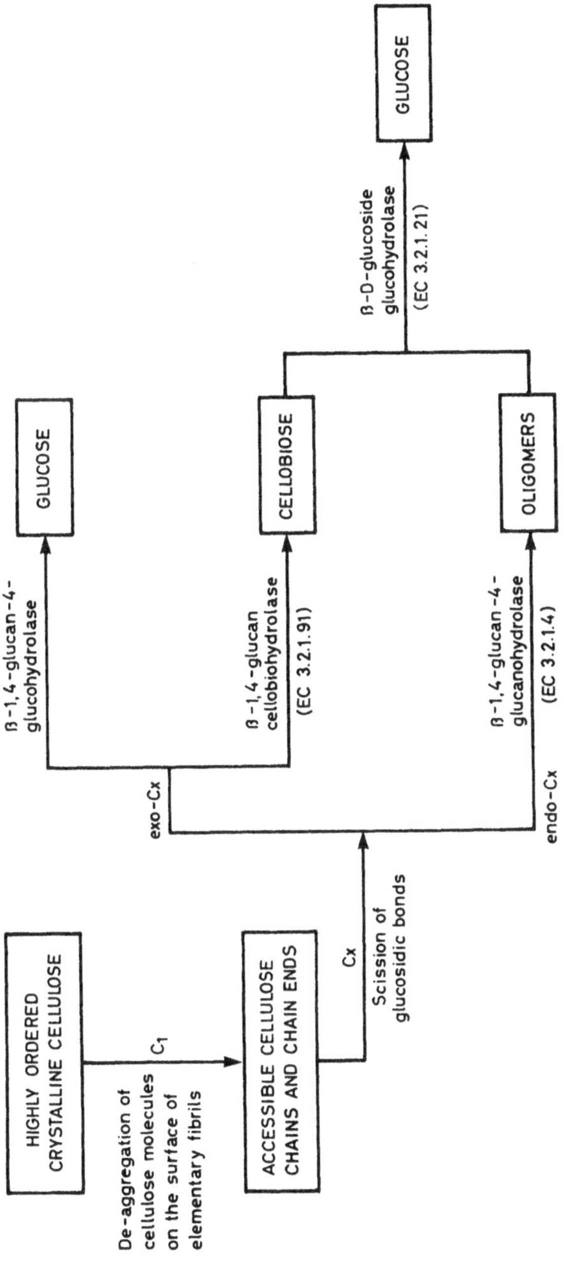

Figure 1. Actions of the various components of the cellulase complex on highly ordered crystalline cellulose.

cellobiohydrolase are one and the same enzyme activity and gave pertinent reasons for rejecting this suggestion. In the light of the new evidence, however, he did modify the original concept to include the possibility that C_1 randomly splits the glucosidic bonds of cellulose chains situated in the crystallite surface. Wood *et al.* (1980) have confirmed that the cellulase complex of *P. funiculosum* contains an enzyme component that attacks H_3PO_4-swollen cellulose at a low rate, is not inhibited by low concentrations of D-glucono-1,5-lactone, shows no transferase activity, is highly specific for the β-1,4-linkage, and attacks cellooligosaccharides at a rate that increases as the degree of polymerization increases. These observations indicate that this enzyme is an exo-β-1,4-glucanase and, since cellobiose is the principal product of hydrolysis, that the enzyme is a β-1,4-glucan cellobiohydrolase.

Exhaustive attempts (Sagar, 1978) failed to detect any significant changes in the fine structure of native cotton cellulose following treatment with the individual C_1 or C_x components, confirming that highly ordered cellulose is totally unaffected unless both these enzyme components are present (Fig. 2). The observation that the combined action of the C_1 and C_x components appears to be confined to the pair of faces of the crystalline elementary fibrils containing the 2,3 and 2,3,6 hydroxyl groups, accessible on alternate anhydroglucose units along the cellulose chains, led Sagar (1978, 1985) to suggest that a component of the cellulase system has a specific affinity for the cellulose molecules in this pair of surfaces (associated with the unique spatial configuration of their free hydroxyl groups), leading to chain scission and production of free cellulose chain ends accessible to the C_x exoglucanases. The possibility must be considered that this unique stereospecific endoglucanase activity (which we shall continue to call C_1) resides in the same protein molecule as that which contains the cellobiohydrolase activity. Alternatively, it is possible that the C_1 activity is mediated by an enzyme–enzyme complex formed between the β-1,4-glucan cellobiohydrolase and one of the endo-β-1,4-glucanases (Sagar, 1978, 1985; Wood and McCrae, 1978). The observation that the synergism between the C_1 and C_x components of some fungi is more effective than that between others would support such a hypothesis (Wood *et al.*, 1980).

Another exo-β-1,4-glucanase has been characterized from the cellulase of *P. funiculosum* (Wood and McCrae, 1982). This enzyme has the activity of an exo-β-1,4-glucan-4-glucohydrolase, but, unlike the cellobiohydrolase, does not act in synergism with the C_x components to hydrolyze crystalline cellulose.

2.2. Regulation of Cellulase Activity

The cellulase complex from *Penicillia* can be regulated in either or both of two ways: (1) inhibition of synthesis of one or more of the components of

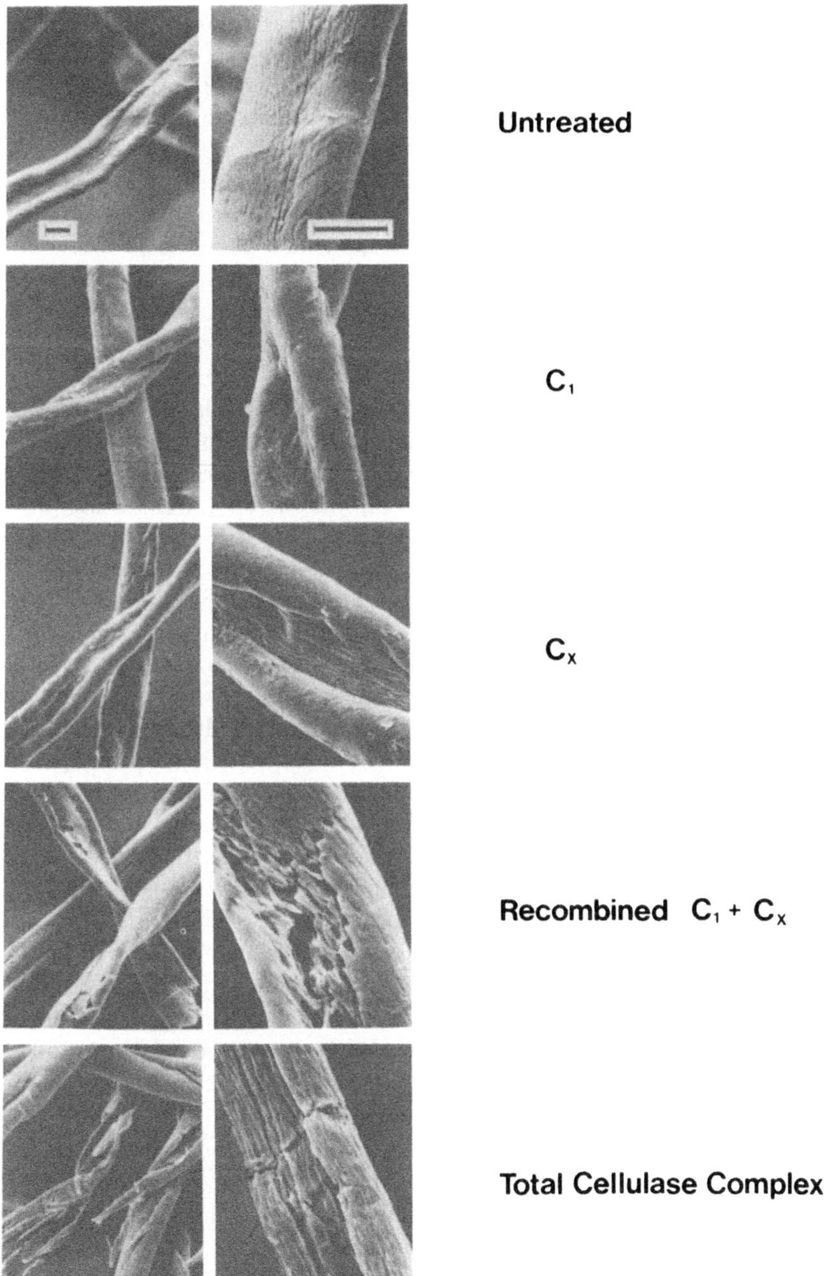

Figure 2. Appearance of cotton fiber under the scanning electron microscope before and after treatment with various components of the cellulase complex. Scale bars: 10 μm.

the complex and (2) inhibition of the activity of any single component of the complex.

Spalding *et al.* (1973) showed that C_x production by *P. expansum* (as measured viscometrically with carboxymethylcellulose as substrate) could be inhibited by several sugars and an amino acid. When 0.1 M concentrations of arabinose, mannose, galactose, glucose, sucrose, raffinose, galacturonic acid, and glutamic acid were added to the growth medium, C_x production was inhibited, even when the sugars and the amino acid were added 24–48 hr after inoculation of the medium.

In the case of raffinose, the enzyme repression may have been associated with one or more of its hydrolysis products (glucose, galactose, fructose), rather than with the parent compound.

Production of endo-β-1,4-glucanases by *P. janthinellum* is induced by cellobiose but not by sophorose (Rapp *et al.*, 1981), unlike the case of *T. viride*, in which both these sugars act as inducers (Sternberg and Mandels, 1979). Indeed, sophorose was found to repress endo-β-1,4-glucanase production by *P. janthinellum* (Rapp *et al.*, 1982), as did glucose and glycerol. The β-1,4-glucanases, once induced, were also subject to product inhibition. They were inhibited competitively by glucose (enzyme inhibitor constant $K_i = 2.75$ mM) and cellobiose ($K_i = 1.6$ mM), although the value of K_i for cellobiose may be artificially high, due to the formation of glucose by cellobiase associated with the β-1,4-glucanases.

Two cellobiases are produced by *P. janthinellum*; although they did not differ very much with respect to the values of their Michaelis constants (K_m) when cellobiose was used as substrate, they differed significantly in the extent of their competitive inhibition by glucose and D-glucono-1,5-lactone (Rapp *et al.*, 1982). Rapp and colleagues concluded that the unusually strong inhibition (K_i for glucose was 0.075 mM and for D-glucono-1,5-lactone 0.035 mM) of one of the cellobiases might indicate that it is not extracellular, but is located in the cell wall, periplasmic space, or interior of the cell. The reason for the differences in inhibition of the two cellobiases is not clear. If D-glucono-1,5-lactone is produced at all in *P. janthinellum*, it is produced intracellularly from glucose by glucose oxidase, whereas glucose is formed mainly outside the cell by the combined action of the various extracellular components of the cellulase complex. The lactone could therefore be part of the system that regulates intracellular glucose levels by preventing the cellobiases from hydrolyzing cellobiose to glucose. This in turn could regulate endo-β-1,4-glucanase biosynthesis through an induction-repression mechanism.

2.3. Influence of Substrate on Cellulase Action

The efficiency of the action of any cellulase complex depends markedly on the source of the cellulose substrate. Betrabet *et al.* (1980) examined the action of the cellulase produced by *P. funiculosum* on cotton, bagasse pulp, wheat-straw pulp, and chlorite-bleached sawdust. The authors found cotton to be the most resistant of these substrates to enzymatic hydrolysis (as determined by the amount of reducing sugar produced after 6 hr of enzyme treatment), this resistance being attributable to its high degree of crystallinity and its relatively large crystallite dimensions. The remaining three substrates had similar degrees of crystallinity and crystallite dimensions, although less reducing sugar was produced from the sawdust after 6 hr of enzyme treatment than from the bagasse pulp and wheat-straw pulp, the difference being attributed to the relatively low α-cellulose and high lignin contents of the sawdust. Dighe *et al.* (1981) investigated the susceptibility of a variety of bleached cellulosic substrates to hydrolysis by the cellulase of *P. funiculosum.* Wheat straw, jute stalk, bagasse, and cottonseed hulls were found to be most susceptible to enzymatic hydrolysis, newspaper waste was moderately susceptible, and rice husk, sawdust, and cotton stalk were comparatively resistant. The susceptibility of the substrates did not appear to be dependent on their α-cellulose content; cottonseed hulls and bagasse contained 54 and 81% of α-cellulose, respectively, but yielded almost equal amounts of reducing sugar after enzymatic hydrolysis. These authors also suggested that a combination of factors—viz., degree of crystallinity, unit cell dimension of the crystallite, degree of polymerization of the cellulose, and the presence of extraneous material—could be responsible for the differences in susceptibility.

2.4. Cellulase-Producing Species of *Penicillium*

Several species of *Penicillium* have been reported to have cellulolytic activity (Table I).

Sidhu and Sandhu (1979) considered a number of fungi, including a *Penicillium* sp. The organism was shown to possess C_1 activity (as measured by the ability of a culture filtrate to hydrolyze filter-paper cellulose), C_x activity (using carboxymethylcellulose as substrate), and β-glucosidase activity (using salicin as substrate). Lynch *et al.* (1981) isolated a variety of cellulolytic microorganisms from soil using two isolation techniques. In the first method, soil was added to an enrichment medium in a batch fermenter, and samples were removed at 1-week intervals and screened for the presence of cellulolytic microorganisms. In the second method, soil was distributed evenly among glass beads contained in a column, and enrichment medium from a reservoir was circulated continuously through the column; samples were again screened for cellulolytic microorganisms. These two isolation methods yielded different populations; *P. chrysogenum* was among those isolated by the batch fermenter method, and *P. cyclopium* was among those

Table I. Production of Cellulase Enzymes by Various Species of *Penicillium*

Species	References
Penicillium sp.	Sidhu and Sandhu (1979), Durand *et al.* (1984)
P. chrysogenum	Lynch *et al.* (1981), B. S. Singh and Saksena (1981)
P. citrinum	B. S. Singh and Saksena (1981)
P. commune	B. S. Singh and Saksena (1981)
P. cyclopium	Lynch *et al.* (1981)
P. dupontii	Qureshi *et al.* (1980a)
P. funiculosum	Selby (1968), Wood and McCrae (1977), Somani and Wangikar (1979), Betrabet *et al.* (1980), Dighe *et al.* (1981), Joglekar *et al.* (1983), Mishra *et al.* (1983)
P. iriensis	Enari (1983)
P. janthinellum	Rapp *et al.* (1981)
P. nigricans	B. S. Singh and Saksena (1981)
P. notatum	D. P. Sharma and Saksena (1979), B. S. Singh and Saksena (1981)
P. variabile	Szakacs *et al.* (1981)
P. vermiculatum	Szakacs *et al.* (1981)
P. verruculosum	Szakacs *et al.* (1981)

isolated from the column. The culture filtrates of both species contained cellulases active against Avicel, carboxymethylcellulose, and cellobiose, although the cellulase from *P. chrysogenum* was more active in every case than the cellulase from *P. cyclopium*. D. P. Sharma and Saksena (1979) examined the cellulolytic ability of 58 species of *Penicillium* isolated from various sources including soil, humus, litter, and composts of paddy straw, leaf, and fruit. Cellulase activity was assayed using cellulose powder, carboxymethylcellulose, and filter paper as substrates. All but three of the species were found to produce cellulases, of which the enzyme from *P. notatum* was most active. B. S. Singh and Saksena (1981) also demonstrated cellulolytic activity from five species of *Penicillium* (*P. chrysogenum*, *P. nigricans*, *P. notatum*, *P. citrinum*, and *P. commune*), and Somani and Wangikar (1979) showed *P. funiculosum* to be a good cellulase producer.

Qureshi *et al.* (1980a) examined the cellulolytic activity of a number of thermotolerant and thermophilic fungi, including *P. dupontii* grown at 45°C. Cellulolytic activity was assessed by three methods: loss in weight of filter paper, reducing sugars produced with carboxymethylcellulose as substrate, and an agar plate clearance technique, again using carboxymethylcellulose and Schultze's stain. All assays were carried out at 45°C. Under the fermentation conditions used, the cellulase from *P. dupontii* was found to produce reducing groups from carboxymethylcellulose but not to be active against filter paper, indicating that this microorganism may not elaborate the C_1 component required for activity against crystalline cellulose.

2.5. Bioconversion of Cellulose as a Renewable Resource

In a world searching for renewable resources, interest in the possibility of using cellulose as an alternative source of food, chemical feedstocks, and fuel has increased greatly in recent years. A very significant part of the extensive research effort on this topic has been directed toward the bioconversion of lignocellulosic materials to produce glucose, which can be further fermented anaerobically to a variety of products such as ethanol, butanol, or acetone or aerobically to microbal protein as an animal feed or for direct human consumption. Traditionally, species of *Trichoderma* have generally been adjudged to be the most suitable organisms for the production of cellulolytic enzymes for this purpose, especially *T. reesei* and its mutants (Mandels *et al.*, 1971, 1974). However, various species of *Penicillium* have also been considered to be suitable cellulolytic organisms for cellulose bioconversion.

As long ago as 1968, a provisional patent was filed (Selby *et al.*, 1968) for the use of *P. funiculosum* in the utilization of bagasse. Szakacs *et al.* (1981) considered the cellulolytic ability of a number of microorganisms, including three species of *Penicillium* (*P. variabile*, *P. vermiculatum*, and *P. verruculosum*). Of all the fungi examined, the cellulase complex from *P. verruculosum* was judged to be the most active against cotton cellulose (as measured by the amount of reducing sugar liberated from the cotton substrate). The enzyme complex had pH and temperature optima of 4.2 and 60°C, and this organism was considered to be a possible alternative to *T. reesei* and meritorious of further investigation.

Durand *et al.* (1984) compared the cellulases produced by four fungi, including *T. reesei* and a *Penicillium* sp. The authors considered the polysaccharolytic spectrum (activities of β-glucosidase and carboxymethylcellulase and activities of the cellulase against filter paper and cotton), heat and pH effects on stability and activity, and the ability of the enzymes to hydrolyze natural substrates. They found the polysaccharolytic spectrum of the *Penicillium* sp. to be similar to that of *T. reesei*, and the optimum temperature of enzymatic activity was similar, although the *T. reesei* cellulase appeared more stable at higher temperatures. The pH stability and ability of the two cellulases to hydrolyze natural substrates were also similar.

Penicillium funiculosum has also been reinvestigated recently and shown to produce a cellulase that is potentially suitable for the commercial hydrolysis of cellulose; the high activity of the β-1,4-glucosidase component of the cellulase complex facilitates the production of glucose as the major end product (Mishra *et al.*, 1983). Joglekar *et al.* (1983) also examined cellulase production by *P. funiculosum*. They found that the organism produced high activities of exo-β-1,4-glucanase, endo-β-1,4-glucanase, and β-glucosidase and concluded that *P. funiculosum* is an attractive alternative to *T. reesei* and its mutants for the commercial saccharification of lignocellulosic substrates.

3. POLYSACCHARASES OTHER THAN CELLULASES

In addition to cellulase, various other extracellular polysaccharases are produced by members of the *Penicillia* (Table II) and are directed toward the mobilization of potential carbon sources in the environment. Some of

Table II. Production of Extracellular Polysaccharases by Various Species of *Penicillium*

Enzyme	Species	References
Glucanases		
Amylase	*P. dupontii*	Qureshi *et al.* (1980b)
	P. expansum	Belloc *et al.* (1975)
	P. italicum, *P. spiculisporum*	K. V. Singh and Agrawal (1981)
Mycodextranase	Several species	Reese and Mandels (1964)
Dextranase	*P. lilacinum*	Charles and Farrell (1957), Tsuchiya *et al.* (1952)
	P. funiculosum, *P. verruculosum*	Tsuchiya *et al.* (1952)
Barley β-glucanase	*P. emersonii*	James *et al.* (1976)
β-D-1,2-Glucanase	Several species	Reese *et al.* (1961)
β-D-1,3-Glucanase	*P. italicum*	Santos *et al.* (1978, 1979), Sánchez *et al.* (1982)
	Several species	Reese and Mandels (1959)
β-D-1,6-Glucanase	Several species	Reese *et al.* (1962)
Pectic enzymes	*P. chrysogenum*	Phaff (1947)
	P. citrinum	Olutiola and Akintunde (1979)
	P. crustosum	Awasthi and Mishra (1982)
	P. digitatum	Bush and Codner (1970), Lobanok *et al.* (1977)
	P. expansum	Spalding and Abdul-Baki (1973), Spalding *et al.* (1973), Swinburne and Corden (1969)
	P. italicum	Bush and Codner (1970)
	P. steckii	Olutiola (1983)
	Penicillium sp.	Durand *et al.* (1984)
Hemicellulases	*Penicillium* sp.	Durand *et al.* (1984)
	Several species	Reese *et al.* (1973)
	P. wortmanni	Deleyn *et al.* (1978)

these enzymes have found industrial application and are described in more detail in Section 5.1.

3.1. Glucanases

The largest group of extracellular polysaccharases elaborated by various species of *Penicillium* to have been studied are the glucanases. These enzymes show a high degree of substrate specificity. They can be conveniently divided into α- and β-glucanases depending on their ability to hydrolyze α- or β-glucans, respectively.

3.1.1. α-Glucanases

Extracellular enzymes have been isolated from species of *Penicillium* that are capable of degrading α-1,4-glucans, dextran (which is predominantly an α-1,6-linked glucan), and mycodextran (the mixed α-1,3-/α-1,4-linked glucan, also known as nigeran) (Table II).

Starch-degrading enzymes or amylases are widely distributed in microorganisms, and the fungal enzymes fall into two distinct classes (Fogarty and Kelly, 1980). Amyloglucosidases or glucoamylases are exo-acting amylases that hydrolyze α-1,4- and α-1,6-linkages to produce glucose as the sole end product from the chain ends of starch and related polymers. In contrast, α-amylases are endo-acting enzymes that randomly hydrolyze α-1,4-bonds but bypass α-1,6-linkages in amylopectin and glycogen. So far as production of amylases by species of *Penicillium* is concerned, two isoenzymes of glucoamylase have been characterized from *P. oxalicum* (Yamasaki *et al.*, 1977a, b), but these enzymes were extracted from the mycelia of this species by urea treatment, and the authors did not present any evidence that they were produced as extracellular enzymes. K. V. Singh and Agrawal (1981) investigated both intracellular and extracellular amylase production by *P. italicum* and *P. spiculisporum* grown on media containing starch as carbon source and either $NaNO_3$ or $(NH_4)_2SO_4$ as nitrogen source. Amylolytic activity was determined using a plate-clearance method. Interestingly, with *P. italicum*, extracellular enzyme production was observed only in the medium that contained $(NH_4)_2SO_4$, whereas with *P. spiculisporum*, it was detected only in the medium that contained $NaNO_3$. Intracellular amylolytic activity was observed with both organisms and both sources of nitrogen and was generally higher than the extracellular activity. Extracellular amylolytic activity has also been detected in the thermophilic fungus *P. dupontii* (Qureshi *et al.*, 1980b), and a process for the production of extracellular α-amylase from a strain of *P. expansum* has been patented by Belloc *et al.* (1975).

An enzyme capable of hydrolyzing α-1,6-glucosidic linkages in dextran has been isolated as an extracellular enzyme from *P. lilacinum* by Charles and Farrell (1957), and the authors claim to have used the enzyme to prepare clinical dextran. This endo-acting enzyme has also found application in the sugar industry to break down dextran formed by *Leuconostoc* bacteria, which can infect damaged sugar beet and sugarcane tissue (Section 5.1). Other species capable of producing extracellular dextran-degrading enzymes include *P. funiculosum* and *P. verruculosum* (Tsuchiya *et al.*, 1952).

Reese and Mandels (1964) discovered a new enzyme, mycodextranase, that splits only the α-1,4-linkages in mycodextran (nigeran), an unbranched glucan that has alternating α-1,3- and α-1,4-links. Of 166 microbial cultures grown on mycodextran plus glucose, only 6 produced mycodextranase, and 4 of these were species of *Penicillium*, namely, *P. funiculosum*, *P. melinii*, *P. spinulosum*, and *P. thomii*. When mycodextran was omitted from the growth medium, the enzyme yields were very low. The fungal mycodextranases are extracellular enzymes and have no action against α-1,4-, α-1,6-, or β-linked glucans. They act in a random manner on mycodextran to produce nigerose and a tetramer, which are subsequently converted into glucose by an oligosaccharase, α-glucosidase.

3.1.2. β-Glucanases

Reese and co-workers have carried out an extensive investigation into the production of fungal β-D-glucanases, and the *Penicillia* feature prominently among the most active organisms (Reese and Mandels, 1959; Reese *et al.*, 1961, 1962). In addition to cellulase, many fungi produce extracellular enzymes that are active against β-1,2-glucans, β-1,3-glucans, and β-1,6-glucans. β-1,2-Glucanase has been found to be an adaptive enzyme like cellulase, whereas β-1,3- and β-1,6-glucanases are produced constitutively. Thus, while it is possible by controlling the conditions of growth to produce the latter enzymes free of β-1,2- and β-1,4-glucanase activities, it is not easy to find β-1,2-glucanase free of β-1,3- or β-1,6-glucanase activities. However, these enzymes can be separated from each other by zone electrophoresis, confirming that each of these β-glucanase activities resides on distinctly different protein molecules (Reese *et al.*, 1961). Each of the four β-glucanases moves independently under the influence of the applied potential, and each acts only on its own substrate. However, more than one active component may be present for particular activity with some of these fungi.

Of the species of *Penicillium* that are active in the elaboration of the various β-glucanases, many fall into closely related taxonomic groups. Thus, β-1,2-glucanase-producing organisms include all members of the *P. funiculosum* series and some members of the *P. javanicum* series (Reese *et al.*, 1961). *Penicillia*, especially of the *P. luteum* series, produced high yields of β-1,3-glucanase, although the vast majority of fungi tested produced detectable amounts of this activity (Reese and Mandels, 1959). Enzymatic hydrolysis of β-1,2- and β-1,6-glucans occurred in a random manner, with the production of a series of oligosaccharides including sophorose in the former case and gentiobiose in the latter. In contrast, two distinct types of β-1,3-glucanase activities were found. An exo-acting enzyme, which produced glucose as the sole product, appeared to be more common than the endo-acting enzyme, which hydrolyzed β-1,3-glucans to produce laminaribiose and higher oligosaccharides. Most of the β-glucanase preparations contained β-glucosidase activity that hydrolyzed sophorose, gentiobiose, and laminaribiose to glucose.

More recent studies have focused on the β-1,3-glucanases, which are of particular interest because β-1,3-glucans are important structural components of fungal cell walls. *Penicillium italicum* can produce at least three enzymes with β-1,3-glucanase activity (Santos *et al.*, 1978). A tentative characterization of these enzymes, which were named β-1,3-glucanase-I, -II, and -III in order of their elution from a DEAE–Sephadex column, indicated that they had different modes of action; the first was an endoglucanase, the second was an exoglucanase, and the third appeared to be capable of both modes of action. During active growth when an excess of glucose is present, only small amounts of β-1,3-glucanase-II and -III are produced, whereas the absence of glucose not only results in a greater rate of synthesis of these two enzymes, but also triggers the production of β-1,3-glucanase-I, which is undetectable in actively growing mycelium. The location of the various enzymes was studied under conditions of β-1,3-glucanase derepression in a low-glucose medium (Santos *et al.*, 1979). After 24 hr of incubation in this medium, 45% of the total activity was located in the cytoplasm, in which the most abundant of the three enzymes was β-1,3-glucanase-II. Another 40% was found to be located in the periplasmic space, where only β-1,3-glucanase-II and -III appeared to be present. The remaining 15% was extracellular and made up of similar concentrations of each of the three enzymes. These results indicate important differences in the roles of these enzymes. β-1,3-Glucanase-I had a very low affinity for the cell walls of *P. italicum*; it neither bound to the walls nor was capable of releasing reducing sugar groups from them *in vitro*. Enzymes II and III are the only ones produced during active growth and can bind to the cell walls and hydrolyze their components *in vitro*. Thus, β-1,3-glucanase-II and -III must be the enzymes involved in the modification of cell-wall glucan during active growth for morphogenesis and cell-wall extension; they are also responsible for the mobilization of cell-wall glucan under conditions of derepression either to supply energy or to induce a morphogenetic event such as conidiation. In contrast, the production of β-1,3-glucanase-I is directed toward the degradation of extracellular glucan for metabolic purposes in glucose-starved cells. Further characterization of the latter enzyme (Sánchez *et al.*, 1982) indicates that it is an acidic glycoprotein with a molecular weight (65,000–68,000) and isoelectric point within the range found for other β-glucanases (Villanueva *et al.*, 1976). In this regard, β-1,3-glucanase-I does not differ grossly from enzymes II and III, which are also acidic glycoproteins with similar molecular weights. However, whereas the latter two enzymes bind to concanavalin A, β-1,3-glucanase-I does not. The most significant properties of enzyme I are its uncommonly low Michaelis constant (K_m) on laminarin (0.04 mg/ml), which is at least 10 times lower than the average K_m for other glucanases (Villanueva *et al.*, 1976), and its minimal activity against heavily branched substrates such as yeast cell-wall β-1,3-glucan.

A concerted effort by Glaxo Laboratories Ltd. and the Shirley Institute set out to isolate a thermostable enzyme capable of degrading the β-1,4/β-1,3 mixed-linked glucan found in varying amounts in barley. Barley is commonly used as an animal feed, particularly for chickens, but the digestive system of the chicken, in common with that of other nonruminants, is not capable of degrading the viscous aqueous solution of this β-glucan, which passes through the bird's digestive system, lowering the total metabolizable energy obtained from the feed and resulting in the production of sticky droppings. Live steam is used in the production of feed pellets, hence the requirement for a thermostable enzyme. This thermostable barley β-glucanase has subsequently been used to improve the filtration efficiency in brewing processes in which raw barley is used to replace malted barley in mashing, by reducing the viscosity of the hot wort (Section 5.1). A wide range of microorganisms were screened for the production of a suitable β-glucanase, and the best enzyme was isolated from a strain of *P. emersonii* (James *et al.*, 1976). This extracellular thermostable enzyme was characterized by retention of at least 75% of its original activity on being heated in aqueous solution (pH 6.0) at 95°C for 10 min.

3.2. Pectic Enzymes

Pectic substances are complex heteropolysaccharides, and a whole range of extracellular enzymes are involved in their degradation. These enzymes, which play an important role in many natural processes (e.g., plant pathogenesis, food spoilage) and in the flax-retting process, are produced in varying proportion by many different microorganisms; details concerning the distribution of microbial pectolytic enzymes have been adequately covered in other review articles (Rombouts and Pilnik, 1980; Fogarty and Kelly, 1983). Phaff (1947) investigated the production of extracellular pectin-esterase and polygalacturonase by *P. chrysogenum* and demonstrated that these enzymes are strongly induced by pectin, pectic acid, D-galacturonate, mucate, and L-galactonate, but not by closely related compounds such as D-galactose, D-galactonate, and L-galacturonate. He suggested the possibility that mucate is the only true stimulatory compound and that L-galactonate and D-galacturonate undergo oxidative dissimilation by the mold via mucate. *Penicillium expansum* causes blue mold rot of apples, and a comparison is available between the production of polygalacturonases (Swinburne and Corden, 1969) and pectin lyase (Spalding and Abdul-Baki, 1973) in both apple tissue and artificial media. Formation of both enzymes is repressed by a variety of sugars and also by galacturonate and glutamate, which could explain why their production in artificial culture is greatly increased using dialyzed rather than nondialyzed apple medium (Spalding *et al.*, 1973). *Penicillium digitatum* and *P. italicum* are responsible for postharvest decay of citrus fruits, and both species produce an endo-acting pectin transeliminase that, in the absence of other known pectic enzymes, is capable of causing maceration of orange rind (Bush and Codner, 1970). This enzyme is produced constitutively by *P. digitatum* independent of growth rate and conditions of culture (Lobanok *et al.*, 1977). Cocoa beans are susceptible to serious biodeterioration by *P. steckii* and *P. citrinum* during storage, and both species produce a complex of enzymes that have pectolytic activity (Olutiola and Akintunde, 1979; Olutiola, 1983). When the latter species was grown in liquid culture media, production of pectin methylesterase appeared, at least in part, to be constitutive, while pectic lyase activity was inducible. *Penicillium crustosum* is also an active producer of pectolytic enzymes that, together with its cellulolytic enzymes, largely contribute to the competitive saprophytic ability of this species (Awasthi and Mishra, 1982).

3.3. Hemicellulases

There are only a few reports documenting the production of hemicellulases by species of *Penicillium*. The spectrum of extracellular polysaccharases produced by *Penicillium* sp. PD20 (probably related to *P. funiculosum*) included endo-β-1,4-D-xylanase and β-D-xylosidase in addition to pectolytic and cellulolytic activities (Durand *et al.*, 1984). Reese *et al.* (1973) screened 250 microorganisms for their ability to produce extracellular β-D-xylosidases, and *P. wortmanni* consistently gave the highest yields. To attain these values, xylan was used as the major carbon source that induced β-D-xylosidase production, probably via the oligosaccharides resulting from hydrolysis by xylanase, although the latter activity was not measured. The extracellular β-xylosidase of *P. wortmanni* has been purified and shown to have a pH optimum of 3.3–4.0 and a molecular weight of about 100,000 (Deleyn *et al.*, 1978).

4. LIPASES AND PROTEOLYTIC ENZYMES

Lipolytic and proteolytic enzymes are found in all microorganisms. In the case of the *Penicillia*, interest has focused on the probable role of these enzymes in the development of characteristic flavors and the ripening of cheeses. Although the species of *Penicillium* used in cheese manufacture would appear to be a useful source of lipolytic and proteolytic preparations for other branches of the food industry, to date they do not appear to have found any commercial application.

4.1. Lipases

4.1.1. Occurrence of Lipases in the *Penicillia*

Lipases are enzymes that can hydrolyze a wide range of fatty acid esters, although glycerides are normally the preferred substrates, and they are frequently produced as extracellular enzymes during the growth of many different species of fungi. Extracellular lipases have been isolated from various species of *Penicillium* including the cheese molds *P. camemberti*, *P. candidum*, *P. caseicolum*, and *P. roqueforti* (Table III).

The lipolytic system of *P. roqueforti* has been extensively investigated. Early studies demonstrated that *P. roqueforti* can elaborate lipase when grown on a fat-free medium (Shipe, 1951), and Morris and Jezeski (1953) proposed that more than one activity is produced by this mold. Imamura and Kataoka (1963a) examined the lipase-producing ability of a particular

Table III. Production of Extracellular Lipolytic Enzymes by Various Species of *Penicillium*

Species	References
P. camemberti	Stepaniak *et al.* (1980)
P. candidum	Kornacki *et al.* (1980), Stepaniak *et al.* (1980)
P. caseicolum	Lamberet and Lenoir (1976a)
P. chrysogenum	Chander *et al.* (1977, 1980)
P. cyclopium	El-Gendy and Marth (1980), Iwai *et al.* (1975)
P. funiculosum	Chander and Klostermeyer (1982)
P. patulum	El-Gendy and Marth (1980)
P. puberulum	El-Gendy and Marth (1980)
P. roqueforti	Eitenmiller *et al.* (1970), Imamura and Kataoka (1963a), Kornacki *et al.* (1980), Lobyreva and Marchenkova (1980), Menassa and Lamberet (1982), Morris and Jezeski (1953), Shipe (1951), Stepaniak *et al.* (1980)
Penicillium sp	Kazanina *et al.* (1981)
Penicillium sp. strain 14-3	Tokiwa and Suzuki (1980)

strain of *P. roqueforti*, which was the best producer of lipase of the five strains tested. Lipase production was inhibited by the presence of sugars (lactose, glucose, or galatose) in the fermentation medium, but stimulated by butterfat or tributyrin and when the inorganic nitrogen source was replaced by casein or peptone. Maximum lipase production was achieved at the lowest incubation temperature tested (7°C) and was hardly detectable at 37°C. These optimum conditions for lipase production are similar to those for the ripening of Roquefort-type cheeses. Further studies have shown that *P. roqueforti* can produce at least two extracellular lipases that are distinguished by acid and alkaline pH optima (Eitenmiller *et al.*, 1970; Menassa and Lamberet, 1982). Lobyreva and Marchenkova (1980) have reported the isolation from a strain of *P. roqueforti* of three lipases that differ in molecular weight and substrate specificity. A lipase with an alkaline pH optimum has also been isolated from several strains of *P. caseicolum*, but there was no evidence for the production of an acid lipase by this species (Lamberet and Lenoir, 1976a).

Stepaniak *et al.* (1980) have examined the ability of *P. roqueforti*, *P. candidum* (syn. *P. caseicolum*), and *P. camemberti* to produce extracellular lipases and proteinases in identical cultivation conditions. They concluded that there could be selected from each species strains that show relatively high lipolytic and low proteolytic activity or vice versa. These strains could be suitable for the commercial production of either lipolytic or proteolytic preparations for use in cheese manufacture and other branches of the food industry. Further investigations (Kornacki *et al.*, 1980) indicated that the relative proportions of lipases and proteinases produced by *P. roqueforti* and *P. candidum* could be controlled to some extent by the type of medium and cultivation procedure. The maximum amounts of extracellular lipases were obtained by surface rather than submerged cultivation.

Several other species of *Penicillium* have also been examined for extracellular lipolytic activity. These include *P. chrysogenum*, which showed maximum lipase production in shaken rather than stationary cultures (Chander *et al.*, 1977). The optimum conditions for lipase production were 30°C at pH 6.0 in a medium containing glucose and peptone as carbohydrate and nitrogen source, respectively. Lipase biosynthesis appeared to be constitutive in this strain of *P. chrysogenum*, but the addition of various natural and synthetic triglycerides to the medium caused a decrease in growth and lipase production (Chander *et al.*, 1980). Similar results were obtained with *P. funiculosum* (Chander and Klostermeyer, 1982), which showed maximum lipase production in shaken cultures at 20°C using a growth medium containing glucose and peptone at pH 6.0. Although tributyrin stimulated lipase biosynthesis by 20%, other lipids suppressed both growth and enzyme levels. A possible explanation of the inhibition of growth and biosynthesis of lipase is the production of free fatty acids, which may be toxic to these organisms. *Penicillium cyclopium* produces at least two extracellular lipases with different pH optima and molecular weights when grown on a medium containing rice bran and corn steep liquor (Iwai *et al.*, 1975). Two active components were also detected in a lipolytic complex isolated from the culture fluid of an unidentified species of *Penicillium* (Kazanina *et al.*, 1981), and an enzyme capable of catalyzing the hydrolysis of poly(ethylene glycol adipate), assumed by the authors to be a type of lipase, was produced by another unidentified *Penicillium* species (Tokiwa and Suzuki, 1980). In a study of the proteolytic and lipolytic activities of several species of *Penicillium*, the greatest lipolytic activity was associated with *P. cyclopium*, *P. patulum*, *P. puberulum*, and *P. roqueforti* (El-Gendy and Marth, 1980).

4.1.2. Properties of Lipases Isolated from the *Penicillia*

Various groups of workers have purified and characterized extracellular lipolytic enzymes produced by *P. roqueforti* (Table IV). An alkaline lipase was partly purified from a culture filtrate of *P. roqueforti* by ammonium sulfate precipitation (Eitenmiller *et al.*, 1970) and was quite active against butter oil over a pH range of 7.5–9.0 with an optimum at pH 8.0. It was a thermolabile enzyme, being inactivated completely within 10 min at 50°C. The substrate specificity of this lipase was studied using several triglycerides. It was most specific for tributyrin and hydrolyzed tributyrin, tricaprylin, tricaprin, tripropionin, and triolein in decreasing order, the latter two substrates being hydrolyzed at much lower rates than the other triglycerides. Menassa and Lamberet (1982) have compared the properties of both the alkaline and acid lipases produced by the same strain of *P. roqueforti* under different culture conditions. With the use of an acid broth culture, the main lipolytic activity had an optimum pH at 6.0, while the partially purified extract from a basic broth culture was most active at pH 9.0. Further characterization of the lipase obtained by growth in an acid medium was undertaken, following precipitation with 50% ethanol and purification using DEAE–Sephadex A50 ion-exchange chromatography and two successive treatments with SP Sephadex C50 (Lamberet and Menassa, 1983a). The enzyme was active from pH 1.7 to 10.5 with an optimum activity at pH 6.5 at 20°C and at pH 6.0 at 30°C. The authors suggested that this apparent shift in the pH optimum could be a result of changes in enzyme stability at the different temperatures. In terms of substrate specificity, the maximum activity was observed with tricaproin, while tributyrin, tricaprylin, butter oil, and triolein were hydrolyzed with decreasing efficiency. The very high molecular weight reported for this enzyme suggests that it occurred in an aggregated form in the culture filtrate. Lobyreva and Marchenkova (1980) isolated from a culture of *P. roqueforti* three lipolytically

Table IV. Properties of Some Extracellular Lipolytic Enzymes

Species	Enzyme	pH optimum	Temperature optimum (°C)	Molecular weight	References
P. caseicolum	Alkaline lipase	9.0–9.6	35	23,000–25,000	Lamberet and Lenoir (1976b)
P. cyclopium	A-lipase	7.5	35	27,000	Iwai *et al.* (1975), Tsujisaka and Iwai (1984)
	B-lipase	5.8	40	36,000	
	Partial glyceride hydrolase	6.0	40	32,000–35,000	Okumura *et al.* (1980)
P. roqueforti	Acid lipase	6.0–6.5	30–40	700,000	Lamberet and Menassa (1983a)
	Alkaline lipase	8.0	37	—	Eitenmiller *et al.* (1970)
	Alkaline lipase	9.0	35	—	Menassa and Lamberet (1982)
	Lipase I	7.0	45	7,930	Lobyreva and Marchenkova (1980, 1981)
	Lipase II	6.0	45	9,100	
	Lipase III	7.0	45	11,420	
	Acid lipase	6.5	—	—	Imamura and Kataoka (1963b)
	Neutral lipase	7.5	—	—	
Penicillium sp.	Lipase 2	7.5–8.0	37–40	29,000–30,000	Kazanina *et al.* (1981)
Penicillium sp. strain 14-3	Poly(ethylene glycol adipate)-degrading enzyme	4.5	45	25,000	Tokiwa and Suzuki (1980)

active proteins (Table IV) that differed in molecular weight and substrate specificity. Lipase III was most active in the hydrolysis of plant oils containing mainly unsaturated fatty acids. Lipase II most effectively hydrolyzed synthetic triglycerides containing saturated fatty acids, in particular tricaproin, tricaprylin, and trimyristin. Tributyrin was more actively hydrolyzed with lipase I than with lipases II and III. Further characterization of the purified lipases demonstrated that optimum activity of lipases I and III occurred at pH 7.0 and 45 °C and of lipase II at pH 6.0 and 45°C (Lobyreva and Marchenkova, 1981). These lipases were found to be stable on being heated to 55°C over a wide range of pH and contained small amounts of carbohydrate (1.4–3.3%). Japanese workers have isolated two types of lipase from the culture medium of *P. roqueforti* (Imamura and Kataoka, 1963b). The acid lipase had an optimum pH for hydrolysis of tributyrin at about 6.5, while a neutral lipase had an optimum pH of 7.5. Their results suggested that the acid lipase, which selectively liberated caproic, capric, and butyric acids from butterfat, may be the more important activity in cheese manufacture. In comparing the characteristics of the lipases of *P. roqueforti* reported in the literature, it is clear that although there are similarities, for example, among the acid lipases of Imamura and Kataoka (1963b), Lamberet and Menassa (1983a), and lipase II (Lobyreva and Marchenkova, 1980, 1981), some differences remain to be resolved, and the actual number of extracellular lipolytic enzymes produced by this species still needs to be established.

The characteristics of the lipolytic system of *P. caseicolum* have been studied using crude enzyme preparations obtained from the cultures of eight strains that showed different abilities to produce extracellular lipases (Lamberet and Lenoir, 1976a). Lipases produced by these cultures were all very similar, having an alkaline pH optimum and hydrolyzing tributyrin at a rate about 5 times higher than that of triolein. There was no evidence for the production of an acid lipase from any of these strains. The lipase from one of these cultures was purified by means of ammonium sulfate precipitation, gel-filtration chromatography on Sephadex G75, and cation-exchange chromatography on DEAE–Sephadex A50 and was shown to have an optimum activity at pH 9.0–9.6 (Lamberet and Lenoir, 1976b). *Penicillium cyclopium* produces at least two extracellular lipases with different pH optima and molecular weights (Table IV). In addition, the substrate specificities of the two lipases are significantly different from each other (Iwai *et al.*, 1975). Only moderate differences were observed between the rates of hydrolysis of various long-chain saturated and unsaturated fatty acid methyl esters. However, the B-lipase was much more active against these esters than the A-lipase. The A-lipase was generally found to be more active on short-chain triglycerides than on long-chain triglycerides, and the best substrate was tributyrin followed by tripropionin (Tsujisaka and Iwai,

1984). This enzyme also hydrolyzed methyl butyrate, albeit at a rate 33 times less than that for tributyrin. Another strain of *P. cyclopium* has been found to produce a novel lipase characterized as being a partial glyceride hydrolase (Okumura *et al.*, 1980). This enzyme hydrolyzed the partial glycerides diolein and monoolein much more rapidly than it did the triglyceride triolein. When the enzyme was used in conjunction with a conventional lipase, the hydrolysis of a triglyceride was much more rapid than that obtained with either of the enzymes alone. Such a cooperative reaction between a conventional lipase and a partial glyceride hydrolase could have industrial application for ensuring the complete hydrolysis of glycerides.

Two active components were found in a lipolytic complex isolated from an unidentified species of *Penicillium* (Kazanina *et al.*, 1981). Both components had an optimum pH range of 7.5–8.0 and a temperature range of 37–40°C. These results suggested that the high-molecular-weight component of the complex was an aggregate form of the low-molecular-weight component. A poly(ethylene glycol adipate)-degrading enzyme isolated from another unidentified species of *Penicillium* had an optimum pH and temperature of 4.5 and 45°C, respectively (Tokiwa and Suzuki, 1980). This enzyme hydrolyzed various plant oils, triglycerides, and methyl esters of fatty acids in addition to various kinds of saturated and unsaturated aliphatic polyesters.

4.2. Proteolytic Enzymes

4.2.1. Occurrence of Proteinases in the *Penicillia*

The production of extracellular proteolytic enzymes has been documented for various species of *Penicillium* and, as for lipolytic enzymes, the cheese molds again figure prominently in this work (Table V). Indeed, the complexity of fungal proteolytic systems is exemplified by *P. roqueforti*, which elaborates two endopeptidases (Zevaco *et al.*, 1973; Gripon and Hermier, 1974; Modler *et al.*, 1974a) and three exopeptidases (Gripon and Debest, 1976) when grown in culture. One of the endopeptidases is an acid (aspartyl) proteinase and is produced mainly in acid conditions (pH 4.0), while the highest yields of the other endopeptidase, a metalloproteinase, are obtained in pH 6.0 buffered medium. A similar range of proteolytic activities are produced by *P. caseicolum*, *P. cyclopium*, and *Penicillium frequentans*, but they differ with respect to electrophoretic mobility and behavior with inhibitors to the enzymes from *P. roqueforti* (Gripon and Debest, 1976). The main component of the extracellular proteolytic system produced by *P. caseicolum* on a neutral medium is also a metalloproteinase, while an acid

Table V. Production of Extracellular Proteolytic Enzymes by Various Species of *Penicillium*

Species	References
P. camemberti	Stepaniak *et al.* (1980)
P. candidum	Kornacki *et al.* (1980), Stepaniak *et al.* (1980)
P. caseicolum	Gripon and Debest (1976), Lenoir and Auberger (1977a)
P. citrinum	O. P. Sharma and K. D. Sharma (1980)
P. cyaneo-fulvum	K. Singh and Martin (1960)
P. cyclopium	El-Gendy and Marth (1980), Gripon and Debest (1976)
P. dupontii	Hashimoto *et al.* (1973a)
P. frequentans	Gripon and Debest (1976)
P. funiculosum	O. P. Sharma and K. D. Sharma (1980)
P. janthinellum	Ghosh and Thangamani (1973), Hofmann and Shaw (1964), Jones and Hofmann (1972), Sodek and Hofmann (1970a)
P. notatum	Ashy *et al.* (1981), Satyanarayana and Jain (1979), D. P. Sharma and Saksena (1981)
P. patulum	El-Gendy and Marth (1980)
P. purpurogenum	O. P. Sharma and K. D. Sharma (1980)
P. roqueforti	Gripon and Debest (1976), Gripon and Hermier (1974), Kornacki *et al.* (1980), Modler *et al.* (1974a), Stepaniak *et al.* (1980), Zevaco *et al.* (1973)

proteinase produced at low levels on this medium is elaborated at much higher levels on an acid medium at pH 4.0 (Lenoir and Auberger, 1977a).

An acid proteinase capable of activating trypsinogen has been isolated from a culture of *P. janthinellum* grown on a medium kept below pH 4.5 by repeated addition of lactic acid; the enzyme is unstable above pH 6.0 (Hofmann and Shaw, 1964). This enzyme, named "penicillopepsin" (see Section 4.2.2), has been produced in larger quantities by Sodek and Hofmann (1970a), using a 1500 liter fermenter. The enzyme appears only when the stationary phase of growth is reached, but the time required to reach maximal levels of penicillopepsin in the fermenter was only 6 days. The enzyme from the fermentation filtrate was concentrated by passage through a DEAE–cellulose column and purified by chromatography on aminoethylcellulose and phosphocellulose. Ghosh and Thangamani (1973) observed that replacement of nitrate as the sole source of nitrogen by ammonium salts, urea, or bacto-peptone in liquid synthetic medium reduced the delay in the formation of penicillopepsin together with an extracellular acid ribonuclease also produced by *P. janthinellum.* An extracellular carboxypeptidase has also been isolated from the culture medium of *P. janthinellum*, and like penicillopepsin, this enzyme is released only when the stationary phase of growth is reached (Jones and Hofmann, 1972).

A thermostable acid proteinase is produced by the thermophilic fungus

P. dupontii (Hashimoto *et al.*, 1973a), and its dual properties of acid and thermal stability make this enzyme a potentially useful candidate for the industrial production of protein hydrolysates (Emi *et al.*, 1976). In contrast, an alkaline proteinase has been isolated from the culture filtrate of *P. cyaneofulvum* grown on a beef heart infusion–peptone–glucose medium (K. Singh and Martin, 1960). *Penicillium notatum* has been reported to produce proteolytic enzymes constitutively on a glucose–gelatin broth (Satyanarayana and Jain, 1979; D. P. Sharma and Saksena, 1981), and Ashy *et al.* (1981) have produced a rennetlike preparation with milk-clotting activity from a culture of *P. notatum* grown on a medium containing whey, glucose, and inorganic salts. In a study on the proteolytic and lipolytic activity of several species of *Penicillium*, the greatest proteolytic activity was associated with strains of *P. patulum* and *P. cyclopium*, and these cultures also showed marked lipolytic activity (El-Gendy and Marth, 1980). O. P. Sharma and K. D. Sharma (1980) have investigated the production of proteolytic enzymes by cultures of *P. citrinum*, *P. purpurogenum*, and *P. funiculosum* isolated from deteriorated leather. Optimum pH and temperature for maximum enzyme production were found to be 7.0 and 30°C, respectively, on a gelatin broth medium.

4.2.2. Properties of Proteinases Isolated from the *Penicillia*

The acid proteinases liberated by several species of *Penicillium* have been purified and partly characterized (Table VI). Optimal activity of these enzymes occurs within the pH range 2.5–5.5. In the case of the acid proteinase produced by *P. roqueforti*, Zevaco *et al.* (1973) reported an optimum pH of 3.5 both for casein or hemoglobin hydrolysis and for bovine trypsinogen activation, but noted a second minor optimum at pH 5.5 with casein. The interval between these two optima coincides with the zone of precipitation of casein. Modler *et al.* (1974a) found that a crude preparation of the same enzyme had pH optima of 3.0 and 5.5 for bovine serum albumin and casein, respectively, but these workers incubated the enzyme with casein only over the pH range 5.0–7.0. The enzyme showed maxium stability to pH in the range 3.0–6.0, and thermal inactivation commenced at 48°C. Milk-clotting studies revealed that the proteolytic activity of the acid proteinase of *P. roqueforti* was 14 times greater than that of calf rennet for identical clotting times. However, because of its general proteolytic activity, this enzyme was found not to be an acceptable substitute for calf rennet in small cheese-making trials (Modler *et al.*, 1974b). The acid proteinase of *P. caseicolum* has many similarities to the enzyme from *P. roqueforti*. It has an optimal pH of about 3.5–4.0 with hemoglobin and 5.0 with casein, but shows a second minor optimum at pH 3.5 with the latter substrate (Lenoir *et al.*, 1979). Maximal stability of this enzyme is found between pH 3.5 and 5.5.

Table VI. Properties of Some Extracellular Proteolytic Enzymes

Species	Enzyme	pH optimum	Temperature optimum (°C)	Molecular weight	References
P. caseicolum	Acid proteinase	5.0[a] 3.5–4.0[b]	45	35,000	Lenoir *et al.* (1979)
	Metalloproteinase	6.0[a] 5.0[b]	50	20,000	Lenoir and Auberger (1977b)
P. cyaneo-fulvum	Alkaline proteinase	9.5–11.0[a]	—	45,000	K. Singh and Martin (1960)
P. dupontii	Acid proteinase	2.5[a] 3.0[b]	55 at pH 2.5 70 at pH 3.6 75 at pH 4.6	41,000	Hashimoto *et al.* (1973b)
	Acid proteinase	2.5[a] 3.0–3.5[b]	60 at pH 2.5 75 at pH 3.7	41,590	Emi *et al.* (1976)
P. janthinellum	Acid proteinase	3.0–4.0[c]	—	32,000	Hofmann and Shaw (1964)
	Carboxypeptidase	4.0–5.0	—	48,000	Jones and Hofmann (1972)
P. roqueforti	Acid proteinase	3.5[a,b]	50	—	Zevaco *et al.* (1973)
	Acid proteinase	3.0[d] 5.5[a]	46	45,000–49,000	Modler *et al.* (1974a, b)
	Metalloproteinase	5.5[a] 4.2[b]	—	20,000	Gripon and Hermier (1974)
	Acid carboxypeptidase	3.5–4.0	40	110,000	Gripon (1977a)
	Alkaline carboxypeptidase	—	—	—	Gripon and Debest (1976)
	Alkaline aminopeptidase	8.0	—	35,000	Gripon (1977b)

[a] Casein. [b] Hemoglobin. [c] Trypsinogen activation. [d] Bovine serum albumin.

The trypsinogen-activating acid proteinase isolated from *P. janthinellum* has been named penicillopepsin because a number of properties of the enzyme, including the amino acid sequence around its active site, indicate that it may be homologous to pepsin (Sodek and Hofmann, 1970b). A comparison of the N-terminal amino acid sequence of penicillopepsin, the acid proteinases from *P. roqueforti* and *Rhizopus chinensis*, and several mammalian pepsins has provided further evidence for the evolutionary homology of these acid proteinases (Gripon *et al.*, 1977a). The enzyme has a pH optimum for trypsinogen activation of 3.0–4.0 and is most stable between pH 3.0 and 5.5 (Hofmann and Shaw, 1964). An acid proteinase produced by the thermophilic fungus *P. dupontii* has been characterized by Hashimoto *et al.* (1973b) and Emi *et al.* (1976), and the results obtained by both groups of workers are in good agreement (Table VI). The temperature optima for this enzyme, as well as its thermal stability, are dependent on pH. This thermostable acid proteinase retains full activity after 1 hr at 60°C in the pH range between 3.5 and 5.5 and at its most stable pH of 4.5 retains more than 65% of its activity after 1 hr at 70°C (Hashimoto *et al.*, 1973b). The acid proteinases from *P. dupontii* (Emi *et al.*, 1976) and *P. roqueforti* (Houmard and Raymond, 1979) are both reported to be glycoproteins.

In addition to an acid proteinase, *P. caseicolum* and *P. roqueforti* also produce an extracellular endopeptidase that is inhibited by chelating agents such as EDTA. Metal determination has confirmed that both enzymes are metalloproteinases, each containing one atom of zinc per molecule that is essential for enzymatic activity (Gripon *et al.*, 1980). These two enzymes have an acid pH optimum (5.5–6.0 on casein, 4.2–5.0 on hemoglobin) and a molecular weight of about 20,000 (Gripon and Hermier, 1974; Lenoir and Auberger, 1977b). Determination of amino acid compositions and N-terminal sequences showed a strong homology between the two enzymes, which have properties similar to those of the metalloproteinases of *Aspergillus sojae* and *Aspergillus oryzae*, and it has been suggested that all these enzymes constitute a homogeneous group to be called "acid metalloproteases" (Gripon *et al.*, 1980).

Penicillium cyaneo-fulvum produces an extracellular alkaline proteinase with an optimum pH for caseolysis of 9.5–11.0 and a molecular weight of about 45,000 (K. Singh and Martin, 1960). This enzyme was shown to be an endopeptidase of low specificity (Martin *et al.*, 1962) that hydrolyzed a variety of substrates including casein, denatured hemoglobin, insulin, and gelatin, but not short-chain peptides (dimers to tetramers).

The extracellular proteolytic system of *P. roqueforti* includes three exopeptidases, two of which have been partly characterized. Gripon (1977a) isolated an acid carboxypeptidase with a molecular weight of about 110,000 and an optimum pH of 3.5–4.0 for hydrolysis of *N*-carbobenzoxy-L-glutamyl-L-tyrosine. This enzyme was stable in the pH range of 5.0–5.5 at

35°C. In contrast, an alkaline aminopeptidase had an optimum pH for L-leucine-*p*-nitroanalide of 8.0 and a molecular weight of 35,000 (Gripon, 1977b). At 35°C, the latter enzyme was stable between pH 6.0 and 7.0. An extracellular carboxypeptidase has also been isolated from the culture medium of *P. janthinellum*. This enzyme, named "penicillocarboxypeptidase-S," is a nonspecific, SH-dependent exopeptidase and has a pH optimum between 4.0 and 5.0 and a molecular weight of about 48,000 (Jones and Hofmann, 1972).

4.3. Role of *Penicillium* during Cheese Ripening

The growth of the mold *P. roqueforti* in the body of blue-veined cheeses and the surface growth of the molds *P. candidum*, *P. caseicolum*, or *P. camemberti* on Camembert, Brie, and related types of cheeses play an important role in the development of the characteristic flavors of these cheeses. In practice, considerable proteolysis and lipolysis occurs during cheese ripening, but these are complex phenomena because of the diversity of enzymes involved, which can include native milk enzymes, the milk-clotting enzymes present in rennet (bovine pepsin and chymosin), and enzymes of the microflora. Because these enzyme systems work together, it is difficult to elucidate the role of individual enzymes during traditional cheese-making. Therefore, most investigations have involved either *in vitro* studies (e.g., the hydrolysis of isolated caseins) or the use of aseptic model curds free of contaminating microorganisms.

It is generally considered that species of *Penicillium* are the main agents responsible for proteolysis is these various cheeses. Supporting evidence has come from studies on model curds in which it was demonstrated that *P. caseicolum* or *P. roqueforti* alone could produce large quantities of soluble nitrogen and amino acids, comparable to those observed in commercial cheeses (Desmazeaud *et al.*, 1976). Purified preparations of the acid proteinase from *P. roqueforti* and neutral proteinase from *P. caseicolum* added to aseptic model curds also induced large increases in soluble nitrogen (pH 4.6) and nonprotein nitrogen, but had little effect on the production of free amino acids (Gripon *et al.*, 1977b). The latter result can be explained by the absence of carboxypeptidase and aminopeptidase in these purified enzyme preparations. Thus, it would appear that these two endopeptidases play a fundamental role in the proteolysis induced by their respective species during cheese ripening. Investigations have also been carried out to compare the *in vitro* activity of the two acid proteinases and the two metalloproteinases of *P. caseicolum* and *P. roqueforti* in the hydrolysis of caseins (Trieu-Cuot and Gripon, 1981; Trieu-Cuot *et al.*, 1982a, b). The action of the metalloproteinases of both species on α_{S1}- and β-caseins is similar, although there are some differences in the rate of hydrolysis and the nature of the bonds that are cleaved. The action of the two acid proteinases is even more similar, the only observed differences being in the rate of hydrolysis of some bonds of α_{S1}-casein. Using the same techniques of isoelectric focusing and two-dimensional electrophoresis developed in these studies, Trieu-Cuot and Gripon (1982) have examined proteolysis due to the action of rennet, plasmin (endogenous alkaline milk proteinase), and the acid and metalloproteinases of *P. caseicolum* during Camembert cheese ripening. The respective activities of the acid proteinase and metalloproteinase were characterized and followed using β-casein degradation products as markers. Peptides resulting from the action of the metalloproteinase on β-casein in the surface of the cheese were detectable immediately after the development of *P. caseicolum*. Subsequently, their level decreased while that resulting from the action of the acid proteinase increased, suggesting a more important role for the latter enzyme. Nevertheless, it is difficult to account for the softening of the inner parts of a Camembert cheese due to the action of the surface flora, because these enzymes are not good candidates for diffusion on the basis of their molecular weights. Using aseptic rennet-free cheeses inoculated with a strain of *P. caseicolum*, Noomen (1983) demonstrated that the proteolytic enzymes penetrated only just over 6 mm into the cheese. Noomen (1983) suggested that with respect to the development of consistency in this type of cheese, the role of the surface flora is predominantly to establish a pH gradient by consumption of lactic acid diffusing toward the cheese surface and by the formation of alkaline breakdown products such as NH_3 diffusing from the surface toward the center. Combined with the breakdown of α_{S1}-casein by rennet, this can explain the visible softening that starts at the outside and gradually extends into the interior of the cheese. Of course, the surface flora is also of crucial importance in the formation of taste and flavor substances characteristic of a particular variety of cheese.

The unique peppery or piquant flavor and aroma of good-quality blue-veined cheese are thought to depend mainly on the production of fatty acids by *P. roqueforti* and their subsequent bioconversion to methyl ketones, secondary alcohols, and other flavor compounds (Bakalor, 1962). In this regard, the lipolytic system of *P. roqueforti* is of particular importance. Lamberet and Menassa (1983b) examined the lipolytic activity of seven samples of blue-veined cheeses using suspensions of cheese and mycelium of *P. roqueforti* as the source of extracellular enzyme and synthetic triglycerides as substrate. They concluded that the total lipolytic activity of the cheeses could be considered as the resultant of activities of the acid and alkaline lipases of *P. roqueforti* previously characterized. Using a sterile blue-cheese slurry system inoculated with *P. roqueforti*, King and Clegg (1979) demonstrated that under the correct conditions, *P. roqueforti* is capable of causing extensive lipolysis of cheese fat. Fatty acids are toxic to *P. roqueforti* and inhibit lipolysis, and it has been suggested that the bioconversion of fatty acids to methyl ketones is a detoxifying mechanism.

5. INDUSTRIAL APPLICATIONS

5.1. Cellulase and Other Polysaccharases

Several of the commercially available polysaccharases are obtained from species of *Penicillium* (Table VII). Although industrial enzymes are characterized by their principal activity, many contain side enzymatic activities that may also be useful for some applications. Thus, Cellulase CP (Sturge Enzymes) was found to contain chitinase and β-1,3-glucanase activities, making it a useful lytic preparation for the isolation of protoplasts from various species of fungi (Hamlyn *et al.*, 1981). Bamforth (1983) observed that the β-glucanase in Cellulase CP is more stable to heating than a heat-labile amyloglucosidase found in the same preparation. Heat-treated Cellulase CP can therefore be used in the enzymatic method for measuring the β-glucan content of cereals without interference by glucose derived from starch. The principal activities of Cellulase CP are C_1, endo-β-glucanase, exo-β-glucanase, and β-glucosidase, which together are capable of converting crystalline cellulose to glucose. Use of cellulolytic enzymes to produce glucose from wood, peat, or cellulosic waste materials is a topical and challenging area. Our knowledge of the enzymatic breakdown of cellulose has reached a point where all the basic information needed for the design of an industrial process is available (Enari, 1983). However, the economics of such a system are still unfavorable, and the industrial use of cellulase enzymes has been limited mainly to food processing (e.g., in the extraction of soybean protein and increasing the digestibility of animal feeds). Bhatawdekar (1983) has investigated the effectiveness of a cellulase preparation from *P. funiculosum* in treating fabrics to remove size derived from tamarind kernel powder in place of starch. More than 90% of the size was removed by steeping the sized fabric in a solution of the cellulase at 50°C for 2 hr. This enzyme preparation was also effective in reducing the turbidity of lime juice cordial to give a clear product that retained its original flavor (Bhatawdekar, 1981).

Industrial dextranases are obtained from strains of *P. lilacinum* and used in sugar processing to degrade dextran. Dextran, the α-1,6-linked glucan produced from sucrose by *Leuconostoc* bacteria, is found as a slimy substance in damaged cane and beet tissue. This polysaccharide exhibits a high viscosity in solution, thereby causing trouble in both clarification and evaporation of the sugar liquors and hampering the sucrose-crystallization process. When added to contaminated sugar juice or syrup, the endo-acting dextranase reduces the viscosity by breaking down the large dextran

Table VII. Some Commercially Available Polysaccharases Produced by Various Species of *Penicillium*

Trade or general name	Source	Product form	Optimum pH (or range)	Optimum temperature (or range) (°C)	Activity (mfr's. own units)	Application and special features	Supplier[a]
β-Glucanase[b]	*P. emersonii*	Liquid/powder	4.0	30–90	200 and 750 BGU/g	Brewing and food-processing industries	ABM
Barley β-glucanase[b]	*P. emersonii*	Liquid	4.0	65–95	200 Glaxo units/ml	Brewing, distilling, feeds	Glaxo
Pectinase	*P. simplicissium*	Powder	4.0–5.0	15–60	15,000 AJDU/g		Kali
Dextranase-Novo	*P. lilacinum*	Liquid	5.0–6.0	50–60	25 and 50 KDU/g	Sugar-processing industry	Novo
Cellulase CP[b]	*P. funiculosum*	Powder	3.5–6.5	60	10,000 CMCU/g	Active on crystalline cellulose	Sturge
Pectinase[c]	*P. funiculosum*		4.0–5.0	30–50	1000 apple juice units/g	Polygalacturonase, pectin methyl esterase, and apple juice viscometric reducing activity	Sturge

[a] ABM: ABM Brewing and Food Group, Woodley, Stockport, Cheshire SK6 1PQ, England. Glaxo: Glaxo Operations UK Ltd., Ulverston, Cumbria LA12 9DR, England. Kali: Miles Kali Chemie GmbH and Co. KG, Hans-Bockler-Allee 20, Postfach 69 03 07, 3 Hannover-Kleefeld, West Germany. Novo: Novo Industri A/S, Enzymes Division, Novo Alle, DK-2880 Bagsvaerd, Denmark. Sturge: John and E. Sturge Ltd., Denison Road, Selby, North Yorkshire Y08 8EF, England.
[b] After Godfrey and Reichelt (1983).
[c] Under development (Milsom, private communication).

molecules. The optimum operating conditions for Dextranase 25L (Novo Industri A/S) are pH 5–6 at a temperature of 50–60°C using 10–20 g enzyme per ton of cane processed (Godfrey, 1983a).

A thermostable *β*-glucanase from *P. emersonii*, originally developed to improve the digestibility of barley used as an animal feed (Section 3.1), is now commonly employed by the brewing industry when raw barley is used as a substitute for malt (Godfrey, 1983b). This enzyme, which serves to lower the viscosity of the mash caused by the presence of any barley *β*-glucans, is best used in combination with a *β*-glucanase from another species (e.g., *Bacillus subtilis*) because there are some differences in the specificity of the various enzymes that combine to give a far higher rate of degradation of the most viscous materials. These enzymes may also be added at the fermentation stage for the complete elimination of any haze caused by the *β*-glucans.

Pectolytic enzymes are widely used in food processing—for example, to assist in the extraction of fruit juices and in the clarification of fruit juices or wines. These and other applications have been described in some detail in recent reviews (Rombouts and Pilnik, 1980; Fogarty and Kelly, 1983). Commercial preparations of pectolytic enzymes are available from several species of fungi, including *P. simplicissium* and *P. funiculosum.*

5.2. Glucose Oxidase

Glucose oxidase is produced by various species of fungi including *P. glaucum*, *P. notatum*, *P. purpurogenum*, *P. amagasakiense*, and *Aspergillus niger*, but commercial preparations are normally obtained only from the latter two species (Richter, 1983). The enzyme from *P. amagasakiense* is excreted into the culture medium, whereas that from *A. niger* remains cell-bound in the fungal mycelium, which must be mechanically broken down to liberate the intracellular enzyme. Kusai *et al.* (1960) have purified the glucose oxidase of *P. amagasakiense* from a commercially available extract named "Deoxine" (Nagase Sangyo Co. Ltd.). This enzyme has a molecular weight of about 154,000 and contains two FAD moieties per molecule of enzyme. Further characterization of the enzyme supplied by Nagase Sangyo Co. Ltd. has shown that it consists of four polypeptide chains of equal molecular size, two of which are held together by a disulfide bond to form a dimer; two of these dimeric units associate noncovalently to form a tetramer of molecular weight 160,000 (Yoshimura and Isemura, 1971). Hayashi and Nakamura (1976) have carried out a useful comparison of three commercial preparations of glucose oxidase from *A. niger* with the *Penicillium* preparation from Nagase Sangyo Co. Ltd. The enzymes from both species are glycoproteins with mannose as the main carbohydrate component, and other similarities suggest that they might have evolved from a common ancestral precursor.

However, the value of the Michaelis constant for glucose was considerably lower for the *Penicillium* enzyme as compared with the constants for the *Aspergillus* enzyme, indicating some difference between the enzymes from the two species. The optimum pH range is from 4.0 to 5.5 for the *Penicillium* enzyme and from 3.5 to 6.5 for the enzyme from *A. niger* (Nakamura and Fujiki, 1968). Differences in absorption spectra due to the FAD moiety indicate that these enzymes differ in protein structure in the vicinity of the prosthetic group.

Production of extracellular glucose oxidase has also been documented for other species of *Penicillium.* It may appear unusual that an enzyme involved in the energy-producing mechanism of the cells should be produced extracellularly, but Scott (1975a) has put forward the view that it is the result of the tendency for *Penicillium* species to autolyze very readily, as compared to *Aspergillus,* which leads to the excretion of glucose oxidase into the fermentation medium. Nakamatsu *et al.* (1975) reported that an isolate of *P. purpurogenum* produced a high yield of the enzyme (32,000 units/ml of broth, where one unit of enzyme results in the consumption of 1 μliter of oxygen per hr) when grown on a simple medium containing beet molasses, $NaNO_3$, and KH_2PO_4 in submerged culture for 3 days at 30°C. The purified enzyme had an optimum pH and temperature of 5.0, and 35°C, respectively. It was stable at pH 5.0–7.0 when incubated at 40°C for 2 hr and at temperatures lower than 50°C when incubated at pH 5.6 for 15 min. When the glucose oxidase activity of this species was measured manometrically using the culture broth, the oxygen uptake was not affected by the addition of catalase and H_2O_2 was not detected in the reaction mixture. These observations suggest that *P. purpurogenum* is also a good producer of catalase and could have commercial application for the production of both enzymes. Górniak and Kaczkowski (1973) showed that $NaNO_3$ was the best nitrogen source for the production and subsequent elaboration of glucose oxidase by *P. notatum,* whereas ammonium ions appeared to have an inhibitory effect on both these processes.

The industrial applications of glucose oxidase have been reviewed by Scott (1975b) and, more recently, by Richter (1983). Commercial products used in food processing do not consist of a single enzyme, but of an enzyme system comprising glucose oxidase plus catalase and often other components, including the coenzyme FAD. A major application of this system is its use as an antioxidant in the food industry, primarily to prevent changes in the color and flavor of food products (e.g., canned soft drinks, beer, wine, salad dressings, and various dried foods), both during processing and in storage. The second most important market for glucose oxidase and catalase in the food industry is for the removal of glucose from egg whites and whole eggs to prevent browning and development of off-flavors, especially in the manufacture of dried eggs. Glucose oxidase (without catalase) is also widely used as a diagnostic agent in medicine in the measurement of glucose levels in patients. The method, which is completely specific for glucose, is available in the form of test strips impregnated with glucose oxidase, peroxidase, and a dye and is widely used for the detection of glucose in urine. The presence of the peroxidase is necessary to oxidize the colorless reduced chromogen in the presence of H_2O_2 to produce a colored dye species.

5.3. Future Developments

Microbial enzymes have become an indispensable tool to industry, and their use will continue to expand by the development or modification of existing applications. The development of several novel applications for enzymes currently in commercial production can also be expected in the near future. For example, the addition of glucose oxidase and amyloglucosidase to toothpastes could grow to a market of unexpected commercial significance (Richter, 1983). The generation of hydrogen peroxide by the action of these enzymes appears to be especially effective against lactobacilli and streptococci present in dental plaque, and a new toothpaste, based on these results, is now available in several countries under the trade name "Zendium."

Until recently, industrial enzyme production has been largely restricted to relatively simple hydrolytic enzymes, such as proteinases, pectinases, and amylases. However, with increasing numbers of enzymes being identified from microorganisms, several other types of enzymes are now being prepared on an industrial scale for use in the pharmaceutical industry, as well as for analytical purposes and in medical research (Atkinson and Mavituna, 1983). Kobayashi and Horikoshi (1982) have isolated an extracellular polyamine oxidase from *Penicillium* sp. No. PO-1 cultured on a medium containing 1,3-propanediamine as the inducer, and this enzyme could prove useful in the field of diagnostics for determining the amount of polyamine in body fluids in relation to cancer.

In the food industry, the identification of the substances responsible for color and flavor will allow the selection of enzymes better adapted to release or produce them from economical sources (Godfrey, 1983c). Thus, the extracellular 5′-phosphodiesterase from *P. citrinum*, an enzyme formerly known to occur only in such exotic materials as snake venom, has been used to degrade isolated pure RNA obtained from yeast grown on sulfite waste liquor from the pulp industry in order to produce the flavor-enhancing nucleotides inosinic acid and guanylic acid (Arima, 1964).

Recent studies involving the enzymatic conversion of cellulosic wastes have included the development of a one-step process for the production of ethyl alcohol, utilizing both cellulase enzymes and yeast (Hodge, 1977). The alcohol formed is removed continuously by vacuum distillation.

However, not all new applications need involve the use of isolated enzymes, and selected cultures of cellulolytic fungi could prove useful in the composting of surplus straw, which typically reduces plant yields if incorporated directly into soil because of toxins produced as it decomposes. Inoculation of nonsterile straw with *P. corylophilum* and a nitrogen-fixing anaerobic bacterium, *Clostridium butyricum*, results in a significant increase in the decomposition rate under laboratory conditions compared with uninoculated straw (Lynch and Harper, 1983).

With the notable exception of starch-size removal by amylases, scant attention has been given to the application of enzymes in the textile industry. The preparation of certain textile fibers such as flax, hemp, and jute by dew-retting involves the action of pectolytic enzymes from various species of microorganisms including *Penicillium*, which degrade pectin in the middle lamella of these plant fibers. To date, no reports on the use of isolated enzyme preparations to produce the desired effects have been published (Fogarty and Kelly, 1983). However, the use of isolated enzymes to remove sizes, fats and waxes, pectins, seed-coat material, and colored impurities from loomstate cotton and cotton/polyester fabrics, leading to a novel, low-energy fabric-preparation process, is currently being investigated at the Shirley Institute.

Undoubtedly, the use of microbial enzymes will expand into many other areas than have been indicated here. Certainly, the *Penicillia* are likely to feature prominently among the organisms selected to produce these enzymes. Although this review has been concerned with extracellular enzymes, the industrial production of intracellular enzymes is also likely to expand considerably in the future. However, much new knowledge will be needed if the energy-dependent and cofactor-using enzymes are to be made commercially effective (Godfrey and Reichelt, 1983). A large increase in the use of immobilized enzymes can also be anticipated, particularly for the processing of large volumes of dilute solutions where recovery and re-use of the enzyme catalyst would be extremely cost-effective. The development of the genetics and genetic engineering of fungi will also serve to encourage the economical production of useful new enzymes.

REFERENCES

Arima, K., 1964, Microbial enzyme production, in: *Global Impacts of Applied Microbiology* (M. P. Starr, ed.), John Wiley, New York, pp. 277–294.

Ashy, M. A., Khalil, A. E. G. M., Abou-Zeid, A. Z. A., and Abou-Alnaga, K. O., 1981, Utilisation of cheese whey in production of milk-clotting enzymes, *Agric. Wastes* **3**:277–284.

Atkinson, B., and Mavituna, F., 1983, *Biochemical Engineering and Biotechnology Handbook*, Macmillan, London; Nature Press, New York.

Awasthi, P. B., and Mishra, U. S., 1982, Enzymatic studies on *Penicillium crustosum* Thom, *Sci. Cult.* **48**(3):114–115.

Bakalor, S., 1962, Research related to the manufacture of blue-veined cheese—part 1, *Dairy Sci. Abstr.* **24**(II):529–535.

Bamforth, C. W., 1983, *Penicillium funiculosum* as a source of β-glucanase for the estimation of barley β-glucan, *J. Inst. Brew. London* **89**:391–392.

Belloc, A., Florent, J., Mancy, D., and Verrier, J., 1975, New amylase and its preparation by fermentation of a *Penicillium*, *U.S. Patent* 3,906,113.

Betrabet, S. M., Paralikar, K. M., and Patil, N. E., 1980, Effect of cellulase on the morphology and fine structure of cellulosic substrates. 3. SEM and X-ray diffraction studies, *Cellul. Chem. Technol.* **14**:811–820.

Bhatawdekar, S. P., 1981, Clarification of lime juice by cellulase of *Penicillium funiculosum*, *J. Food Sci. Technol.* **18**:207–208.

Bhatawdekar, S. P., 1983, Studies on optimum conditions of enzymatic desizing of LTKP sized fabric by cellulase-steeping and cellulase-padding methods, *J. Text. Assoc.* **44**:83–86.

Bush, D. A., and Codner, R. C., 1970, Comparison of the properties of the pectin trans-eliminases of *Penicillium digitatum* and *Penicillium italicum*, *Phytochemistry* **9**:87–97.

Chander, H. von, and Klostermeyer, H., 1982, Production of lipase by *Penicillium funiculosum* under various growth conditions, *Arch. Lebensmittelhyg.* **33**(2):42–44.

Chander, H., Sannabhadti, S. S., Elias, J., and Ranganathan, B., 1977, Factors affecting lipase production by *Penicillium chrysogenum*, *J. Food Sci.* **42**(6):1677.

Chander, H., Singh, J., and Khanna, A., 1980, Role of lipids on the growth and lipase production by *Penicillium chrysogenum*, *Milchwissenschaft* **35**(3):153.

Charles, A. F., and Farrell, L. N., 1957, Preparation and use of enzymatic material from *P. lilacinum* to yield clinical dextran, *Can. J. Microbiol.* **3**:239–247.

Deleyn, F., Claeyssens, M., Van Beeumen, J., and DeBruyne, C. K., 1978, Purification and properties of β-xylosidase from *Penicillium wortmanni*, *Can. J. Biochem.* **56**:43–50.

Desmazeaud, M. J., Gripon, J. C., Bars, D. Le., and Bergere, J. L., 1976, Étude du rôle des micro-organismes et des enzymes au cours de la maturation des fromages. III. Influence des micro-organismes, *Lait* **56**(557):379–396.

Dighe, A. S., Khandeparkar, V. G., and Betrabet, S. M., 1981, Enzymatic saccharification of cellulosic materials, *Indian J. Microbiol.* **21**(2):126–130.

Durand, H., Soucaille, P., and Tiraby, G., 1984, Comparative study of cellulases and hemicellulases from four fungi: Mesophiles *Trichoderma reesei* and *Penicillium* sp. and thermophiles *Thielavia terrestris* and *Sporotrichum cellulophilum*, *Enzyme Microb. Technol.* **6**:175–180.

Eitenmiller, R. R., Vakil, J. R., and Shahani, K. M., 1970, Production and properties of *Penicillium roqueforti* lipase, *J. Food Sci.* **35**:130–133.

El-Gendy, S. M., and Marth, E. H., 1980, Proteolytic and lipolytic activities of some toxigenic and nontoxigenic *Aspergilli* and *Penicillia*, *J. Food Protection* **43**(5):354–355.

Emi, S., Myers, D. V., and Iacobucci, G. A., 1976, Purification and properties of the thermostable acid protease of *Penicillium duponti*, *Biochemistry* **15**(4):842–848.

Enari, T.-M., 1983, Microbial cellulases, in: *Microbial Enzymes and Biotechnology* (W. M. Fogarty, ed.), Applied Science Publishers, London and New York, pp. 183–223.

Fogarty, W. M., and Kelly, C. T., 1980, Amylases, amyloglucosidases and related glucanases, in: *Economic Microbiology*, Vol. 5, *Microbial Enzymes and Bioconversions* (A. H. Rose, ed.), Academic Press, New York, pp. 115–170.

Fogarty, W. M., and Kelly, C. T., 1983, Pectic enzymes, in: *Microbial Enzymes and Biotechnology* (W. M. Fogarty, ed.), Applied Science Publishers, London and New York, pp. 131–182.

Ghosh, R. K., and Thangamani, A., 1973, Influence of inorganic nitrate on the formation of extracellular protease and ribonuclease by *Penicillium janthinellum*, *Can. J. Microbiol.* **19**:1219–1223.

Godfrey, T., 1983a, Dextranase and sugar processing, in: *Industrial Enzymology—The Application of Enzymes in Industry* (T. Godfrey and J. R. Reichelt, eds.), Macmillan, London; Nature Press, New York, pp. 422–423.

Godfrey, T., 1983b, Brewing, in: *Industrial Enzymology—The Application of Enzymes in Industry* (T. Godfrey and J. R. Reichelt, eds.), Macmillan, London; Nature Press, New York, pp. 221–259.

Godfrey, T., 1983c, Flavouring and colouring, in: *Industrial Enzymology—The Application of Enzymes in Industry* (T. Godfrey and J. R. Reichelt, eds.), Macmillan. London; Nature Press, New York, pp. 305–314.

Godfrey, T., and Reichelt, J. R., 1983, Introduction to industrial enzymology, in: *Industrial Enzymology—The Application of Enzymes in Industry*, Macmillan, London, Nature Press, New York, pp. 1–7.

Górniak, H., and Kaczkowski, J., 1973, The influence of nitrogen source on the biosynthesis and properties of glucose oxidase in cultures of *Penicillium notatum*, *Bull. Acad. Polish Sci. Ser. Biol.* **21:**571–576.

Gripon, J. C., 1977a, Le système protéolytique de *Penicillium roqueforti*. IV. Propriétés d'une carboxypeptidase acide, *Ann. Biol. Anim. Biochem. Biophys.* **17**(3A)**:**283–298.

Gripon, J. C., 1977b, The proteolytic system of *Penicillium roqueforti*. V. Purification and properties of an alkaline aminopeptidase, *Biochimie* **59:**679–686.

Gripon, J. C., and Debest, B., 1976, Étude électrophorétique du système protéolytique exocellulaire de *Penicillium roqueforti*, *Lait* **56**(557)**:**423–438.

Gripon, J. C., and Hermier, J., 1974, Le système protéolytique de *Penicillium roqueforti*. III. Purification, propriétés et spécificité d'une protéase inhibée par l'E.D.T.A., *Biochimie* **56:**1324–1332.

Gripon, J. C., Rhee, S. H., and Hofmann, T., 1977a, N-Terminal amino acid sequences of acid proteases: Acid proteases from *Penicillium roqueforti* and *Rhizopus chinensis* and alignment with penicillopepsin and mammalian proteases, *Can. J. Biochem.* **55:**504–506.

Gripon, J. C., Desmazeaud, M. J., Bars, D. Le., and Bergere, J. L., 1977b, Role of proteolytic enzymes of *Streptococcus lactis*, *Penicillium roqueforti*, and *Penicillium caseicolum* during cheese ripening, *J. Dairy Sci.* **60:**1532–1538.

Gripon, J. C., Auberger, B., and Lenoir, J., 1980, Metalloproteases from *Penicillium caseicolum* and *P. roqueforti*: Comparison of specificity and chemical characterization, *Int. J. Biochem.* **12:**451–455.

Hamlyn, P. F., Bradshaw, R. E., Mellon, F. M., Santiago, C. M., Wilson, J. M., and Peberdy, J. F., 1981, Efficient protoplast isolation from fungi using commercial enzymes, *Enzyme Microbiol. Technol.* **3:**321–325.

Hashimoto, H., Kaneko, Y., Iwaasa, T., and Yokotsuka, T., 1973a, Production and purification of acid protease from thermophilic fungus *Penicillium duponti* K1014, *Appl. Microbiol.* **25**(4)**:**584–588.

Hashimoto, H., Iwaasa, T., and Yokotsuka, T., 1973b, Some properties of acid protease from the thermophilic fungus, *Penicillium duponti* K1014, *Appl. Microbiol.* **25**(4)**:**578–583.

Hayashi, S., and Nakamura, S., 1976, Comparison of fungal glucose oxidases—chemical, physicochemical and immunological studies, *Biochim. Biophys. Acta* **438:**37–48.

Hodge, W. H., 1977, Process for making alcohol from cellulosic material using plural ferments, *U.S. Patent* 4,009,075.

Hofmann, T., and Shaw, R., 1964, Proteolytic enzymes of *Penicillium janthinellum*. 1. Purification and properties of a trypsinogen-activating enzyme (peptidase A), *Biochim. Biophys. Acta* **92:**543–557.

Houmard, J., and Raymond, M. N., 1979, Further characterization of the *Penicillium roqueforti* acid protease, *Biochimie* **61:**979–982.

Imamura, T., and Kataoka, K., 1963a, Biochemical studies on the manufacturing of Roquefort type cheese. I. Lipase-producing ability of *Penicillium roqueforti*, *Jpn. J. Zootech. Sci.* **34**:344–348.

Imamura, T., and Kataoka, K., 1963b, Biochemical studies on the manufacturing of Roquefort type cheese. II. Characteristics of lipases produced by *Penicillium roqueforti*, *Jpn. J. Zootech. Sci.* **34**:349–353.

Iwai, M., Okumura, S., and Tsujisaka, Y., 1975, The comparison of the properties of two lipases from *Penicillium cyclopium* Westring, *Agric. Biol. Chem.* **39**(5):1063–1070.

James, A. E., Fare, G., Sagar, B. F., Lucas, F., and Mitchell, I. D. G., 1976, Improvements in or relating to enzymes, *U.K. Patent* 1,421,127.

Joglekar, A. V., Srinivasan, M. C., Manchanda, A. C., Jogdand, V. V., and Karanth, N. G., 1983, Studies on cellulase production by a *Penicillium funiculosum* strain in an instrumented fermentor, *Enzyme Microbiol. Technol.* **5**:22–24.

Jones, S. R., and Hofmann, T., 1972, Penicillocarboxypeptidase-S, a nonspecific SH-dependent exopeptidase, *Can. J. Biochem.* **50**:1297–1310.

Kazanina, G. A., Selezneva, A. A., and Tereshin, I. M., 1981, Lipolytic complex from *Penicillium* sp., *Appl. Biochem. Microbiol*, **17**:657–661.

King, R., and Clegg, G. H., 1979, The metabolism of fatty acids, methyl ketones and secondary alcohols by *Penicillium roqueforti* in blue cheese slurries, *J. Sci. Food Agric.* **30**:197–202.

Kobayashi, Y., and Horikoshi, K., 1982, Purification and characterization of extracellular polyamine oxidase produced by *Penicillium* sp. no. PO-1, *Biochim. Biophys. Acta* **705**:133–138.

Kornacki, K., Stepaniak, L., Adamiec, I., Grabska, J., and Wrona, K., 1980, Production of lipases and proteases by moulds of *Penicillium roqueforti* and *Penicillium candidum* under selected conditions of surface and submerged cultivation, *Acta Aliment. Pol.* **6**(4):281–288.

Kusai, K., Sekuzu, I., Hagihara, B., Okunuki, K., Yamauchi, S., and Nakai, M., 1960, Crystallization of glucose oxidase from *Penicillium amagasakiense*, *Biochim. Biophys. Acta* **40**:555–557.

Lamberet, G., and Lenoir, J., 1976a, Les caractères du système lipolytique de l'espèce *Penicillium caseicolum*: Nature du système, *Lait* **56**(553-554):119–134.

Lamberet, G., and Lenoir, J., 1976b, Les caractères du système lipolytique de l'espèce *Penicillium caseicolumn*: Purification et propriétiés de la lipase majeure, *Lait* **56**(559-560):622–644.

Lamberet, G., and Menassa, A., 1983a, Purification and properties of an acid lipase from *Penicillium roqueforti*, *J. Dairy Res.* **50**:459–468.

Lamberet, G., and Menassa, A., 1983b, Détermination et niveau des activités lipolytiques dans les fromages à pâte persillée, *Lait* **63**(629-630):333–344.

Lenoir, J., and Auberger, B., 1977a, Les caractères du système protéolytique de *Penicillium caseicolumn.* I. Précaractérisation de l'activité exocellulaire, *Lait* **57**(563-564):164–183.

Lenoir, J., and Auberger, B., 1977b, Les caractères du système protéolytique de *Penicillium caseicolum.* II. Caracterisation d'une protéase neutre, *Lait* **57**(568):471–491.

Lenoir, J., Auberger, B., and Gripon, J. C., 1979, Les caractères du système protéolytique de *Penicillium caseicolumn.* III. Caractérisation d'une protéase acide, *Lait* **59**(585-586):244–268.

Lobanok, A. G., Mikhailova, R. V., and Sapunova, L. I., 1977, Kinetics of constitutive synthesis of polymethylgalacturonases by *Penicillium digitatum* as function of the carbon source and pH, *Mikrobiologiya* **46**:920–925.

Lobyreva, L. B., and Marchenkova, A. I., 1980, Isolation and characteristics of lipases from *Penicillium roqueforti*, *Mikrobiologiya* **49**:924–930.

Lobyreva, L. B., and Marchenkova, A. I., 1981, Characteristics of lipases in the culture liquid of *Penicillium roqueforti*, *Mikrobiologiya* **50**:64–68.

Lynch, J. M., and Harper, S. H. T., 1983, Straw as a substrate for co-operative nitrogen fixation, *J. Gen. Microbiol.* **129**:251–253.

Lynch, J. M., Slater, J. H., Bennett, J. A., and Harper, S. H. T., 1981, Cellulase activities of some aerobic micro-organisms isolated from soil, *J. Gen. Microbiol.* **127**:231–236.

Mandels, M., Weber, J., and Parizek, R., 1971, Enhanced cellulase produced by a mutant of *Trichoderma viride*, *Appl. Microbiol.* **21**:152–154.

Mandels, M., Hontz, L., and Nystrom, J., 1974, Enzymatic hydrolysis of waste cellulose, *Biotechnol. Bioeng.* **16**:1471–1493.

Martin, S. M., Singh, K., Ankel, H., and Khan, A. H., 1962, The specificity of a protease from *Penicillium cyaneo-fulvum*, *Can. J. Biochem. Physiol.* **40**:237–246.

Menassa, A., and Lamberet, G., 1982, Contribution à l'étude du système lipolytique de *Penicillium roqueforti*, *Lait* **62**:32–43.

Mishra, C., Rao, M., Seeta, R., Srinivasan, M. C., and Deshpande, V., 1983, Hydrolysis of lignocelluloses by *Penicillium funiculosum* cellulase, *Biotechnol. Bioeng.* **26**:370–373.

Modler, H. W., Brunner, J. R., and Stine, C. M., 1974a, Extracellular protease of *Penicillium roqueforti*. I. Production and characteristics of crude enzyme preparation, *J. Dairy Sci.* **57**(5):523–527.

Modler, H. W., Brunner, J. R., and Stine, C. M., 1974b, Extracellular protease of *Penicillium roqueforti*. II. Characterization of a purified enzyme preparation, *J. Dairy Sci.* **57**(5):528–534.

Morris, H. A., and Jezeski, J. J., 1953, The action of micro-organisms on fats. II. Some characteristics of the lipase system of *Penicillium roqueforti*, *J. Dairy Sci.* **36**:1285–1298.

Nakamatsu, T., Akamatsu, T., Miyajima, R., and Shiio, I., 1975, Microbial production of glucose oxidase, *Agric. Biol. Chem.* **39**(9):1803–1811.

Nakamura, S., and Fujiki, S., 1968, Comparative studies on the glucose oxidases of *Apergillus niger* and *Penicillium amagasakiense*, *J. Biochem.* **63**(1):51–58.

Noomen, A., 1983, The role of the surface flora in the softening of cheeses with a low initial pH, *Neth. Milk Dairy J.* **37**:229–232.

Okumura, S., Iwai, M., and Tsujisaka, Y., 1980, Purification and properties of partial glyceride hydrolase of *Penicillium cyclopium* M1, *J. Biochem.* **87**:205–211.

Olutiola, P. O., 1983, Cell wall degrading enzymes associated with deterioration of cocoa beans by *Penicillium steckii*, *Int. Biodeterior. Bull.* **19**(1):27–36.

Olutiola, P. O., and Akintunde, O. A., 1979, Pectin lyase and pectin methylesterase production by *Penicillium citrinum*, *Trans. Br. Mycol. Soc.* **72**(1):49–55.

Phaff, H. J., 1947, The production of exocellular pectic enzymes by *Penicillium chrysogenum*. I. On the formation and adaptive nature of polygalacturonase and pectinesterase, *Arch. Biochem.* **13**:67–81.

Qureshi, M. S. A., Mizra, J. H., and Malik, K. A., 1980a, Cellulolytic activity of some thermophilic and thermotolerant fungi of Pakistan, *Biologia* **26**:201–217.

Qureshi, M. S. A., Mizra, J. H., and Malik, K. A., 1980b, Amylolytic activity of some thermophilic and thermotolerant fungi of Pakistan, *Biologia* **26**(1-2):225–237.

Rapp, P., Grote, E., and Wagner, F., 1981, Formation and location of 1,4-β-glucanases and 1,4-β-glucosidases from *Penicillium janthinellum*, *Appl. Environ. Microbiol.* **41**(4):857–866.

Rapp, P., Kmobloch, U., and Wagner, F., 1982, Repression of endo-1,4-β-glucanase formation in *Penicillium janthinellum* and product inhibition of its 1,4-β-glucanases and cellobiases, *J. Bacteriol.* **149**(2):783–786.

Reese, E. T., 1976, History of the cellulase program at the U.S. Army Natick Development Center, *Biotechnol. Bioeng. Symp.* **6**:9–24.

Reese, E. T., and Mandels, M., 1959, β-D-1,3-Glucanases in fungi, *Can. J. Microbiol.* **5**:173–185.

Reese, E. T., and Mandels, M., 1964, A new α-glucanase: Mycodextranase, *Can. J. Microbiol.* **10**(2):103–114.

Reese, E. T., Siu, R. G. H., and Levinson, H. S., 1950, The biological degradation of soluble cellulose derivatives and its relationship to the mechanism of cellulose hydrolysis, *J. Bacteriol.* **59**:485–497.

Reese, E. T., Parrish, F. W., and Mandels, M., 1961, β-D-1,2-Glucanases in fungi, *Can. J. Microbiol.* **7**:309–317.

Reese, E. T., Parrish, F. W., and Mandels, M., 1962, β-D-1,6-Glucaneses in fungi, *Can. J. Microbiol.* **8**:327–334.

Reese, E. T., Maguire, A., and Parrish, F. W., 1973, Production of β-D-xylopyranosidases by fungi, *Can. J. Microbiol.* **19**:1065–1074.

Richter, G., 1983, Glucose oxidase, in: *Industrial Enzymology—The Aplication of Enzymes in Industry* (T. Godfrey and J. R. Reichelt, eds.), Macmillan, London; Nature Press, New York, pp. 428–436.

Rombouts, F. M., and Pilnik, W., 1980, Pectic enzymes, in: *Economic Microbiology*, Vol. 5, *Microbial Enzymes and Bioconversions* (A. H. Rose, ed.), Academic Press, New York, pp. 227–282.

Sagar, B. F., 1978, Investigation of assessibility changes in crystalline cellulose, in: *Shirley Institute Report*, Publ. European Res. Office, U.S. Army; available NTIS (AD-A058080).

Sagar, B. F., 1985, Mechanism of cellulase action, in: *Cellulose and Its Derivatives* (J. F. Kennedy, G. O. Phillips, D. J. Wedlock, and P. A. Williams, eds.), Ellis Horwood, Chichester, U.K., pp. 199–207.

Sánchez, M., Nombela, C., Villanueva, J. R., and Santos, T., 1982, Purification and partial characterization of a developmentally regulated 1,3-β-glucanase from *Penicillium italicum*, *J. Gen. Microbiol.* **128**:2047–2053.

Santos, T., Sánchez, M., Villanueva, J. R., and Nombela, C., 1978, Regulation of the β-1,3-glucanase system in *Penicillium italicum*: Glucose repression of the various enzymes, *J. Bacteriol.* **133**(2):465–471.

Santos, T., Sánchez, M., Villanueva, J. R., and Nombela, C., 1979, Derepression of β-1,3-glucanases in *Penicillium italicum*: Localization of the various enzymes and correlation with cell wall glucan mobilization and autolysis, *J. Bacteriol.* **137**(1):6–12.

Satyanarayana, T., and Jain, S., 1979, Studies on extra-cellular protease production by seed-borne fungi of jowar, *Curr. Trends Life Sci.* **7**:189–195.

Scott, D., 1975a, Oxidoreductases, in: *Enzymes in Food Processing* (G. Reed, ed.), Academic Press, New York, pp. 219–254.

Scott, D., 1975b, Applications of glucose oxidase, in: *Enzymes in Food Processing* (G. Reed, ed.), Academic Press, New York, pp. 519–547.

Selby, K., 1968, Mechanism of biodegradation of cellulose, in: *Biodeterioration of Materials*, Vol. 1 (A. H. Walters and J. J. Elphick, eds.), Elsevier, London, pp. 62–77.

Selby, K., Maitland, C. C., and Thompson, K., 1968, Biochemical process, British Patent Application 42663/68.

Sharma, D. P., and Saksena, S. B., 1979, Production of cellulose degrading enzymes by fungi, *Current Trends Life Sci.* **7**:121–132.

Sharma, D. P., and Saksena, S. B., 1981, Production of proteases by some saprophytic fungi, *J. Indian Bot. Soc.* **60**:330–333.

Sharma, O. P., and Sharma, K. D., 1980, Studies on *in vitro* production of proteolytic enzymes by some leather deteriorating fungi, *Rev. Roum. Biochim.* **17**(3):209–215.

Shipe, W. F., Jr., 1951, A study of the relative specificity of lipases produced by *Penicillium roqueforti* and *Aspergillus niger*, *Arch. Biochem. Biophys.* **30**:165–179.

Sidhu, M. S., and Sandhu, D. K., 1979, Cellulolytic enzymes of seven fungi isolated from decomposing sugar-cane bagasse, *Indian J. Microbiol.* **19**(4):204–208.

Singh, B. S., and Saksena, S. B., 1981, Estimation of extracellular C_1 and C_x cellulases produced by certain species of *Penicillium*, *Natl. Acad. Sci. Lett.* **4**:431–432.

Singh, K., and Martin, S. M., 1960, Purification and properties of a protease from *Penicillium cyaneo-fulvum*, *Can. J. Biochem. Physiol.* **38**:969–980.

Singh, K. V., and Agrawal, S. C., 1981, Amylase production by some dermatophytes, *Acta Bot. Indica* **9**:32–34.

Sodek, J., and Hofmann, T., 1970a, Large-scale preparation and some properties of penicillopepsin, the acid proteinase of *Penicillium janthinellum*, *Can. J. Biochem.* **48**:425–431.

Sodek, J., and Hofmann, T., 1970b, Amino acid sequence around the active site aspartic acid in penicillopepsin, *Can. J. Biochem.* **48**:1014–1016.

Somani, R. B., and Wangikar, P. D., 1979, Cellulolytic activity of some soil fungi, *Food Farming Agric.* **12**(4):96–98.

Spalding, D. H., and Abdul-Baki, A. A., 1973, *In vitro* and *in vivo* production of pectin lyase by *Penicillium expansum*, *Phytopathology* **63**:231–235.

Spalding, D. H., Wells, J. M., and Allison, D. W., 1973, Catabolite repression of polygalacturonase, pectin lyase, and cellulase synthesis in *Penicillium expansum*, *Phytopathology* **63**:840–844.

Stepaniak, L., Kornacki, K., Grabska, J., Rymaszewski, J., and Cichosz, G., 1980, Lipolytic and proteolytic activity of *Penicillium roqueforti*, *Penicillium candidum* and *Penicillium camemberti* strains, *Acta Aliment. Pol.* **6**(3):155–164.

Sternberg, D., and Mandels, G. R., 1979, Induction of cellulolytic enzymes in *Trichoderma reesei* by sophorose, *J. Bacteriol.* **139**:761–769.

Swinburne, T. R., and Corden, M. E., 1969, A comparison of the polygalacturonases produced *in vivo* and *in vitro* by *Penicillium expansum* Thom, *J. Gen. Microbiol.* **55**:75–87.

Szakacs, G., Reczey, K., Hernadi, P., and Dobozi, M., 1981, *Penicillium verruculosum* WA 30: A new source of cellulase, *Eur. J. Apl. Microbiol. Biotechnol.* **11**:120–124.

Tokiwa, Y., and Suzuki, T., 1980, Purification and some properties of polyethylene adipate-degrading enzyme produced by *Penicillium* sp. strain 14-3, *Rep. Ferment. Res. Inst. (Yatabe)* **55**:21–34.

Trieu-Cuot, P., and Gripon, J. C., 1981, Casein hydrolysis by *Penicillium caseicolum* and *P. roqueforti* proteinases: A study with isoelectric focusing and two-dimensional electrophoresis, *Neth. Milk Dairy J.* **35**:353–357.

Trieu-Cuot, P., and Gripon, J. C., 1982, A study of proteolysis during Camembert cheese ripening using isoelectric focusing and two-dimensional electrophoresis, *J. Dairy Res.* **49**:501–510.

Trieu-Cuot, P., Archieri-Haze, M. J., and Gripon, J. C., 1982a, Effect of aspartyl proteinases of *Penicillium caseicolum* and *Penicillium roqueforti* on caseins, *J. Dairy Res.* **49**:487–500.

Trieu-Cuot, P., Archieri-Haze, M. J., and Gripon, J. C., 1982b, Étude comparative de l'action des métalloprotéases de *Penicillium caseicolum* et *Penicillium roqueforti* sur les caséines alpha S1 et bêta, *Lait* **62**(615–616):234–249.

Tsuchiya, H. M., Jeanes, A., Bricker, H. M., and Wilham, C. A., 1952, Dextran-degrading enzymes from moulds, *J. Bacteriol.* **64**:513–519.

Tsujisaka, Y., and Iwai, M., 1984, Comparative study on microbial lipases, *Kagaku To Kogyo (Osaka)* **58**(2):60–69.

Villanueva, J. R., Notario, V., Santos, T., and Villa, T. G., 1976, β-Glucanases in nature, in: *Microbial and Plant Protoplasts* (J. F. Peberdy, A. H. Rose, H. J. Rogers, and E. C. Cocking, eds.), Academic Press, London, pp. 323–355.

Wood, T. M., and McCrae, S. I., 1977, The mechanism of cellulase action with particular reference to the C_1 component, in: *Bioconversion of Cellulosic Substances into Energy, Chemicals and Microbial Protein* (T. K. Ghose, ed.), IIT, Delhi, pp. 111–141.

Wood, T. M., and McCrae, S. I., 1978, The cellulase of *Trichoderma koningii*, *Biochem. J.* **171**:61–72.

Wood, T. M., and McCrae, S. I., 1982, Purification and some properties of a (1 → 4)-β-D-glucan glucohydrolase associated with the cellulase from the fungus *Penicillium funiculosum*, *Carbohydr. Res.* **110**:291–303.

Wood, T. M., McCrae, S. I., and Macfarlane, C. C., 1980, The isolation, purification and properties of the cellobiohydrolase component of *Penicillium funiculosum* cellulase, *Biochem. J.* **189**:51–65.

Yamasaki, Y., Suzuki, Y., and Ozawa, J., 1977a, Purification and properties of two forms of glucoamylase from *Penicillium oxalicum*, *Argic. Biol. Chem.* **41**(5):755–762.

Yamasaki, Y., Suzuki, Y., and Ozawa, J., 1977b, Properties of two forms of glucoamylase from *Penicillium oxalicum*, *Agric. Biol. Chem.* **41**(8):1443–1449.

Yoshimura, T., and Isemura, T., 1971, Subunit structure of glucose oxidase from *Penicillium amagasakiense*, *J. Biochem.* **69**:839–846.

Zevaco, C., Hermier, J., and Gripon, J. C., 1973, Le système protéolytique de *Penicillium roqueforti*. II. Purification et propriétés de la protéase acide, *Biochimie* **55**:1353–1360.

Suggested Additional Reading

Bajpai, R. K., and Ruess, M., 1981, A mechanistic model for penicillin production, *J. Chem. Technol. Biotechnol.* **30:**332–344.

Bernard, A., and Cooney, C. L., 1981, Studies on oxygen limitations in the penicillin fermentation, in: *2nd European Congress of Biotechnology, Eastbourne, England. Abstracts of Communications.* Soc. Chem. Ind., London, p. 81.

Calam, C. T., 1982, Factors governing the production of penicillin by *Penicillium chrysogenum,* in: *Overproduction of Microbial Metabolites* (V. Krumphanzl, B. Sikyta, and Z. Vanek, eds.), Academic Press, New York, pp. 98–95.

Cooney, C. L., 1979, Conversion yields in penicillin production: Theory vs practice, *Proc. Biochem.* **May:**31–33.

Court, J. R., and Pirt, S. J., 1977, Fed batch culture of *Penicillium chrysogenum* for penicillin production, *Proc. Soc. Gen. Microbiol.* **4:**139–140.

Court, J. R., and Pirt, S. J., 1981, Carbon and nitrogen limited growth of *Penicillium chrysogenum* in fed batch culture: The optimal ammonium ion concentration for penicillin production, *J. Chem. Technol. Biotechnol.* **31:**235–240.

Demain, A. L., 1974, Biochemistry of penicillin and cephalosporin fermentations, *Lloydia* **37:**147–167.

Enfers, S. O., and Nilsson, H., 1979, Design and response characteristics of an enzyme electrode for measurement of penicillin in fermentation broth, *Enzyme Microbiol. Technol.* **1:**260–264.

Hegewald, E., Wolleschensky, B., Guthke, R., Neubert, M., and Knorre, W. A., 1981, Instabilities of product formation in a fed-batch culture of *Penicillium chrysogenum, Biotechnol. Bioeng.* **21:**1563–1572.

Hersbach, G. J. M., van der Beck, C. P., and van Dijk, P. W. W., 1984, The penicillins: properties, biosynthesis and fermentation, in: *Biotechnology of Industrial Antibiotics* (E. J. Van damne, ed.), Marcel Dekker, New York and Basel, pp. 46–140.

Jensen, E. B., Nielson, R., and Emborg, C., 1981, The influence of acetic acid on penicillin production, *Eur. J. Appl. Microbiol. Biotechnol.* **13:**29–33.

Kennel, Y. M., and Demain, A. L., 1978, Effect of carbon sources on β-lactam antibiotic formation by *Cephalosporium acremonium, Expt. Mycol.* **2:**234–238.

Konig, B., Seewald, C., and Schugeol, K., 1981, Process engineering investigations on penicillin production, *Eur. J. Appl. Microbiol. Biotechnol.* **12:**205–211.

Kuenzi, M. T., 1979, Comparison of the fermentation kinetics of *Cephalosporium acremonium* in shake flasks and fermenters by means of ribonucleic acid measurements, *Biotechnol. Lett.* **1:**127–132.

Kuenzi, M. T., 1980, Regulation of cephalosporin synthesis in *Cephalosporin acremonium* by phosphate and glucose, *Arch. Microbiol.* **128:**78–83.

Lara, F., Mateos, R. C., Vasquez, G., and Sanchez, S., 1982, Induction of penicillin biosynthesis by L-glutamate in *Penicillium chrysogenum, Biochem. Biophys. Res. Commun.* **105:**172–178.

Lee, J. S., Shin, K. C., Yang H. S., and Ryn, D. D. Y., (1978, Effects of carbon sources and other process variables in fed-batch fermentation of penicillin, *Korean J. Microbiol.* **16**:21–29.

Luengo, J. M., Revilla, G., Lopez, M. J., Villanueva, J. R., and Martin, J. F., 1979, Lysine regulation of penicillin biosynthesis in low producing and industrial strains of *Penicillium chrysogenum*, *J. Gen. Microbiol.* **115**:207–211.

Lurie, L. M., Stepanova, N. E., Bartoshevich, Y. E., and Levitov, M. M., 1979, Study of physiological role of oils in penicillin biosynthesis, *Antibiotiki (Moscow)* **24**:86–92.

Mou, D.-G., and Cooney, C. L., 1983, Growth monitoring and control through computer aided on-line mass balancing in a fed batch penicillin fermentation, *Biotechnol. Bioeng.* **25**:225–255.

Matsumura, M., Imanaka, T., Yoshida, T., and Taguchi, H., 1978, Effect of glucose and methionine consumption rates on cephalosporin C production by *Cephalosporium acremonium*, *J. Ferm. Technol.* **56**:345–353.

Mou, D.-G., and Cooney, C. L., 1983, Growth monitoring and control in a complex medium. A case study employing fed-batch penicillin fermentation and computer-aided on-line mass balancing, *Biotechnol. Bioeng.* **25**:257–269.

Ott, J. L., Godzoski, C. W., Pavey, D., Farron, J. D., and Horton, D. R., 1962, Biosynthesis of cephalosporin C. I. Factors affecting the fermentation, *Appl. Microbiol.* **10**:515–523.

Ryu, D. D. Y., and Hospodka, J., 1980, Quantitative physiology of *Penicillium chrysogenum* in penicillin fermentation, *Biotechnol. Bioeng.* **22**:289–298.

Sanchez, S., Paniagua, L., Mateos, R. C., Lara, F., and Mara, J., 1982, Nitrogen regulation of penicillin G synthesis in *Penicillium chrysogenum*, in: *Advances in Biotechnology*, vol. 3, (M. Moo-Young, ed.), Pergamon, London, pp. 147–154.

Shen, Y. Q., Hein, J., Solomon, N. A., Wolfe, S., and Demain, A. L., 1984, Repression of β-lactam production in *Cephalosporium acremonium* by nitrogen sources, *J. Antibiotics* **37**:503–511.

Vardar, F., and Lilly, M. D., 1982, Effect of cycling dissolved oxygen concentrations on product formation in penicillin fermentation, *Eur. J. Appl. Microbiol. Biotechnol.* **14**:203–211.

Witter, R., and Schugel, K., 1985, Interrelation between penicillin productivity and growth rate, *Appl. Microbiol. Biotechnol.* **21**:348–356.

Species Index

Subject Index

Numbers in parentheses refer to chemical structures described in Chapter 6.

GPSR Compliance

The European Union's (EU) General Product Safety Regulation (GPSR) is a set of rules that requires consumer products to be safe and our obligations to ensure this.

If you have any concerns about our products, you can contact us on ProductSafety@springernature.com

In case Publisher is established outside the EU, the EU authorized representative is:

Springer Nature Customer Service Center GmbH
Europaplatz 3
69115 Heidelberg, Germany

Batch number: 10761599

Printed by Printforce, the Netherlands